Technology and Management Applied in Construction Engineering Projects

Technology and Management Applied in Construction Engineering Projects

Editors

Mariusz Szóstak
Marek Sawicki
Jarosław Konior

MDPI • Basel • Beijing • Wuhan • Barcelona • Belgrade • Manchester • Tokyo • Cluj • Tianjin

Editors
Mariusz Szóstak
Wrocław University of
Science and Technology
Poland

Marek Sawicki
Wrocław University of
Science and Technology
Poland

Jarosław Konior
Wrocław University of
Science and Technology
Poland

Editorial Office
MDPI
St. Alban-Anlage 66
4052 Basel, Switzerland

This is a reprint of articles from the Special Issue published online in the open access journal *Applied Sciences* (ISSN 2076-3417) (available at: https://www.mdpi.com/journal/applsci/special_issues/Engineering_Technology_Management).

For citation purposes, cite each article independently as indicated on the article page online and as indicated below:

LastName, A.A.; LastName, B.B.; LastName, C.C. Article Title. *Journal Name* **Year**, *Volume Number*, Page Range.

ISBN 978-3-0365-6039-7 (Hbk)
ISBN 978-3-0365-6040-3 (PDF)

© 2022 by the authors. Articles in this book are Open Access and distributed under the Creative Commons Attribution (CC BY) license, which allows users to download, copy and build upon published articles, as long as the author and publisher are properly credited, which ensures maximum dissemination and a wider impact of our publications.

The book as a whole is distributed by MDPI under the terms and conditions of the Creative Commons license CC BY-NC-ND.

Contents

About the Editors vii

Preface to "Technology and Management Applied in Construction Engineering Projects" . . . ix

Mariusz Szóstak, Jarosław Konior and Marek Sawicki
Technology and Management Applied in Construction Engineering Projects
Reprinted from: *Appl. Sci.* **2022**, *12*, 11823, doi:10.3390/app122211823 1

Daniel Przywara and Adam Rak
Monitoring of Time and Cost Variances of Schedule Using Simple Earned Value Method Indicators
Reprinted from: *Appl. Sci.* **2021**, *11*, 1357, doi:10.3390/app11041357 7

Xun Liu, Le Shen and Kun Zhang
Estimating the Probability Distribution of Construction Project Completion Times Based on Drum-Buffer-Rope Theory
Reprinted from: *Appl. Sci.* **2021**, *11*, 7150, doi:10.3390/app11157150 19

Hubert Anysz, Jerzy Rosłon and Andrzej Foremny
7-Score Function for Assessing the Strength of Association Rules Applied for Construction Risk Quantifying
Reprinted from: *Appl. Sci.* **2022**, *12*, 844, doi:10.3390/app12020844 37

Tadeusz Kasprowicz, Anna Starczyk-Kołbyk and Robert Wójcik
Randomized Estimation of the Net Present Value of a Residential Housing Development
Reprinted from: *Appl. Sci.* **2022**, *12*, 124, doi:10.3390/app12010124 61

Janusz Sobieraj and Dominik Metelski
Project Risk in the Context of Construction Schedules—Combined Monte Carlo Simulation and Time at Risk (TaR) Approach: Insights from the Fort Bema Housing Estate Complex
Reprinted from: *Appl. Sci.* **2022**, *12*, 1044, doi:10.3390/app12031044 81

Martina Milat, Snježana Knezić and Jelena Sedlar
Resilient Scheduling as a Response to Uncertainty in Construction Projects
Reprinted from: *Appl. Sci.* **2021**, *11*, 6493, doi:10.3390/app11146493 117

Parinaz Jafari, Malak Al Hattab, Emad Mohamed and Simaan AbouRizk
Automated Extraction and Time-Cost Prediction of Contractual Reporting Requirements in Construction Using Natural Language Processing and Simulation
Reprinted from: *Appl. Sci.* **2021**, *11*, 6188, doi:10.3390/app11136188 137

Luz Duarte-Vidal, Rodrigo F. Herrera, Edison Atencio and Felipe Muñoz-La Rivera
Interoperability of Digital Tools for the Monitoring and Control of Construction Projects
Reprinted from: *Appl. Sci.* **2021**, *11*, 10370, doi:10.3390/app112110370 161

Edison Atencio, Pablo Araya, Francisco Oyarce, Rodrigo F. Herrera, Felipe Muñoz-La Rivera and Fidel Lozano-Galant
Towards the Integration and Automation of the Design Process for Domestic Drinking-Water and Sewerage Systems with BIM
Reprinted from: *Appl. Sci.* **2022**, *12*, 9063, doi:10.3390/app12189063 193

Paul Sestras
Methodological and On-Site Applied Construction Layout Plan with Batter Boards Stake-Out Methods Comparison: A Case Study of Romania
Reprinted from: *Appl. Sci.* **2021**, *11*, 4331, doi:10.3390/app11104331 . . . . **225**

Zian Xu and Minshui Huang
Improving Bridge Expansion and Contraction Installation Replacement Decision System Using Hybrid Chaotic Whale Optimization Algorithm
Reprinted from: *Appl. Sci.* **2021**, *11*, 6222, doi:10.3390/app11136222 . . . . **245**

Elżbieta Szafranko and Jolanta Harasymiuk
Modelling of Decision Processes in Construction Activity
Reprinted from: *Appl. Sci.* **2022**, *12*, 3797, doi:10.3390/app12083797 . . . . **271**

Yingjie Zhao, Fan Yang and Yijiang Ma
Experimental Method for Flow Calibration of the Aircraft Liquid Cooling System
Reprinted from: *Appl. Sci.* **2022**, *12*, 5056, doi:10.3390/app12105056 . . . . **293**

About the Editors

Mariusz Szóstak

Mariusz Szóstak employed at the Department of Building Engineering, Faculty of Civil Engineering, at Wroclaw University of Science and Technology (WUST), Poland. He obtained an MSc in Civil Engineering from WUST in 2013, a PhD (Hons) in Civil Engineering in April 2018. Since 2021, he has been the Deputy Head of the Department of Building Engineering. He is the author of more than 70 scientific papers, including papers on work safety in construction and articles in journals in the Journal Citation Reports database. He developed several dozens of reviews of articles in journals from the JCR list such as: Applied Sciences, Buildings, International Journal of Environmental Research and Public Health, Sustainability, etc. His research interests concern issues related to safety and health protection in construction processes, particularly the modeling of accidents in the construction industry, and an analysis of the causes of accidents in construction and construction project management, such as a cost analysis of construction projects, value engineering, and BIM technology. He has been Project Manager of the Wroclaw University of Science and Technology of the project Erasmus+ Strategic Partnerships for vocational educational and training. Innovation "SafeCRobot Virtual Reality Immersive Safety Training Environment for Robotised and Automated Construction Sites" 2020-1-UK01-KA202-079176, 2020-2022 and is a member of the research team of the project "Model of the assessment of risk of the occurrence of building catastrophes, accidents, and dangerous events at workplaces with the use of scaffolding" supported by Polish National Centre of Research and Development within Project PBS3/A02/19/2015. Since 2022 Member of the Academy of Young Scholars and Artists.

Marek Sawicki

Marek Sawicki is employed at the Department of Building Engineering, Faculty of Civil Engineering, at Wroclaw University of Science and Technology (WUST), Poland. He received his Master's degree in civil engineering from WUST in 1988, his Ph. (Hons) in Civil Engineering in July 1997. He is the author of more than 90 scientific papers, including works on building renovation, organization of works and work safety in the construction industry and articles in journals listed in the Journal Citation Reports database. He is a member of the Program Council and the Board of Reviewers of *Builder* journal. He has actively participated in two research grants on knowledge maps and scaffolding safety. He was a member and Chairman of the Supervisory Board of a housing cooperative from 2001 to 2022. His research interests include: issues regarding the recycling of construction materials, renovation of balconies in residential buildings, organization and technology of works, assembly and issues related to safety and health in construction processes, particularly the modeling of accident phenomena in construction, analysis of causes of accidents in construction, and management of construction projects with a cost analysis of construction projects.

Jarosław Konior

Jarosław Konior is employed at the Department of Building Engineering, Faculty of Civil Engineering, at Wroclaw University of Science and Technology (WUST), Poland. He received his master's degree in civil engineering from WUST in 1987, his Ph. (Hons) in Civil Engineering in 1997 and University Professor degree in April 2022. He is the author of more than 50 scientific papers, including works on buildings' technical depreciation and construction project management in conditions of uncertainty, and articles in journals listed in the Journal Citation

Reports database. Jarosław Konior has over 30 years of professional experience in Polish companies and multi-international, global enterprises providing a variety of services for investment process: technical consultancy and expertise, design and engineering, site supervision and co-ordination, project and construction management. His area of professional and scientific interests is construction project management and quality management in construction engineering and supervision, cost discount and financial control of construction projects, multi-criteria feasibility studies of investment enterprises, qualitative and quantitative research of engineering buildings by probabilistic and fuzzy sets approach, multidiscipline expertise investigations of non-typological building samples.

Preface to "Technology and Management Applied in Construction Engineering Projects"

Construction project management is a process that includes a number of operations, activities, and decisions that are closely related to an executed investment and that aim to create new, or increase, existing fixed assets in order to achieve utility effects. The utility effect of the construction process may be the construction of a new building, or the renovation or modernization of an existing building. In each construction process, according to the definition of the building object's life cycle, the following four basic phases are distinguished: the programming/planning phase, the implementation phase, the operation/use/maintenance phase, and the phase of decommissioning or demolition. Appropriate planning of the entire construction process is a very important operation, which has a direct impact on whether success is achieved when implementing an investment project.

The construction industry is characterized by the high complexity of implemented construction processes, variability of implementation conditions, diversity of facilities, applied technologies, and methods of work organization. The execution of construction projects is specific and particularly difficult because each implementation is a unique, complex, and dynamic process that consists of several or more subprocesses that are related to each other, in which various participants of the investment process participate.

Therefore, there is still a vital need to study, and research engineering technology and management, as applied in construction projects.

Mariusz Szóstak, Marek Sawicki, and Jarosław Konior
Editors

Editorial

Technology and Management Applied in Construction Engineering Projects

Mariusz Szóstak *, Jarosław Konior * and Marek Sawicki

Department of Building Engineering, Faculty of Civil Engineering, Wrocław University of Science and Technology, 50-370 Wrocław, Poland
* Correspondence: mariusz.szostak@pwr.edu.pl (M.S.); jaroslaw.konior@pwr.edu.pl (J.K.)

Abstract: The current Special Issue is a digest of 13 published articles that referred to the following scientific and professional areas: construction project management and quality management in construction engineering and supervision; cost discount and the financial control of construction projects; multi-criteria feasibility studies of investment enterprises; the qualitative and quantitative research of engineering buildings by probabilistic and fuzzy sets approach; multidiscipline expertise investigations of buildings that significantly differ in structure and use.

Keywords: management; project cost; investment schedule; risk mitigation; uncertainty; randomness; fuzziness; health and safety control

Citation: Szóstak, M.; Konior, J.; Sawicki, M. Technology and Management Applied in Construction Engineering Projects. *Appl. Sci.* **2022**, *12*, 11823. https://doi.org/10.3390/app122211823

Received: 19 October 2022
Accepted: 17 November 2022
Published: 21 November 2022

Publisher's Note: MDPI stays neutral with regard to jurisdictional claims in published maps and institutional affiliations.

Copyright: © 2022 by the authors. Licensee MDPI, Basel, Switzerland. This article is an open access article distributed under the terms and conditions of the Creative Commons Attribution (CC BY) license (https://creativecommons.org/licenses/by/4.0/).

1. Introduction

Construction project management is a process that includes several operations, activities, and decisions that are closely related to carried-out enterprises, which aim to increase existing or create new fixed assets to achieve utility effects. The utility effect of the construction process may be the construction of a new building or the renovation or modernization of an existing building. In each construction process, according to the definition of the building's life cycle, the following four basic phases are distinguished: the programming/planning phase, the implementation phase, the operation/use/maintenance phase, and the decommissioning or demolition phase. The appropriate planning of the entire construction process is an important operation that has a direct impact on the success achieved while implementing an investment project.

The construction industry is characterized by a high complexity of implemented construction processes, variability of implementation conditions, and diversity of facilities, applied technologies, and methods of work organization. The execution of construction projects is specific and difficult because each implementation is a unique, complex, and dynamic process that consists of several subprocesses related to each other in which various participants of the investment process take part.

Therefore, there is still a vital need to study, research, and conclude engineering technology and management applied in construction projects.

2. Contributions

The current Special Issue includes 13 research articles, presented in Table 1. A total of 35 authors or co-authors took part. The authors originate from seven countries: Canada, Chile, China, Croatia, Poland, Romania, and Spain. Each of the published papers represents valuable and novel research works on state-of-the-art applications.

Table 1. The content of the Special Issue "Technology and Management Applied in Construction Engineering Projects".

No.	Author	Country	Title	Keywords
1	Daniel Przywara, Adam Rak [1]	Poland	Monitoring of Time and Cost Variances of Schedule Using Simple Earned Value Method Indicators	earned value method—EVM; time variances; cost variances; schedule
2	Xun Liu, Le Shen, Kun Zhang [2]	China	Estimating the Probability Distribution of Construction Project Completion Times Based on Drum-Buffer-Rope Theory	construction project; PERT; theory of constraint (TOC); drum-buffer-rope (DBR); construction schedule
3	Hubert Anysz, Jerzy Rosłon, Andrzej Foremny [3]	Poland	7-Score Function for Assessing the Strength of Association Rules Applied for Construction Risk Quantifying	association analysis; tabu search; delay; risk; construction project
4	Tadeusz Kasprowicz, Anna Starczyk-Kołbyk, Robert Wójcik [4]	Poland	Randomized Estimation of the Net Present Value of a Residential Housing Development	efficiency; risk; randomization; construction project
5	Janusz Sobieraj, Dominik Metelski [5]	Poland, Spain	Project Risk in the Context of Construction Schedules—Combined Monte Carlo Simulation and Time at Risk (TaR) Approach: Insights from the Fort Bema Housing Estate Complex	time schedules; project risk; construction project management; time-at-risk (TaR); investment-construction process model; Monte Carlo simulation
6	Martina Milat, Snježana Knezić, Jelena Sedlar [6]	Croatia	Resilient Scheduling as a Response to Uncertainty in Construction Projects	resilience; baseline schedule; uncertainty; taxonomy; construction project
7	Parinaz Jafari, Malak Al Hattab, Emad Mohamed, Simaan AbouRizk [7]	Canada	Automated Extraction and Time-Cost Prediction of Contractual Reporting Requirements in Construction Using Natural Language Processing and Simulation	construction reports; construction contracts; natural language processing; machine learning; simulation modeling
8	Luz Duarte-Vidal, Rodrigo F. Herrera, Edison Atencio, Felipe Muñoz-La Rivera [8]	Chile, Spain	Interoperability of Digital Tools for the Monitoring and Control of Construction Projects	monitoring progress; construction phase; automated monitoring; digital tools; as-built; as-planned
9	Edison Atencio, Pablo Araya, Francisco Oyarce, Rodrigo F. Herrera, Felipe Muñoz-La Rivera, Fidel Lozano-Galant [9]	Chile, Spain	Towards the Integration and Automation of the Design Process for Domestic Drinking-Water and Sewerage Systems with BIM	building information modelling (BIM); automatization; facilities design; domestic plumbing and sanitation

Table 1. Cont.

No.	Author	Country	Title	Keywords
10	Paul Sestras [10]	Romania	Applied Construction Layout Plan with Batter Boards Stake-Out Methods Comparison: A Case Study of Romania	land surveyor; construction surveying; building layout; polar coordinates; stake-out methods; total station
11	Zian Xu, Minshui Huang [11]	China	Improving Bridge Expansion and Contraction Installation Replacement Decision System Using Hybrid Chaotic Whale Optimization Algorithm	bridge expansion and contraction installation (BECI); decision making (DM); technical condition assessment; analytic hierarchy process (AHP); whale optimization algorithm; tent chaotic mapping; Lévy flight
12	Elżbieta Szafranko, Jolanta Harasymiuk [12]	Poland	Modelling of Decision Processes in Construction Activity	decision-making process; decision modelling in construction activities; decisions in civil engineering
13	Yingije Zhao, Fan Yang, Yijiang Ma [13]	China	Experimental Method for Flow Calibration of the Aircraft Liquid Cooling System	liquid cooling system; flow calibration; differential pressure; experimental method; aircraft

Appropriate planning and effective control with the constant monitoring of construction projects are important for the successful execution of a project within planned initial conditions. Budget planning at the investment preparation stage, as well as the control of cash flows during project implementation, is of key importance for investors, project managers, and construction work contractors [14,15].

The paper of Przywara and Rak [1] attempts to analyze the emerging time and cost deviations using proprietary time variances from the schedule and variances from planned costs based on simple indicators of the Earned Value Method. The construction of a multi-family housing development was used as an example to study the variances between planned and incurred costs.

In turn, the article of Liu et al. [2] applied the "drum-buffer-rope" construction schedule management and control technology into a PERT network to improve the relationship among the activities. The research described that the method derived from the theory of constraints (TOC) attempts to enhance the couplings among tasks to revise and further reduce the uncertainty of construction activities which leads to improving the reliability of project progression. The elements of drum, buffer, and rope (DBR) in TOC are added to the PERT network schedule. To illustrate the impact of DBR applications on improving project schedule reliability, a case of a hydropower station is used as an example to show the enhanced reliability of scheduling.

The planning and implementation of construction projects are difficult processes and are burdened with many risk elements. Over the years, many approaches to construction project risk management have been developed by various researchers.

In the paper of Anysz et al. [3], the 7-Score Function for Assessing the Strength of Association Rules Applied for Construction Risk Quantifying was invented. Based on the 7-score function, the most powerful and the most informative rules can be found, which allows their importance to be ranked. It crucially introduced an innovative method of quantitative risk assessment, not based on the experts' opinions but rather on evidence concerning collected and completed construction contracts of the same kind.

The objective of the research conducted by Kasprowicz et al. [4] was to verify the correct probabilistic method for the analysis and assessment of the Net Present Value of unstable construction projects. The results of the real building investments by the randomized method of the estimation of construction projects' efficiency were presented.

The article of Sobieraj and Metelski [5] proposed a framework for quantifying the risk of time variation in a project based on Monte Carlo simulation and probabilistic Time-at-risk analysis. This is an approach to explicitly quantify the uncertainty in the duration of the whole project as well as its individual stages. The possibilities of the proposed approach are explained using a simple example of the construction of a housing estate in Warsaw-Bemowo, which was carried out in the period 1999–2012.

The aim of the paper of Milat et al. [6] was to structure sources of uncertainty related to complex construction projects. The main objective of the research was to propose a comprehensive mathematical model for resilient scheduling as a trade-off between project robustness, project duration, and contractor profit. The proposed optimization problem is illustrated in the example project network, along which the probabilistic simulation method that was used to validate the results of the scheduling process in uncertain conditions. According to the authors, the proposed resilient scheduling approach leads to more accurate forecasting; therefore, the project planning calculations are accepted with increased confidence levels.

Some problems in the construction industry involve communication and reporting procedures. Construction reports are often required by clients as a way of monitoring project progress, estimating production rates, and resolving disputes and claims. Ineffective reporting systems lead to poor project management. To prevent such drawbacks, Jafari et al. [7] developed an automated reporting requirement identification and time–cost prediction framework.

Monitoring the progress of a construction site during the construction phase is essential. An inadequate understanding of the project status can lead to mistakes and inappropriate actions, causing delays and increased costs. The manuscript of Duarte-Vidal et al. [8] presented a bibliographic synthesis and interpretation of 30 nonconventional digital tools for monitoring a construction project's progress that was achieved by the means of building information modeling (BIM), unmanned aerial vehicles (UAVs), and photogrammetry.

According to Atencio et al. [9], the use of building information modelling (BIM) in construction projects is expanding, and its usability throughout building lifecycles, from planning and construction to operation and maintenance, is gaining increasing evidence.

The issue of planning also concerns the efficient layout planning of a construction site. According to Sestras [10], this is a fundamental task for any project undertaking. It is each survey engineer's responsibility to guide the builders and conduct an accurate, safe, and time- and cost-efficient layout of the designed structure.

Due to the complexity of situations in which construction activity decisions are made, the decision-making process can be supported by different mathematical methods, systems, and models. Xu and Huang [11] proposed the multi-criteria model BECI technical condition assessment approach, which contains specific on-site inspection regulations with both qualitative and quantitative variables. The hybrid chaotic whale optimization algorithm (WOA) was designed and utilized to improve and automate the process of optimal replacement plan selection with the assistance of the analytic hierarchy process (AHP).

The article of Szafranko and Harasymiuk [12] includes descriptions of decision support methods and models, including single-criteria and multi-criteria models, operations research, and fuzzy models. The article contains an analysis of the model approaches proposed in the literature confronted with decision-making processes in engineering practice. The study covered 34 construction projects and 15 companies operating in the construction industry. Several decision situations have been considered. The observations obtained during the research helped the authors to develop the decision support models dedicated to engineering practice while managing construction projects.

3. Conclusions

The presented unanimous research approach is a result of many years of studies conducted by 35 well-experienced authors. The common subject of research concerned the development of methods and tools for modelling multicriteria processes in construction engineering. Real-time decision-making seems to be well-recognized as a deterministic cause–effect event and is supported by many programs/applications in everyday engineering and managerial work. A much more interesting challenge is the analysis of past states, the effects of which are observed in the present, and also the current modelling of continuous processes, the effects of which will be visible in the future. When using two-valued logic, in which the premise results in an arbitrary conclusion, a decision is determined by the inference model. When such logic is insufficient, it has been noticed that decisions cease to be deterministic and instead become multivariate. Actions are also undertaken in a multi-variant way and assume the tracking of circumstances and a possible change during the implementation of the decision. There is no doubt that real situations are more complicated and they, therefore, require a more complex decision-making management process, as presented in the published works.

Author Contributions: Writing—original draft preparation, M.S. (Mariusz Szóstak); writing—review and editing, J.K. and M.S. (Marek Sawicki). All authors have read and agreed to the published version of the manuscript.

Funding: This research received no external funding.

Institutional Review Board Statement: Not applicable.

Informed Consent Statement: Not applicable.

Data Availability Statement: Not applicable.

Acknowledgments: The authors express gratitude to the *Applied Sciences* journal for offering an academic platform for researchers where they can contribute and exchange their recent findings in engineering.

Conflicts of Interest: The authors declare no conflict of interest.

References

1. Przywara, D.; Rak, A. Monitoring of Time and Cost Variances of Schedule Using Simple Earned Value Method Indicators. *Appl. Sci.* **2022**, *12*, 1357. [CrossRef]
2. Liu, X.; Shen, L.; Zhang, K. Estimating the Probability Distribution of Construction Project Completion Times Based on Drum-Buffer-Rope Theory. *Appl. Sci.* **2022**, *12*, 7150. [CrossRef]
3. Anysz, H.; Rosłon, J.; Foremny, A. 7-Score Function for Assessing the Strength of Association Rules Applied for Construction Risk Quantifying. *Appl. Sci.* **2022**, *12*, 844. [CrossRef]
4. Kasprowicz, T.; Starczyk-kołbyk, A.; Wójcik, R. Randomized Estimation of the Net Present Value of a Residential Housing Development. *Appl. Sci.* **2022**, *12*, 124. [CrossRef]
5. Sobieraj, J.; Metelski, D. Project Risk in the Context of Construction Schedules-Combined Monte Carlo Simulation and Time at Risk (TaR) Approach: Insights from the Fort Bema Housing Estate Complex. *Appl. Sci.* **2022**, *12*, 1044. [CrossRef]
6. Milat, M.; Knezić, S.; Sedlar, J. Resilient Scheduling as a Response to Uncertainty in Construction Projects. *Appl. Sci.* **2022**, *12*, 6493. [CrossRef]
7. Jafari, P.; Al Hattab, M.; Mohamed, E.; Abourizk, S. Automated Extraction and Time-Cost Prediction of Contractual Reporting Requirements in Construction Using Natural Language Processing and Simulation. *Appl. Sci.* **2022**, *12*, 6188. [CrossRef]
8. Duarte-Vidal, L.; Herrera, R.F.; Atencio, E.; Rivera, F.M.-L. Interoperability of Digital Tools for the Monitoring and Control of Construction Projects. *Appl. Sci.* **2022**, *12*, 10370. [CrossRef]
9. Atencio, E.; Araya, P.; Oyarce, F.; Herrera, R.F.; Rivera, F.M.-L.; Lozano-Galant, F. Towards the Integration and Automation of the Design Process for Domestic Drinking-Water and Sewerage Systems with BIM. *Appl. Sci.* **2022**, *12*, 9063. [CrossRef]
10. Sestras, P. Methodological and On-Site Applied Construction Layout Plan with Batter Boards Stake-Out Methods Comparison: A Case Study of Romania. *Appl. Sci.* **2022**, *12*, 4331. [CrossRef]
11. Xu, Z.; Huang, M. Improving Bridge Expansion and Contraction Installation Replacement Decision System Using Hybrid Chaotic Whale Optimization Algorithm. *Appl. Sci.* **2022**, *12*, 6222. [CrossRef]
12. Szafranko, E.; Harasymiuk, J. Modelling of Decision Processes in Construction Activity. *Appl. Sci.* **2022**, *12*, 3797. [CrossRef]
13. Zhao, Y.; Yang, F.; Ma, Y. Experimental Method for Flow Calibration of the Aircraft Liquid Cooling System. *Appl. Sci.* **2022**, *12*, 5056. [CrossRef]
14. Konior, J.; Szóstak, M. Methodology of Planning the Course of the Cumulative Cost Curve in Construction Projects. *Sustainability* **2020**, *12*, 2347. [CrossRef]
15. Konior, J.; Szóstak, M. The S-Curve as a Tool for Planning and Controlling of Construction Process-Case Study. *Appl. Sci.* **2020**, *10*, 2071. [CrossRef]

Article

Monitoring of Time and Cost Variances of Schedule Using Simple Earned Value Method Indicators

Daniel Przywara * and Adam Rak *

Department of Civil Engineering and Construction Processes, Faculty of Civil Engineering and Architecture, Opole University of Technology, 45-063 Opole, Poland
* Correspondence: d.przywara@po.edu.pl (D.P.); a.rak@po.edu.pl (A.R.)

Abstract: The Planning and implementation of construction projects are difficult processes and are burdened with many risk elements. The budget spread over time, which is developed on the basis of the schedule, presents the expected distribution of costs throughout the duration of the works, which during the implementation of the project is subject to constant changes resulting from time, cost, and organizational factors. Managing construction contracts requires managers to be able to analyze on an ongoing basis the variances of production costs-from the values calculated in the offer cost estimate and assumed in the Budgeted Cost of Work Scheduled. The article attempts to analyze the emerging time and cost deviations using proprietary time variances from the schedule (T/S) and variances from planned costs (T/C) monitoring, based on simple indicators of the earned value method (EVM). An example of construction of a multi-family housing development was used to study the variances of planned and incurred costs.

Keywords: earned value method—EVM; time variances; cost variances; schedule

---

Citation: Przywara, D.; Rak, A. Monitoring of Time and Cost Variances of Schedule Using Simple Earned Value Method Indicators. *Appl. Sci.* **2021**, *11*, 1357. https://doi.org/10.3390/app11041357

Academic Editor: Mariusz Szóstak
Received: 23 December 2020
Accepted: 28 January 2021
Published: 3 February 2021

Publisher's Note: MDPI stays neutral with regard to jurisdictional claims in published maps and institutional affiliations.

Copyright: © 2021 by the authors. Licensee MDPI, Basel, Switzerland. This article is an open access article distributed under the terms and conditions of the Creative Commons Attribution (CC BY) license (https://creativecommons.org/licenses/by/4.0/).

1. Introduction

The growing number and growing importance of unique, often complex construction projects, results in unflagging interest in project management. Despite the undisputed, more significant than ever, development of effective methods of planning, coordinating, and controlling, the increase in the complexity of operating conditions of the enterprise on the market means that success in project management is not easier. The selection of production control methods depends on many factors-primarily the degree of production repeatability, as well as the degree of details of available data, the details of the records kept, the use of planning documentation, the degree of use of computer techniques and the organizational culture prevailing in the company. Construction production is characterized with significant seasonal fluctuations. They constitute a cycle of repetitive changes, in more or less equal periods of time, with similar intensity. Their identification and inclusion considerably increases the precision of predictions [1–3].

Earned Value Method (EVM) is recognized as an advanced method of controlling production ventures, which provides working results in the form of quantitative and qualitative indicators.

The purpose of developing the EVM method was to link the material progress of works and the costs incurred-against the background of planned values. This tool, "introduced" to construction sites, has been effectively implemented and disseminated. Managers managing complex construction contracts use key words to assess the progress of their schedules, fully describing planned and actual turnover, budget, and timeliness of implementation [4,5]. These include simple control indicators of the earned value method, used in the strategic assessment of the condition of the project-when monitoring the progress of works using the indicator method [6–9]. The EVM method is one of performance-based project management methods and is an effective tool for controlling projects in terms of costs, time, and scope of works done [10]. It consists of controlling the implementation of the project by comparing the scope of works executed and the costs actually incurred against the project schedule and budget adopted in the base plan.

2. Materials and Methods: Description of the Scheduled Examined

Compared to the traditional method of checking the project progress, the EVM method includes the third dimension: earned value, which represents the planned value of the scope of work actually executed, in the assessment of the condition of the project, in addition to planned and actually incurred costs. By developing this method, not only measurement possibilities and assessment of the actual results of operations based on the data found during control periods were obtained, but also analysis of performance trends and forecasts of the future cost of project activities and projections of the final project budget. Appropriate indicators allow at each stage of the implementation of works to assess the current status of the project and the possibility of its implementation within the planned budget and schedule. The indicators used in the EVM method can be calculated for both individual tasks and the entire project, as cumulative earned value [11–16].

The basis of the earned value method is the Budgeted Cost of Work Performed (BCWP). It shows how much the work done as planned costs. If one wants to calculate it, they should have information on the planned cost of work to be incurred by the date of the inspection, and on the amount of work actually executed to that day. The essence of the earned value method is shown in Figure 1. The curve illustrating the course of BCWS- "Budgeted Cost of Work Scheduled" (planned costs) is determined during the project planning phase, its final value is BAC—"Budget at Completion" [17]. Curves illustrating the course of BCWP—"Budgeted Cost of Work Performed" (earned value) and ACWP—"Actual Cost of Work Planned" (costs incurred), are determined during the implementation of the works and can be determined only until the date of inspection.

The course of the curves in Figure 1 also allows to see the variances from the budget and the schedule: both on the day of the inspection and forecast values on the day of completing the project. Data from the BCWS, BCWP, and ACWP curves are the basis for calculating further indicators, to which in the EVM method belong: [10]

- Cost Variance (CV):

$$CV = BCWP - ACWP \qquad (1)$$

- Schedule variance (SV):

$$SV = BCWP - BCWS \qquad (2)$$

where BCWS—planned cost of the planned work until the day of inspection of the quantity of work of a given task, ACWP—actual cost of the work executed up to the date of inspection of the quantity of work of a given task, BCWP—the planned cost of the work executed to the day of work inspection of this task; value checking how much, according to plan, was paid for the work actually executed.

In the study [18], searching for reasons for delayed projects, monotonicity analyzes of the indicator corresponding to these assumptions were carried out, proposing the variance parameter from booked expenditure (AV: Accounting Spending Variance).

This indicator is described in the form:

$$AV = CV - SV \qquad (3)$$

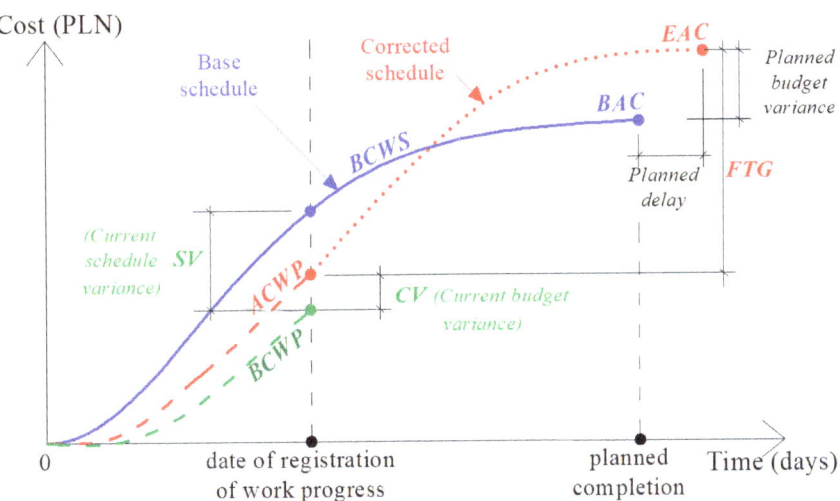

Figure 1. Elements of earned value method (EVM) (description in text). Source: [19].

Thus, we have

$$AV = (BCWP-ACWP)-(BCWP-BCWS) = \\ = BCWP-ACWP-BCWP+BCWS = BCWS-ACWP \quad (4)$$

The proposed AV parameter monitors financial liquidity, which is the difference between planned and actually incurred costs. In this concept, commonly known as cost management, workflow is defined as "data and material movement" through the network of production units.

Current control systems, including EVM, focus on analyzing the "speed" of cost and time increases, ignoring the problem of the flow of production resources, in which the authors see major failures in their implementation processes. These systems are an effective control tool, but they have one serious limitation, treating each schedule process as an independent activity, the effect of which does not translate into the implementation of subsequent processes. In order to demonstrate positive increases in cost variances (CVs), managers strive to level production costs (ACWP) at the expense of continuous increase in process efficiency, which causes unreliability of workflow streams [20].

The conclusion of the study stated that, further research is necessary to develop alternative systems to monitor the progress of projects. A detailed analysis of the time variance of the schedule was undertaken in the study [21–23], proposing a new look at the EVM method, marked as ES (Earned Schedule).

The basic assumption is a comparative analysis of the earned value ratio BCWP and the planned costs BCWS. The algorithm results are the values of time variances at individual moments of the project (t), reflecting the periods in which the equality was to take place: BCWP = BCWS, determined with the indicator of the earned schedule value (ES). This indicator is calculated as follows:

$$SV(t) = ES - AT \quad (5)$$

A graphic interpretation of the simple parameters of the ES method is shown in Figure 2.

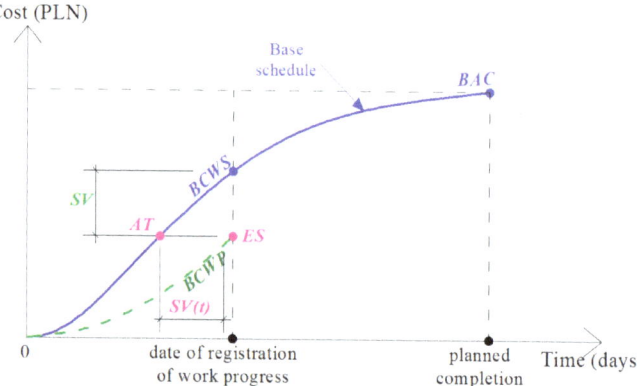

Figure 2. Graphic interpretation of simple parameters of the Earned Schedule (ES) method (description in the text). Source: [19]

In turn, in the study [24], the EVM analysis was extended to include "qualitative" indicators, proposing a qualitative technique in this method—(QEV: Quality Earned Value).

For this purpose, quality assessment parameters (QR: Quality Requirements) have been introduced. The methodology for calculating the QEV assumes formulas for assessing the quality of its components. Its purpose is to measure the project's ability to provide the assumed qualitative levels of the indicated tasks of the examined schedule. An additional research objective was an attempt to link the quality management process with monitoring of works progress. Data was collected on the day each schedule task was completed. The results prove that, taking into account the parameter of assumed quality levels reduces the projected earned value of the project.

In the study [17], in order to increase the accuracy of early forecasting of the Estimated At Completion (EAC) project cost of the EVM method, its analysis was extended with the introduction of an innovative parameter of estimation of non-linear final cost regression (CEAC: nonlinear Cost Estimate at Completion). The methodology introduces the concept of "interim schedule" (ES: Earned Schedule) and ensures that, the CEAC indicator is determined at each stage of project implementation. In the course of the algorithm implementation tests, the equations of time and cost models (logistic, Gompertz, Bass, and Weibull) are built, followed by the selection of one that best describes the implementation of the project, which is determined by the CEAC level.

The proposed methodology allows to determine in a more realistic way the parameter sought for the final cost of the project, by creating a wider field of analysis taking into account extensive mathematical models.

Another approach to expanding EVM analysis is the method of time-cost analysis described in the study [25], assessing the progress of works based on three additional assessment parameters: Early Start Rate (ESR), Early Completion Rate (ECR), and Cost Overrun Rate (COR).

These studies show that, the same orders, which are characterized with for example the purpose of the object, have an impact on their delivery time. A different approach to production monitoring has been proposed in the study [26] in the EDM (Earned Duration Management) method.

The role of this technique is to replace one of the tools of the EVM method, the schedule performance index (SPI) with its own EDI (Earned Duration Index) index in the form of an exponential function with a time argument. It has been shown that, this parameter gives more accurate results in comparison with the SPI. In addition to the purely informational model identifying the variances, the study also contains a method of solving the problem of their leveling, which can also be found in the study [27].

On the basis of simple indicators of the EVM method, a developed time-cost analysis is proposed [19], examining the impact of the project implementation time on the costs incurred and costs on the planned time-by introducing the following assessment parameters:

- Indicator for monitoring changes in the assumed financial liquidity due to arising time variances from the schedule (T/S):

$$T/S = \frac{TV}{SV} = \frac{BCWS - ACWP}{BCWP - BCWS} \qquad (6)$$

- Indicator for monitoring changes in assumed financial liquidity due to arising variances from planned costs (T/C):

$$T/C = \frac{TV}{CV} = \frac{BCWS - ACWP}{BCWP - ACWP} \qquad (7)$$

Their role is to expand the analysis with control indicators to analyze financial liquidity, by taking into account the costs and time variances in the schedule. Equation (6) extends the simple analysis of project liquidity by including the difference of the earned value indicator, corrected by the value of planned costs (BCWS), constituting its budget. In Equation (7), a simple analysis of financial liquidity (difference in the value of planned BCWS costs and costs incurred ACWP) has been expanded to include the planned cost of the actual work done (BCWP earned value index), reduced by the actual cost (ACWP).

In the mathematical structure of indicators, it is unacceptable to equate the values of the BCWS, ACWP, and BCWP parameters-at which the equality of the described tools indicate the implementation of works in accordance with the assumed time and cost. These cases do not require taking remedial actions to improve the monitored schedule. The following tests were performed to calculate the assessment parameters entered for the work schedule model, assuming three basic implementation structures—steady, parallel, and sequence execution methods.

To carry out the analysis of time and cost variances, the schedule of the implemented construction project, consisting of the construction of a multi-family housing development, planned for a period of seven months, with a budget of over PLN 20 million, was used. Figure 3 shows the obtained three scenarios for the implementation of works. The allocation of the means of production was planned using three basic structures of project implementation-methods of steady, parallel, and sequence execution.

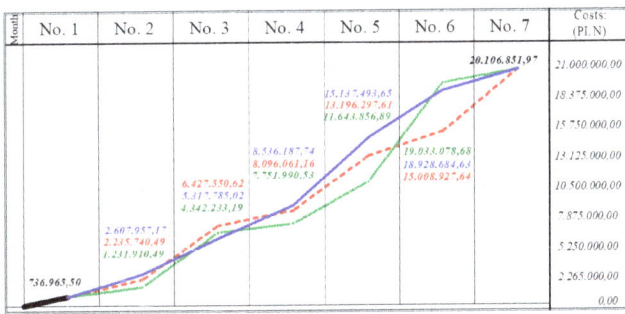

Figure 3. Cumulative values of planned costs (BCWS) of schedules-in implementation methods according to steady (dot chart), parallel (continuous chart), sequence (intermittent chart) execution of works. Source: [19].

3. Results: Application of T/S and T/C Monitoring

T/S and T/C monitoring of variances arising during the implementation of the works covered by the schedule, was divided into three scenarios of its implementation, according to the methods adopted above. The adopted calculation models are characterized by a 10%

increase in the budget of costs (C = 110%, 120%, 130%, and 140%) and a 5% increase in the time of works execution (S = 105%, 110%, 115%, and 120%), respectively. These models are described with symbols A1B1 ÷ A4B4, indicating the increase in variances [8]. For T/S monitoring, the impact of the increasing delay parameter (A models) was adopted, while for T/C monitoring the impact of delays and cost overruns combined (AB models).

The steady execution method assumes a global (overall) impact of variances (A, B)-for all schedule processes. In the methods of parallel and sequence work execution, the local (partial) impact of variances (a, b) was adopted-on selected process sequences.

This is due to the obvious characteristics of these methods, widely described in the literature (delays of the analyzed process sequences do not always affect the delays of their successors).

The analysis of the growth rate of unplanned costs and the work performance deviating from the plan consists in a gradual increase in expenditure on production (C), with a simultaneous systematic reduction of the production rate (S). These phenomena are reflected in the lowering of the value of simple indicators of the earned value method: pool of actual costs (ACWP) and earned value index (BCWP), which determine the monotonicity of the T/S and T/C monitoring indicators.

The results of the analysis for the parallel execution method, with a description of the introduced time and cost disturbances and the division into four variances scenarios, are presented in Table 1.

Table 1. Schedule models according to the method of parallel execution. Source: [19]. Where: BCWP—budget cost of work performed, BCWS—budget cost of work scheduled, ACWP—actual cost of work performed, T/C—Indicator for monitoring changes in assumed financial liquidity due to arising variances from planned costs, T/S—Indicator for monitoring changes in the assumed financial liquidity due to arising time variances from the schedule

Registration Stage/Variance Scenario	BCWP BCWS ACWP (PLN)	BCWP-BCWS (PLN)	BCWP-ACWP (PLN)	T/C T/S (−)
Models a1(II): s = 100%, c = 110% (for T/S)///a1b1(II): s = 105%, c = 110% (for T/C)				
31.03.2017	736.965,50 736.965,50 736.965,50	0.00	0.00	0.000 0.000
30.04.2017	2.578.038,90 2.607.957,17 2.675.075,67	−29.918,27	−97.036,77	0.647 −27.65
31.05.2017	5.249.292,03 5.317.785,02 5.457.917,12	−68.492,99	−208.625,09	0.672 −42.39
30.06.2017	8.433.684,65 8.536.187,74 8.744.085,60	−102.503,09	−310.400,95	0.661 −99.59
31.07.2017	14.919.855,83 15.137.493,65 15.562.631,87	−217.637,82	−642.776,04	0.661 125.81
31.08.2017	18.636.591,14 18.928.684,63 19.362.526,82	−292.093,49	−725.935,68	0.653 −112.89
30.09.2017	19.785.464,08 20.106.851,97 20.749.627,75	−321.387,89	−964.163,67	0.667 0.00

Table 1. *Cont.*

Registration Stage/Variance Scenario	BCWP BCWS ACWP (PLN)	BCWP-BCWS (PLN)	BCWP-ACWP (PLN)	T/C T/S (−)
Models a2(II): s = 100%, c = 120% (for T/S)///a2b2(II): s = 110%, c = 120% (for T/C)				
31.03.2017	736.965,50 736.965,50 736.965,50	0.00	0.00	0.000 0.00
30.04.2017	2.545.693,31 2.607.957,17 2.733.226,91	−62.263,86	−187.533,60	0.647 −51.61
31.05.2017	5.177.570,36 5.317.785,02 5.591.548,02	−140.214,66	−413.977,66	0.669 −81.75
30.06.2017	8.315.325,53 8.536.187,74 8.949.928,73	−220.862,21	−634.603,20	0.675 −198.17
31.07.2017	14.705.597,15 15.137.493,65 15.991.149,23	−431.896,50	−1.285.552,08	0.664 252.62
31.08.2017	18.341.804,13 18.928.684,63 19.952.100,82	−586.880,50	−1.610.296,69	0.631 −331.78
30.09.2017	19.464.076,19 20.106.851,97 21.392.403,53	−642.775,78	−1.928.327,34	0.667 0.00
Models a3(II): s = 100%, c = 130% (for T/S)///a3b3(II): s = 115%, c = 130% (for T/C)				
31.03.2017	736.965,50 736.965,50 736.965,50	0.00	0.00	0.000 0.00
30.04.2017	2.505.066,94 2.607.957,17 2.818.980,75	−102.890,23	−313.913,81	0.672 −75.57
31.05.2017	5.089.287,12 5.317.785,02 5.790.206,78	−228.497,90	−700.919,66	0.665 −123.14
30.06.2017	8.196.966,41 8.536.187,74 9.294.411,49	−339.221,33	−1.097.445,08	0.689 −296.76
31.07.2017	14.494.399,50 15.137.493,65 16.419.666,59	−643.094,15	−1.925.267,09	0.666 379.44
31.08.2017	18.050.078,17 18.928.684,63 20.541.674,82	−878.606,46	−2.491.596,65	0.611 −657.66
30.09.2017	19.145.749,34 20.106.851,97 22.035.179,31	−961.102,63	−2.889.429,97	0.667 0.00

Table 1. Cont.

Registration Stage/Variance Scenario	BCWP BCWS ACWP (PLN)	BCWP-BCWS (PLN)	BCWP-ACWP (PLN)	T/C T/S (−)
Models a4(II): s = 100%, c = 140% (for T/S)///a4b4(II): s = 120%, c = 140% (for T/C)				
31.03.2017	736.965,50 736.965,50 736.965,50	0.00	0.00	0.000 0.00
30.04.2017	2.464.440,57 2.607.957,17 2.882.652,52	−143.516,60	−418.211,95	0.692 −99.52
31.05.2017	5.010.408,67 5.317.785,02 5.934.883,83	−307.376,35	−924.475,16	0.662 −162.50
30.06.2017	8.086.375,80 8.536.187,74 9.537.674,79	−449.811,94	−1.451.298,99	0.695 −395.35
31.07.2017	14.285.360,56 15.137.493,65 16.848.183,95	−852.133,09	−2.562.823,39	0.668 506.25
31.08.2017	17.760.510,91 18.928.684,63 21.131.248,83	−1.168.173,72	−3.370.737,92	0.598 −876.55
30.09.2017	18.829.581,19 20.106.851,97 22.677.955,09	−1.277.270,78	−3.848.373,90	0.667 0.00

Similar calculations were carried out for the steady execution method and the sequence work execution. The picture of emerging disturbances is presented in Figure 4 (monitoring of variances from the assumed time T/S) and Figure 5 (monitoring of variances from the assumed costs T/C).

Monitoring of variances from the assumed duration (T/S) in the schedule planned according to the assumptions of the steady execution method shows the highest values in the fifth control stage.

At this stage, the highest production throughput occurs, reaching 15% of the total costs (Figure 3).

The highest T/S values were recorded in this group of scenarios, which results from cumulative variances related to the production stream of works.

Monitoring (T/S) in the schedule made according to the assumptions of the parallel performance method shows a typical increase in the analyzed values, together with the increase of the stimulus value, in subsequent scenarios. The highest positive T/S monitoring values were also obtained for the fifth stage, in which activities with the largest budget were carried out. During the sixth inspection period, the lowest negative variance levels were recorded to the assumed time, illustrating the losses of the previous, fifth stage.

Monitoring of time variance of the schedule (T/S) in the sequence execution method-subject to the influence of local, gradually increasing disturbances, illustrates the planned course of works until the end of the third month. Then the amplitude of fluctuations in the results reflects the modeled disturbances: twice changing the sign-as in the case of the parallel execution method, although in this case the T/S tool recorded jumps of these values five and ten times higher.

It results from the specifics and assumptions of the analyzed implementation structure of projects. In all cases, the assumed methods of implementation accepted within them, four scenarios of gradual increase in variances illustrate rational, systematic increases in value, demonstrating the proper operation of the tested computational procedure of T/S monitoring (Figure 4).

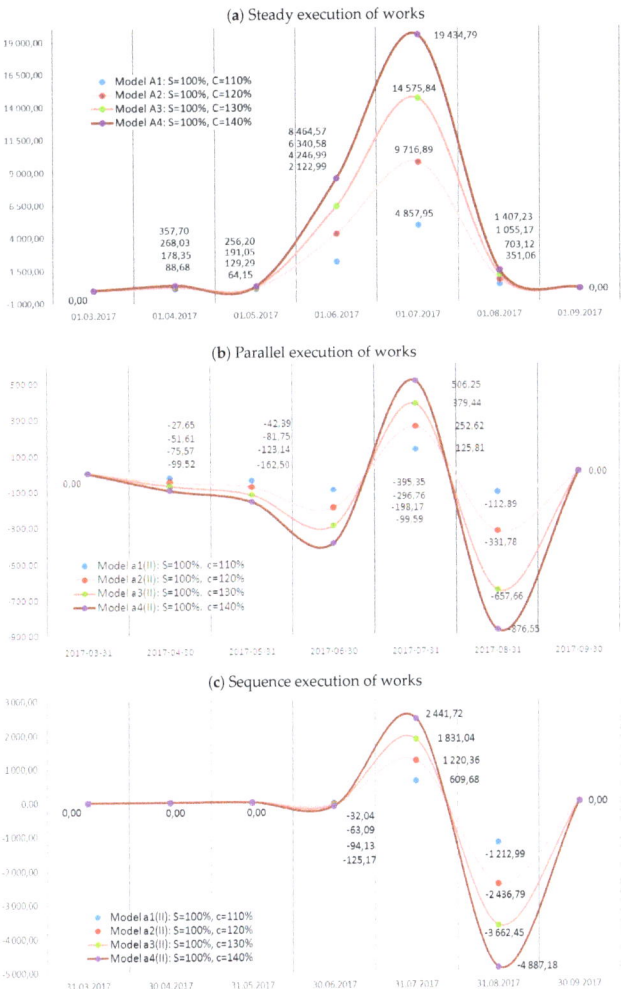

Figure 4. Monitoring indicator charts in the assumed financial liquidity due to the occurring time variances from the schedule (T/S) in models with a gradually increasing time variance-in the implementation methods according to (from above): steady, parallel, sequence execution of works. Source: [19].

In the statement of the second and third stage of registration of works in the schedules planned with the methods of parallel and sequence execution, a clear T/S response to previously modeled variances in the structure of parallel execution of works can be seen.

Monitoring of variances from the assumed costs of the project (T/C) in the steady execution method illustrates, unlike the other two methods, values greater than zero at all stages of recording the progress of works. This is due to the global nature of modeled time and cost distortions in this method.

Similar to T/S monitoring, the largest value amplitudes were noted in the fifth and the sixth stage of registration, in which works with the highest budgets were executed.

The monitoring of cost variances (T/C) in the method of parallel execution records a zero value in the first stage of registration, in which no interference occurs. Then, the T/C monitoring values remain at 60–70% until the end of the project-the disturbances do not accumulate. Contrary to the above conclusion, in the next method-the sequence execution, T/C monitoring presents overlapping losses in sequence recording stages (resulting from modeled disturbances), starting from the fourth control stage. It results from the basic assumption of this method, in which the time delays of the analyzed sequence of processes affect the delays of its successors. In the first three stages of registration, T/C monitoring records zero values because no distortions were modeled in them.

As in the case of T/S monitoring, the definition of four scenarios of gradual increase in interference gives the systematic monotonicity of the graphs of their functions in the project implementation calendar, in each of the structures analyzed (Figure 5).

Figure 5. Monitoring indicator charts in the assumed financial liquidity due to arising variances from planned costs (T/C) in models with a gradually increasing time and cost variance-in the implementation methods according to (from above): steady, parallel, sequence execution of works. Source: [19].

4. Conclusions

The role of additional tools of the EVM method-T/S and T/C monitoring is to take into account the financial liquidity of the project in the analysis of time and cost variances of the assumed schedule. These tools are an important complement to the workshop of the earned value method, fully compatible with its simple indicators.

Under the influence of the budget's underestimation and time slips, the planned cost and time distribution changes, expressed in the course of the financial liquidity (TV) parameter. Planned cost parameter (BCWS), depending on the situation in the project monitoring, should therefore be flexible-already at the budgeting stage.

In the case of modeling total (global) time and cost variances, the largest values from the T/S and T/C monitoring calculations were recorded for periods containing the "longest" and "most expensive" schedule activities. Subjecting the schedule to partial (local) time and cost variances eliminated these values-while maintaining the proportion of their results-in relation to total (global) distortions.

In practice of management of long-term construction projects, these tools may contribute to the identification of variances from the assumed schedules, describing-in contrast to the classical EVM method-their actual values.

Author Contributions: A.R. designed the study. D.P. performed the data analysis and contributed to data interpretation, writing of the manuscript. D.P. contributed to data collection, data interpretation, editing of text, and performed the statistical analysis. D.P. and A.R. drafted the manuscript. D.P. and A.R. participated in the interpretation of the results. All authors have read and agreed to the published version of the manuscript.

Funding: The research was prepared by the authors without the support of external institutions.

Institutional Review Board Statement: The research was approved by the Dean of the Faculty of Civil Engineering and Architecture of the Opole University of Technology, and informed consent was obtained before data collection.

Informed Consent Statement: Informed consent was obtained before data collection.

Data Availability Statement: Data available on request due to restrictions eg privacy or ethical.

Acknowledgments: We thank the Dean of the Faculty of Civil Engineering and Architecture of the Opole University of Technology for supporting our research.

Conflicts of Interest: The authors declare no conflict of interest.

References

1. Czyżewski, A. *Economic Analysis in the Implementation of Investment Projects*; Publishing house of the Poznań University of Economics: Poznań, Poland, 2011.
2. Project Management Institute. *A Guide to the Project Management Body of Knowledge (PMBOK GUIDE)*, 6th ed.; Project Management Institute (PMI): Newtown Square, PA, USA, 2017.
3. Połoński, M. *Management of Construction Investment Process*; Wydawnictwo SGGW: Warszawa, Poland, 2018. (In Polish)
4. Konior, J.; Szóstak, M. The S-Curve as a Tool for Planning and Controlling of Construction Process—Case Study. *Appl. Sci.* **2020**, *10*, 2071. [CrossRef]
5. Konior, J.; Szóstak, M. Methodology of Planning the Course of the Cumulative Cost Curve in Construction Projects. *Sustainability* **2020**, *12*, 2347. [CrossRef]
6. Al-Jibouri, S.H. Monitoring systems and their effectiveness for project cost control in construction. *Int. J. Proj. Manag.* **2003**, *21*, 145–154. [CrossRef]
7. Lo, W.; Chen, Y.-T. Optimization of Contractor's S-Curve. In Proceedings of the 24th International Symposium on Automation & Robotics in Construction (ISARC 2007), Kochi, India, 19–21 September 2007; pp. 417–420. Available online: https://www.irbnet.de/daten/iconda/CIB11259.pdf (accessed on 18 March 2020).
8. Hsieh, T.-Y.; Hsiao-Lung Wang, M.; Chen, C.-W. A Case Study of S-Curve Regression Method to Project Control of Construction Management via T-S Fuzzy Model. *J. Mar. Sci. Technol.* **2004**, *12*, 209–216.
9. Chen, H.L.; Chen, W.T.; Lin, Y.L. Earned value project management: Improving the predictive power of planned value. *Int. J. Proj. Manag.* **2016**, *34*, 22–29. [CrossRef]
10. Trocki, M. *Modern Project Management*; P.W.E.: Warsaw, Poland, 2012.

11. Dziadosz, A.; Kapliński, O.; Rejment, M. Usefulness and fields of the application of the Earned Value Management in the implementation of construction projects. *Bud. Archit.* **2014**, *13*, 357–364. [CrossRef]
12. Waris, M.; Khamidi, M.F.; Idrus, A. The Cost Monitoring of Construction Projects through Earned Value Analysis. *J. Constr. Eng. Proj. Manag.* **2012**, *2*, 42–45. [CrossRef]
13. Bhosekar, S.K.; Vyas, G. Cost Controlling Using Earned Value Analysis in Construction Industries. *Int. J. Eng. Innov. Technol.* **2012**, *1*, 324–332.
14. Vandevoorde, S.; Vanhoucke, M. A comparison of different project duration forecasting methods using earned value metrics. *Int. J. Proj. Manag.* **2006**, *24*, 289–302. [CrossRef]
15. Howes, R. Improving the performance of Earned Value Analysis as a construction project management tool. *Eng. Constr. Archit. Manag.* **2000**, *7*, 399–411. [CrossRef]
16. Czemplik, A. Application of earned value method to progress control of construction projects. *Procedia Eng.* **2014**, *91*, 424–428. [CrossRef]
17. Narbaev, T.; De Marco, A. Combination of growth model and earned schedule to forecast project cost at completion method. *J. Constr. Eng. Manag.* **2014**, *140*, 04013038. [CrossRef]
18. Kim, Y.; Ballard, G. Is the earned-value method an enemy of work flow? In Proceedings of the Eighth Annual Conference of the International Group for Lean Construction (IGLC-8), Brighton, UK, 17–19 July 2000; pp. 142–144.
19. Przywara, D. Time-Cost Analysis in Monitoring the Works of the Construction Schedule. Ph.D. Thesis, Opole University of Technology, Opole, Poland, 2019.
20. Maravas, A.; Pantouvakis, J.-P. Project cash flow analysis in the presence of uncertainty in activity duration and cost. *Int. J. Proj. Manag.* **2012**, *30*, 374–384. [CrossRef]
21. De Koning, P.; Vanhoucke, M. Stability of earned value management: Do project characteristics influence the stability moment of the cost and schedule performance index. *J. Mod. Proj. Manag.* **2016**, *4*, 8–25.
22. Vanhoucke, M.; Vereecke, A.; Gemmel, P. The project scheduling game (PSG): Simulating time/cost trade-offs in projects. *Proj. Manag. J.* **2005**, *36*, 51–59. [CrossRef]
23. Przywara, D.; Rak, A. Analysis of time-cost of monitoring schedule by Earned Value Method. *Tech. J.* **2017**, *2-B(6)*, 41–50.
24. Dodson, M.; Defavari, G.; De Carvahlo, V. Quality: The third element of Earned Value Management. *Procedia Comput. Sci.* **2015**, *64*, 932–939. [CrossRef]
25. Chen, Q.; Jin, Z.; Xia, B.; Skitmore, M. Time and cost performance of design build projects. *J. Constr. Eng. Manag.* **2016**, *142*, 162–169. [CrossRef]
26. Khamooshi, H.; Abdi, A. Project duration forecasting using Earned duration management with exponential Smoothing techniques. *J. Manag. Eng.* **2017**, *33*, 04016032. [CrossRef]
27. Przywara, D.; Rak, A. Modeling of optimal timing transition of front production by the two-punctual network note schedule. *Open J. Archit. Des.* **2014**, *2*, 1–5. [CrossRef]

Article

Estimating the Probability Distribution of Construction Project Completion Times Based on Drum-Buffer-Rope Theory

Xun Liu [1,*], Le Shen [2] and Kun Zhang [3]

1. School of Civil Engineering, Suzhou University of Science and Technology, Suzhou 215000, China
2. School of Business, Suzhou University of Science and Technology, Suzhou 215000, China; s793115763@163.com
3. Institute of Engineering Management, Hohai University, Nanjing 211100, China; dreamerzk@126.com
* Correspondence: liuxun8127@usts.edu.cn; Tel.: +86-136-7510-2267

Abstract: Various factors affecting the construction progress are regarded as bottlenecks giving rise to the project duration overrun. The contractor should combine the project schedule with the plan in order to reduce the uncertainty of the project activities. The present research describes the method derived from the theory of constraints (TOC) attempts to enhance the relationship among activities, to revise and further reduce the uncertainty of construction activities to improve the reliability of project progress. The elements of drum, buffer and rope (DBR) in TOC are added to PERT network schedule; through the identification of schedule in the bottleneck process, the implementation plan of the bottleneck is obtained. By measuring buffer time and calculating network schedule buffer time as well as feeding time, the relationship among activities and uncertainty of duration are also improved. To illustrate the impact of DBR applications on improving project schedule reliability, a case of hydropower station as an example is illustrated to show enhanced reliability of scheduling. As compared to program evaluation and review technique network (PERT) simulation, the simulation results showed that the uncertainty of construction progress could be reduced if the DBR are well cooperated mutually.

Keywords: construction project; PERT; theory of constraint (TOC); drum-buffer-rope (DBR); construction schedule

Citation: Liu, X.; Shen, L.; Zhang, K. Estimating the Probability Distribution of Construction Project Completion Times Based on Drum-Buffer-Rope Theory. *Appl. Sci.* **2021**, *11*, 7150. https://doi.org/10.3390/app11157150

Academic Editors: Mariusz Szóstak, Marek Sawicki and Jarosław Konior

Received: 18 June 2021
Accepted: 13 July 2021
Published: 2 August 2021

Publisher's Note: MDPI stays neutral with regard to jurisdictional claims in published maps and institutional affiliations.

Copyright: © 2021 by the authors. Licensee MDPI, Basel, Switzerland. This article is an open access article distributed under the terms and conditions of the Creative Commons Attribution (CC BY) license (https://creativecommons.org/licenses/by/4.0/).

1. Introduction and Literature Review

There are many uncertain factors in the construction process and they often have negative impacts on the project duration, resulting in project duration stipulated in the contract when the project plan does not match the duration of the practice. Just from the standpoint that the implementation process of the construction project requires a stable environment, if the difference between project completion and intended completion period is quite large, there may be several negative results like increasing idle time of intermediate task in construction process and negative influence on resource distribution rationality. Therefore, these uncertain factors would become bottlenecks to reduce project schedule uncertainty. However, owing to the correlation [1,2], transmission [3] and non-superposition [4] among these risk factors, project managers cannot fully take into account the impact of uncertain factors on activities [5]. In particular, the way the uncertainty is managed in the project and risk management havea direct influence on the success of a project. According to a previousreport, only 44% of the projects could catch up with the finishing line, while 70% of the projects reduce the anticipated work amount, and 30% of the projects were just simply terminated [6]. PERT is a traditional method for modeling uncertainties in project networks, in which the effect on project progress could be reduced through deriving from uncertainty based on buffer mechanism [7]. For example, project managers usually make activities duration under uncertainty influence joined to all levelsof process continued time, to ensure a single activity or overall project can be accomplished within time as scheduled [8]. However, project managers are usually too conservative, or in order to ensure the protection engineering progress implemented as planned, the reliability of the project schedule is very low [7,8]. As a consequence, the higher uncertainty of the schedule, the more prejudice appropriate management measures like rational allocation and optimization of resources in the project [9]. Construction projects are usually executed with various resource constraints, which may change the critical activities of the project and change the project completion time [10]. Therefore, it is necessary to develop a technique that is capable of finding the critical activities and the project completion time by considering the activities' resource requirements and predecessors, and the uncertainties in their durations.

In terms of the view that the duration of every activity is stable, all kinds of risk factors that affect the schedule of the project should be regarded as the bottleneck in TOC [11,12]. Traditional PERT can weaken the probability characteristics of each activity that exists in project network [13]. Simulation-based scheduling could enhance the value of traditional scheduling methods by relaxing some of the restrictive assumptions of PERT [5]. The reliability could be enhanced by describing the project completion time as a probability distribution [14]. However, enough attention has not been paid to the normality assumption that is built into these scheduling methods, the opportunity to improve the reliability of these scheduling methods by finding the best fit probability distribution functions of the many activities in a schedule and an exact probability distribution function of project completion times is ignored [15].There is usually an underestimation of the true project meanby the PERT calculated mean project duration [7,13,16]. Moreover, the reliability of the implementation of the project PERT network progress planning is relatively low. Many researchers have investigated the project scheduling efforts, but it is usually hard to achieve their requirements [17–19]. Due to the uncertainty of constraints, the real concern in construction site for a contractor is how to manage the variation of project duration without reducing it [7].

The uncertainty of a project is the key issue and the potential main cause for most problems [20,21]. In response to the bottleneck caused by the uncertainties, project managers often actively increase the buffer time to absorb the delay of activity duration caused by the occurrence of these uncertainties during the process of construction management [6,11]. The constraint theoryaims to solve the bottleneck management problem, and could thus improve the overall operations and achieve maximum benefits [22]. In Goldratt's opinion, the most important factor is that itcan make actual progress, not conform to the planning progress, which is also related to the bottleneck of the project. Later on, he developed the "drum-buffer-rope" (DBR) schedule planning and control techniques using TOC [23]. Production management uses the DBR principle to rectify the uncertainty of activities and enhance the reliability of schedule. Goldratt indicated that the uncertainty of bottleneck activities duration can be improved by the element "drum", and the originally independent activities can be changed into the correlation mutually by the two elements "buffer" and "rope". The application of the three elements "drum", "buffer" and "rope" in the project schedule can make a contribution to reduce the uncertainty of the construction plan.

Compared with the progress management effects of the traditional industrial manufacturing process, JIT system, Cook indicated that the uncertainty can be significantly reduced after using TOC in progress management [24]. Blackstone also pointed out that with Ford Motor Company in the United States, the on-schedule delivery time achievement rate rose after applying the TOC [25]. Gardiner [26], Spencer, Cox [27] and Wu [28] thought that the application of the element "rope" can help to determine the start time of the bottleneck activities, which ensure that predecessor activities can be completed on time through the settings that project resources are totally put into the bottleneck activities and the start time of bottleneck activities are the same as scheduled.Steyn believed that the application of the TOC can harmonize the relationship between risk factors and the project plan organization, and can be more effective in reducing the uncertainty of the project construction schedule [29]. In view of the fact that the buffer mode cannot effectively guarantee the project completed on schedule in some projects, Vonder et al., proposed a distributed buffer setting mode, and thought that the buffer mode had good robustness with the premise of completion on time [30]. Hu et al. [31] described the three components: plan, control and concentration of the DBR management scheduling model method in detail, and pointed out that such a management method is suitable for resources scheduling of construction projects. Zhang et al., proposed the buffer setting method under uncertain project conditions in using TOC [32] that includes several factors such as resource strain, network planning complexity and risk preference of project managers. It solved the problem that resource strain is difficult to be quantified and unified, and also took into account the using of alternative resources to solve the problem of resource shortage. In consideration of the characteristics that there are gradual gaps in activities and many uses of the element "drum" in the project plan, Bie et al., analyzed the weakness of the centralized buffer setting method in coping with some "drum" elements of the activities network, and proposed one setting method for dispersed capacity constraints. Additionally, he also obtained the project duration, which can be reduced to a great extent under the setting method for dispersed capacity constraint through experiments in different networks with "drum" elements [33]. Apparently, the element "drum" of DBR progress management and control technology in TOC can improve the uncertainty of bottleneck operation duration. The two elements "buffer" and "rope" can strengthen the correlation among the operations activities, which is originally independent. As a result, the application of the three types of elements "drum", "buffer" and "rope", combined with traditional PERT progress management technology, can effectively reduce uncertainty of the project construction schedule [15,26,29].

On the basis of previous research, this paper applied the "drum-buffer-rope" construction schedule management and control technology into a PERT network for improving the relationship among activities. The ultimate goal is to reduce the uncertainty of construction project schedules. The present research (1) proposed simulation model under uncertainty duration, which is founded on the definition that the completion period of the project is within the scope of the contract, and the time distribution for the project should be reduced; (2) the DBR in TOC is applied in construction project scheduling and control; (3) combined with the Monte Carlo simulation method to set up new project construction schedule to reduce the uncertainty of the construction project;and (4) thescheduling of a concrete rock fill dam project was applied as a case study to demonstrate and verify the validity of the proposed model.

2. Theory of Constraints

The TOC concept was addressed by Goldratt et al. [34] based on the principle that complex systems exhibit inherited simplicity. The constraints limit the system's ability to generate the system's real goal. TOC aims to increase throughput while simultaneously decreasing inventory and operation cost [27,34]. One of the thinking tools of the TOC is the effect–cause–effect. That is to say, there is a problem for which a cause is hypothesized. If the cause exists in reality, there are other effects one can predict. If the effect is found, the hypothesis will be strengthened. The five main steps of the TOC [23,35] are as follows: (1) identify the system's constraints; (2) exploit the identified constraints; (3) subordinate everything to the identified constraints; (4) make sure that the constraints are worked to the maximum; and (5) if in the previous steps a constraint has been broken, go back to Step 1.

Step (1) and step (4) are critical for an enterprise to apply the TOC methodology successively [36]. The main TOC technique to identify and exploit the constraint resources is named DBR [37]. As revealed in Figure 1, there are three types of buffers used in the DBR [38]: a constraint buffer, an assembly buffer and a shipping buffer. A constraint buffer is used to protect the schedule of the constraint and is inserted just before the choke point. An assembly buffer is used to ensure that parts coming from a constraint resource do not have to wait for parts coming from non-constraints, and it is located in front of an assembly operation that is fed by both constraint and non-constraint parts. A shipping buffer is used to protect the delivery dates of the orders and is, therefore, located at the end of the process.

That is to say, both the constraint buffer and the assembly buffer are closely related to the constraint machine.

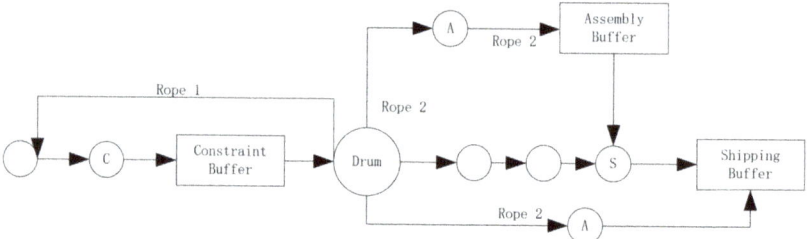

Figure 1. Drum–buffer–rope.

In Figure 1, "Drum" is a control point in the production system associated with the constraint (bottleneck) and its schedule. "Buffer" is time or a time equivalent amount of work in the process, and includes the constraint buffer and the shipping buffer. "Rope" is the term used for the communication feedback to the resources in front of the constraint resources so that each of them produces only the amount that the bottlenecks can complete. Based on this feedback, the entire production of the plan is based on the capability of the bottleneck. In other words, a maximum limit on the number of activities released to the bottleneck but not yet completed is established, and an activity is released whenever the number of activities is below the limit. There are two ropes: Rope 1 determines the schedule at the bottleneck to exploit the constraint according to the organization's goal; Rope 2 then subordinates the system to the bottleneck activity.

The central part of the DBR technique involves inserting buffers in front of the constraint resources and assembly operations to protect the production system from the inevitable fluctuations that are usually caused by the internal disruptions that occur in production of processing time. Contrary to MRP and JIT, in TOC the buffer is considered as a production strategic reserve that can protect the bottleneck from fluctuations in the production process.

3. Method
3.1. Identification and Scheduling of the Bottleneck Operations
3.1.1. Identification the Bottleneck Operations

TOC technology is laid stress on confirming the bottleneck to make non-bottleneck operations fully cooperate with the whole production system [39], so that the whole production system can have a maximum producing capacity without changing the production flow time, and finally it can improve the practical effect of scheduling. Uncertainty on the critical path affects the improvement of the construction schedule. The bottleneck operation in PERT can be regarded as the most influential operation, resulting to the difference between actual and planning construction duration [36,40,41]. The greater the uncertainty or variance of process operation duration is given, the higher possibility that this operation can be a bottleneck in the schedule. On the other hand, the critical path can largely affect the uncertainty, which is also considered as the key chain and affected the reliability of project completion time. Consequently, there are four principles as follow to recognize the bottlenecks:

(1) Project duration is influenced by the major critical path of which the total time difference is zero or minimum. If duration on the critical path does not match the schedule, it makes the start time of subsequent activities and project completion time change.
(2) When the DBR schedule management technique is applied in the process of compiling a project schedule, since the bottleneck operations must limit operation, which reduces the certainty of project completion time, the bottleneck operation must exist

in the critical path. The degree of uncertain operational time can be used as the selection standard of bottleneck operations. In other words, the greater the standard deviation of the operation time that is given, the higher possibility that it may be the bottleneck operation.

(3) Sometimes the critical path is not only one, and each critical path may have a chance to have an effect on distribution of construction, in addition to considering the standard deviation of operation duration, selecting the bottleneck operations also needs to take into account the standard deviation of critical path completion duration at the same time.

(4) When there are several bottlenecks in a project network, the impact of the operation near the convergence point is higher than other operations, and the operation closest to the convergence point should be chosen.

3.1.2. Scheduling of the Bottleneck Operations

Goldratt [23] and Xie et al. [42] pointed out that the drum is determined by backward scheduling from customer orders. Other activity schedules obtain the expected duration of the total processing time from the drum scheduling [29,30]. Since the production schedule is subject to bottleneck, scheduling managers should have to provide sufficient resources including human, machinery and construction materialsin order to ensure the bottleneck activity can be started at the expected time and completed within the scheduled time. Therefore, there should be enough resources in bottleneck activities to reduce the uncertainty of bottleneck activity completion time caused by internal and external risk factors. As a result, activity schedules of bottlenecks should be the most possible operation completion time.

3.2. Determine Buffer Time of Bottleneck Activities

The differences among different projects should be considered. For example, project risk, owners and contractors, and managers often increase duration of activities subjectively. This increased time is named as buffer time or protection time, andis used to prevent the occurrence of activity time uncertainty. However, the activities actually do not have clear and stable buffer times; it means that the buffer times are different because of different characteristics of specific activities.

Usually, duration estimates for individual activities contain some arrangements for possible events or occurrences. DBR scheduling technique considered all possible events or occurrences into a project buffer. This implies that all expected times on individual activities and sub-projects are estimated. Buffers, on the other hand, are calculated to reflect the uncertainty in the estimates of duration of activities [43], and there is no exact constant rate of buffer time in construction activity. For example, the setting out process in surveying engineering may only need half a day sometimes, but floor concrete pouring operation needs to have 14 days for a maintenance period. For each activity to have the same rate of buffer time is simply not practical. If each activity has the same rate of buffer time, it would not match the actual construction situationand notaffect the quality and safety of construction project. This study adopts dynamic buffer, the bigger the variance is, the more the buffer is required. Additionally, the buffer time will be relatively decreased, when the activity completion time falls behind the expected schedule [44].

One of the challenges in DBR is the sufficient sizing of the buffers. If the buffers are estimated more than the necessary size, practical consequences immediately occur. Contrarily, if the buffers are underestimated, they may increase the probability of duration overruns, which can cause financial penalties and a reliable loss on the part of the customers or market. In study of the application of TOC in construction project schedule management, scholars have tried different methods to determine the buffer time of operations. Slusarczyk et al., attempted to apply these concepts and explored the advantages of applying TOC to a complex mega infrastructure project and to compute the buffer size using some of the available methods [45]. In a previous study, software development projects for resource-constrained problems were analyzed and given solutions; an improved root square error was suggested; the setting method of buffer sizes, which is suitable for software development projects, was adopted; and the preemptive scheduling method based on a heuristic algorithm and priority rules was used to plan the scheduling [42]. A buffer sizing method based on comprehensive resource tightness was proposed to better reflect the relationship among activities and to improve the accuracy of project buffer determination [44]. Wei et al., considered the inline mode of security time for each resource conflict activity and proposed to set a reduction ratio to improve the calculation of the buffer time [46]. On the other hand, in the field of scheduling of construction projects, the determination of buffer time was not absolute. For example, Schragengein thought the size of buffersshould be three times the standard deviation of the average bottleneck lead time. He used three times based on relevant work experiences and assumed the reliable lead time complies with the normal distribution [47]. Ronen and Starr thought that the buffer time should be a quarter of the total lead time [48]. Cohen thought that it should accord to the degree of uncertainty level of the target or the sum of 50% of each activity duration [49]. At the same time, the buffer time can be regarded as degree of uncertainty of the activity duration. The higher the degree is, the longer the buffer time is, and also the more possibility that the project duration meet the progress schedule.

Schragenheim [47] and Demmy [44] pointed out three buffer times in manufacturing system constraint buffer, assembly buffer and shipping buffer. The buffer time would be reduced if the operation completion time exceeds predetermined. If the activities can be done as expected and made no idle time in the following activities, it would be helpful to the project managers to arrange as well as the capital input and use of project resources. Additionally, on the other hand, the occurrence of risk has probably a negative impact on the completion as expected of the project. Therefore, impact caused by risk exists in between pessimistic completion time and expected completion time of the activity. Therefore, the buffer time can be calculated according to the following Equations (1)–(3).

$$CB_c = \frac{T_{cb} - T_{ce}}{2}, c \in S_{CB} \quad (1)$$

$$AB_a = \frac{T_{ab} - T_{ae}}{2}, a \in S_{AB} \quad (2)$$

$$SB_s = \frac{T_{sb} - T_{se}}{2}, a \in S_{SB} \quad (3)$$

where

CB_c = Constraint buffer (CB) time;
AB_a = Assembly buffer (AB) time;
SB_s = Shipping buffer (SB) time;
T_{cb} = Pessimistic duration all predecessor activities of constraint buffer;
T_{ce} = Expected duration of all predecessor activities of constraint buffer;
T_{ab} = Pessimistic duration of all predecessor activities of assembly buffer;
T_{ae} = Expected duration of all predecessor activities of assembly buffer;
T_{sb} = Pessimistic duration of all predecessor activities of shipping buffer;
T_{se} = Expected duration of all predecessor activities of shipping buffer;
S_{CB} = Set of predecessor activities of constraint buffer;
S_{AB} = Set of predecessor activities assembly buffer;
S_{SB} = Set of predecessor activities shipping buffer.

Constraint buffer (CB) is a set before the bottleneck operation activities. The purpose is to provide a protective effect produced by the bottleneck, so that the bottleneck resource can reach the goal of predetermined output in terms of progress schedule.The bottleneck buffer must be placed in front of the critical path activity to minimize the resource limit and maximize the duration reliability [15]. Assembly buffer (AB) ensures that the bottleneck is not delayed by postponement of other activities when it is formed by joining components together. The PERT and CPM have a confluence of activities that give rise to duration variation, resulting in an increase in project duration uncertainty. AB is therefore added to the assembly node where the bottleneck and non-bottleneck are merged. Shipping buffer (SB) is established in the product shipping area and aimed to protect production to satisfy the order delivery date, in order to prevent the influence of uncertain factors in the production process that may delay the project delivery date. The traditional production process in the CPM or PERT scheduling belongs to the push system. Once the bottleneck duration is extended, the postponed duration causes a breach of the contract, which is why the shipping buffer occurs after a bottleneck.

The measurements of three buffer times are similar, but have different numbers of predecessor activities of bottleneck, so the buffer time varies. Equations (4)–(6) distribute each activity buffer time in accordance with the ratios of expected duration of predecessor activity of the buffer to expected project duration.

$$BT_c = \frac{b_c}{T_e} \times CB_c, c \in S_{CB} \tag{4}$$

$$BT_a = \frac{b_a}{T_e} \times AB_a, a \in S_{AB} \tag{5}$$

$$BT_s = \frac{b_s}{T_e} \times SB_s, s \in S_{SB} \tag{6}$$

where
BT_c = Buffer time of activity c before the bottleneck buffer;
BT_a = Buffer time of activity a before assembly buffer;
BT_s = Buffer time of predecessor activity s before shipping buffer;
T_e = Expected project duration;
b_c = Pessimistic duration of predecessor activity c before constraint buffer;
b_a = Pessimistic duration of predecessor activity a before assembly buffer;
b_s = Pessimistic duration of predecessor activities of s before shipping buffer.

Given that buffer time is derived from each activity, the redundant safety protection time in each activity is removed to set buffers and centralized as a mechanism to deal with uncertain factors in project implementation process. Equations (7)–(9) remove extra buffer time from each activity. The removed buffer time is then placed in the buffer zone, and can be used to manage the project uncertainty and improve the scheduling reliability.

$$\overline{b_c} = b_c - BT_c \tag{7}$$

$$\overline{b_a} = b_a - BT_a \tag{8}$$

$$\overline{b_s} = b_s - BT_s \tag{9}$$

where
$\overline{b_c}$ = Pessimistic duration without buffer time of predecessor activity c before CB;
$\overline{b_a}$ = Pessimistic duration without buffer time of predecessor activity a before AB;
$\overline{b_s}$ = Pessimistic duration without buffer time of predecessor activity s before SB.

3.3. Constraint Buffer Management of Activities

The settings of buffer time are commonly provided by each activity to remove redundant safety protection time in every activity, which is a centralized mechanism to deal with uncertain factors in the project implementation process [38,43]. Furthermore, project managers can monitor the project progress status and reduce the uncertainty through buffer management.

Nowadays, large projects have complex construction conditions including information, task, techniques, organization, environment, and goal, which determines the dynamic, uncertain and highly interdependent features of the project construction process and system [20,50]. Immediate management of project duration from that fact is of extreme important. Schragenheim suggested that the length of buffer time can be divided into three sections: negligible zone, alert zone and accelerative zone, the length of each zone is equal [47]. The size of each zone is allocated equally according to Equations (1)–(3). The constraint buffer is shown in the following Equation (10).

$$\overline{T_{ce}} = \max\left\{t_i + \frac{a_i + 4m_i + \overline{b_c}}{6}\right\} \tag{10}$$

$\overline{T_{ce}}$ = Expected duration of all predecessor activities of CB without the buffer time;
t_i = Start time of activity i;
a_i = Optimistic duration of activity i;
m_i = Most possible duration of activity i;
$\overline{b_c}$ = Pessimistic duration without buffer time of predecessor activity c before constraint time.

If the actual duration of the predecessor activity of the constraint buffer is longer than that of the expected duration, the project duration is not likely to meet the requirements, therefore, the constraint buffer needs sufficient time to absorb the extra project duration caused by uncertainties.

If the actual duration of the predecessor activity of the constraint buffer is within the negligible section, the buffer time is still sufficient for the project manager to make use of the project duration. If it is located within the alert section, the project manager should watch more closely on the progression of project duration. If it is within the accelerative section or alert section, the project can be expected to complete smoothly. Given the starting time of the bottleneck of adding 2 × (CB/3), the temporal variances are in a range still acceptable for the project manager, thus posing no need of extra overworking resource to start the bottleneck activity ahead of the schedule, since it is more important for the manager keeping the project duration as planned than decreasing it.

While if the actual duration is within the accelerative section, the manager has to increase adequate resources, such as manpower, machinery for construction or working hours, so as to start activities ahead of schedule for earlier completion, allowing bottleneck activity to set in motion as scheduled. When confronting with existing possibility problems of penalty for breach of contract and project delay, it is explicitly necessary to add in a certain amount of resources to shorten the activity time, and that, in the meantime, adds costs to the project [51].

In the field of industrial engineering, the output pace of the bottleneck determines the output efficiency of the system [52]. On the other hand, if the materials were produced too late, activities would delay, which may also affect the arrangement of start time for bottleneck activities, and then influence the output of the whole construction system [51,52]. Therefore, the completion time of activities before bottleneck must be decided appropriately. The element rope was defined as the opposite length of lead time from bottleneck to order starting by Gardiner [26]. The element can make the production speed of all activities in the production system executed according to the production rhythm of the bottleneck. The control function of rope can be attained through the establishment of detailed plan from resources to construction site [26]. Wu thought that the main function of rope is to determine the proper time and correct materials arriving at the construction site [28]. Spencer explained that the purpose of rope control is to ensure the production materials are always enough to support the ongoing bottleneck activities. It means that the rope is used to decide the start time to book materials and resources in addition to make them smoothly go through the non-bottleneck activities and ensure the bottleneck activities to be completed on schedule [27].

It can be summarized that the predecessor activity duration of the bottleneck can be regarded as the scale of rope. The length of the rope is the same as the duration of the predecessor activity of the bottleneck, which uses a push system for synchronous production. The rope scheduling is the milepost for all bottleneck predecessor activities after the start of work. To enable the bottleneck to start as expected in the buffer management, the total completion time of all predecessor activities must not exceed the rope scheduling [26,28,42]. When a constraint is broken, the next constraint needs to be identified and improved.

3.4. Simulation of Construction Project Scheduling in DBR Model

The method proposed in the study is described in the flowchart presented in Figure 2. The DBR model components include the drum, buffer and rope. For example, the schedule model simulations contain two schedule components (C = 2), three possible types of DBR model, the Drum–Buffer schedule, Drum–Rope schedule and Buffer–Rope schedule, which were compared with PERT simulation results. Therefore, the algorithm runs many simulation experiments using activity durations with the eight-schedule model.

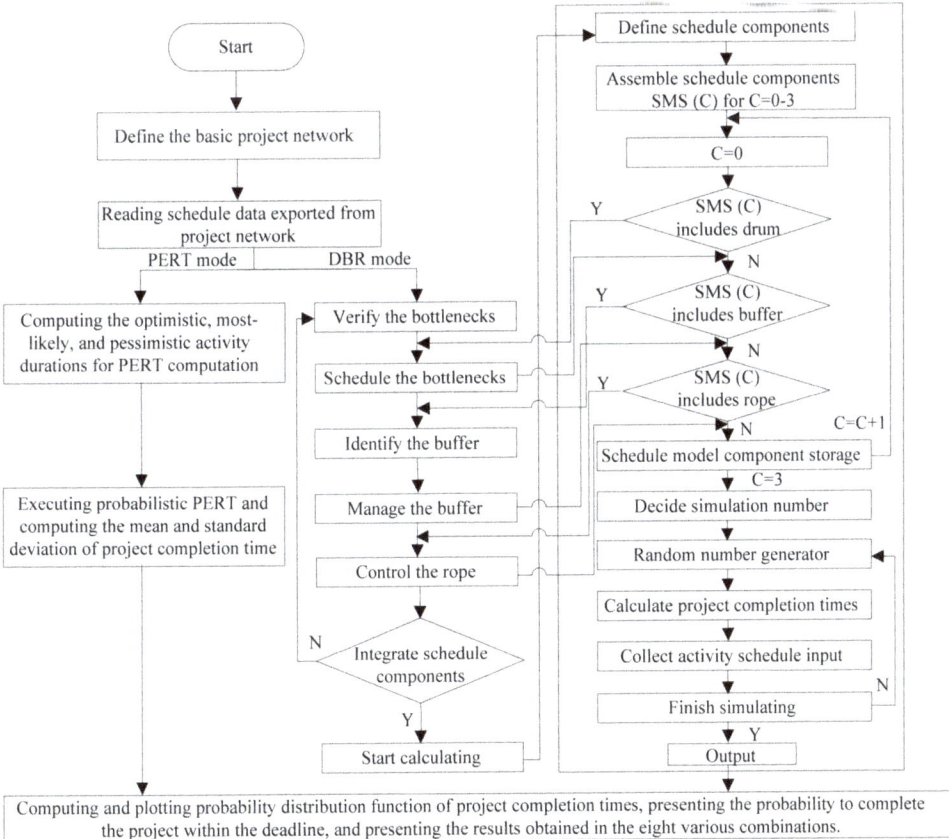

Figure 2. Flowchart of DBR model applied in construction project scheduling.

4. Empirical Examples

4.1. Background

A hydropower station is located on the main stream of the Yellow River, which is a large-scale cascade hydropower station. It is one of China's important power points in the northern channel of the Power Transmission from West to East in the Western Development Strategy. The reservoir design capacity of the hydropower station is 62 million cubic meters, which is a daily regulation reservoir. The main task of this power station is to generate electricity, in addition to the functions of irrigation and water supply. Main hydraulic structures are made up of a concrete face rock fill dam, left bank flood discharging tunnel, left bank spillway, right bank flood discharging tunnel, water diversion and power generation system and powerhouse, with a total investment of 6.6 billion Yuan.

According to the engineering characteristics and the technological requirements of a concrete face rock fill dam and the actual situation of the hydropower station construction, after careful analysis and research, the owner, supervisor, designer and construction contractor reached a consensus that the deployment of the material resources, the arrangement of the construction road and the arrangement of building the dam are the three key factors to ensure the whole of the hydropower station project completed on schedule. For the filling and building construction of the concrete faced rock fill dam, a careful analysis of the relationship between the construction process activities, and an accurate quantitative expression of the correlations, dependencies and constraints among the process activities, are the prerequisite to make rational arrangement of the construction sequence. Additionally, on the foundation of this analysis, the key process and critical path obtained from the network and the problems is reflected, which has its practical guiding significance for managers and construction organizers to better understand the key of construction and to distinguish the primary and secondary points of works, in order to better achieve the project schedule control target.

4.2. Case Analysis

In the present research, a case study was brought to demonstrate the application of conjunctive use of the three elements drum–buffer–rope in practical project management, based on the fundamental hypothesis [53,54], the study in this part is used to show whether the method can effectively reduce the degree of uncertainty of the project construction period. The case in this study is analyzed as follows:

Step 1. Define the basic project network plan

Referring to the logical relations among processes specified in the overall progress of the construction network plan of the concrete panel rock fill dam of the hydropower station project (contract stage research report), the network plan of the general construction schedule can be drawn as shown in Figure 3. As the concrete panel rock fill dam has a long construction period, many processes and complicated relationships are associated with the processes, when drawing the network plan of the overall construction progress, digital as a code name was used for a different construction working procedure. Additionally, complete details of the process (process name, process activity continued time, process engineering quantity, antecedent process and subsequent process), triangular estimate value, expected completion time, variance, and standard deviation of each activity in project network are listed in Table 1.

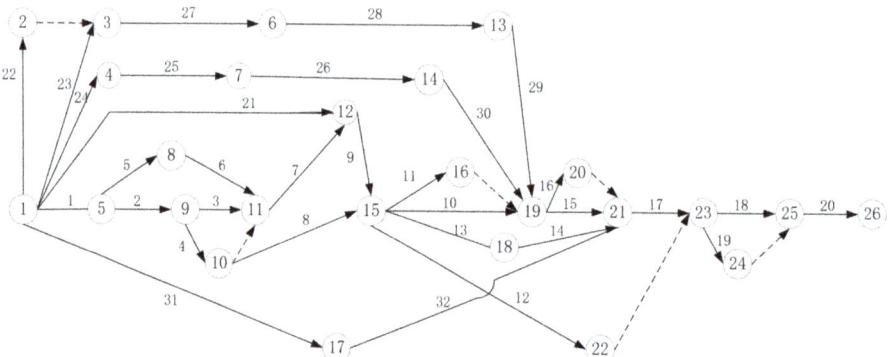

Figure 3. Double code network diagram of the construction of a concrete face rock fill dam.

Table 1. Detail table of engineering construction process information.

Number	Activity Item	Precedence Relation	Duration Estimation (a, m, b)/Day	Activity Expected Duration	Standard Deviation (Day)	Variance
1	Excavation of dam abutment above water on two banks (above ▽1901)	-	170,230,260	225	15	225
2	Closure dike filling and foundation pit drainage	1	40,55,70	55	5	25
3	Foundation pit excavation	2	50,70,90	70	6.67	44.44
4	Foundation excavation of cutoff wall beside dam	2	30,40,55	40.83	4.17	17.36
5	Seepage construction of upstream and downstream enclosing wall	1	40,55,60	53.33	3.33	11.11
6	Filling construction of upstream and downstream cofferdam	5	30,45,50	43.33	3.33	11.11
7	Silicon pouring of toe slab and foundation (below ▽1901)	3,4,6	8,15,25	15.50	2.83	8.03
8	Silicon pouring of cutoff wall beside dam	4	55,70,90	70.83	5.83	34.03
9	Filling construction of temporary section of upstream dam (below ▽1955)	7,21	140,165,185	164.17	7.5	56.25
10	Filling construction of temporary section of down-stream dam (below ▽1955)	8,9	130,150,185	152.5	9.17	84.03
11	Backfilling of gully at axis on left bank	8,9	90,120,145	119.17	9.17	84.03
12	Masonry beside dam	8,9	455,495,560	499.17	17.5	306.25
13	Silicon pouring of panel of first stage	13	45,60,80	60.83	5.83	34.03
14	Water stopping installation on surface	10,11	45,60,75	60	5	25
15	Whole section filling from ▽1955 to dam crest	10,11 29,30	160,195,215	192.5	9.17	84.03
16	Filling of slope body in front of dam (below ▽1940)	10,11 29,30	100,120,145	120.83	7.5	56.25
17	Silicon pouring of panel of second stage	14,1516,32	75,90,115	91.67	6.67	44.44
18	Construction of parapet wall and road on dam crest	12,17	85,120,150	119.17	10.83	117.36
19	Demolition of downstream cofferdam	12,17	70,90,115	90.83	7.5	56.25
20	Filling and masonry on dam crest	18,19	48,60,75	60.5	4.5	20.25
21	Bolt-concrete support of dam abutment on two banks	-	280,330,400	333.33	20	400
22	Construction of helper system in this contract section	-	135,165,195	165	10	100
23	Transformation of machining system of cushion material	-	85,105,115	103.33	5	25
24	Borrow Area Planning and road construction in II zone of water ditch	-	25,30,45	31.67	3.33	11.11
25	Peeling of gravel soil and strong decomposed rock	24	100,120,145	120.83	7.5	56.25
26	Mining and blasting test of transition material	25	12,15,20	15.33	1.33	1.78
27	Machining of test material in cushion	22,23	35,45,50	44.17	2.5	6.25
28	Machining of cushion material in first stage	27	48,60,80	61.33	5.33	28.44
29	Machining of cushion material in second stage	28	105,135,160	134.17	9.17	84.03
30	Mining of transition material and cushion material	26	195,240,280	239.17	14.17	200.69
31	Grouting test	-	72,90,118	91.67	7.67	58.78
32	Grouting and pouring of toe slab on left and right banks	31	300,360,420	360	20	400

Note: in the table, a, m, b, respectively, indicate the optimistic completion time, the most likely completion time and the pessimistic completion time of the process. The value is obtained by modifying original data.

Step 2. Verify the bottlenecks

As shown in Table 1 and Figure 2, the activities on the critical path are 1, 2, 3, 7, 9, 10, 15, 17, 18 and 20. The critical activities are ranked from large to small according to their variation or standard deviation,1,18,10 or 15,9,3 or 17, 2, 20, and 7. As a result, activity 1, 2, 3, 7, 9, 10, 15, 17, 18 and 20 would be thebottleneck activities, and thus, the bottlenecks schedule and the buffer and the rope could be built up.

Step 3. Schedule the bottlenecks

In this paper, using the commercial software Crystal Ball Version, the probability duration distribution of each operation is set as a triangular distribution, which is composed with the most optimistic completion time, the most likely completion time and the most pessimistic completion time [55–57]. The calculations were based on the simulation flowchart shown in Figure 2. By setting a certain number of simulations, for example, setting simulation time up to 20,000, through a random number generator (0–1) and the triangular distribution operations, the completion time of each operation, the project completion time and probability distribution can be obtained. In the progress of planning the model building process, it is necessary to consider the selection and combination of schedule elements, calculate the required data of the schedule elements, and coordinate with the calculation of the PERT schedule. The DBR technical elements include three basic elements of drum, buffer, and rope. It has seven planning patterns in combination. With the traditional PERT network planning model, a total of eight schedule models exist. Through the comparison among the eight schedule models, a suitable project schedule can be found, so as to be able to reduce the completion risk to a minimum. After building the model, the schedule can be placed into the temporary storage. The number of Monte Carlo simulations and operation duration of each simulation can be set according to triangular distribution and random number generator. Once the operation duration of each simulation is determined, the completion time of the project is calculated according to critical path method, until the expected times of simulation are reached.

4.3. Results and Discussion

The simulation results under PERT network schedule of this case and the simulation results including various combinations of different elements under DBR scheduling technique is shown in Figure 4. The data comparison of a variety of simulation results obtained by different progress planning methods are shown in Table 2.

The expected duration in the project progress network diagram is 1211.63 days, but uncertainty of duration in the network schedule in terms of PERT schedule technique is as high as 33.56 days (SD. 33.56 days, Min. 1092.29 days, and Max. 348.92 days). Additionally, when the DBR technology is applied in PERT network diagram, compared to the traditional PERT network technology, drum technology reduces the expected project completion time by 2.68 percent (31.59 days), while the uncertainty of the project completion duration has decreased by 3.8 days (SD. 29.76 days, Min. 1083.51 days, Max. 1288.12 days). Application of buffer technology in projects increases the average completion time by 6.03% (73.01 days), but the degree of uncertainty of the project completion period has decreased by 16.78 days (SD. 16.78 days, Min. 1220.30 days, Max. 1344.39 days). Similarly, the application of rope technology also increases the average completion time of the project, and the rate of increasing was 1.27% (15.44 days), but the uncertainty of project completion period has decreased by 6.97 days (SD. 26.59 days, Min. 1132.12 days, Max. 1322.37 days). When putting two of schedule control elements into the traditional PERT project schedule network diagram, drum-buffer technology increases the expectations of the project completion time, the rate of increasing was 3.59 % (43.52 days), but uncertainty of project completion period has decreased by 22.16 days (SD. 11.40 days, Min. 1213.72 days, Max. 1303.21 days). Drum-rope technology reduces the average completion time of the project 1.27% (15.44 days), while the uncertainty of completion period reduces 10.45 days (SD. 23.11 days, Min. 1117.47 days, Max.1271.83 days). Buffer-rope technology reduces the average completion time of the project 1.91% (23.17 days), while the uncertainty of completion period reduces 24.38 days

(SD. 9.18 days, Min. 1152.89 days, Max. 1216.75 days). As for the comparison between drum-buffer-rope and traditional PERT schedule technology, the DBR reduces the average completion time of the project 2.6% (27.38 days), while the uncertainty of completion period reduces 25.64 days (SD. 7.92 days; Min. 1156.38 days; Max. 1212.46 days).

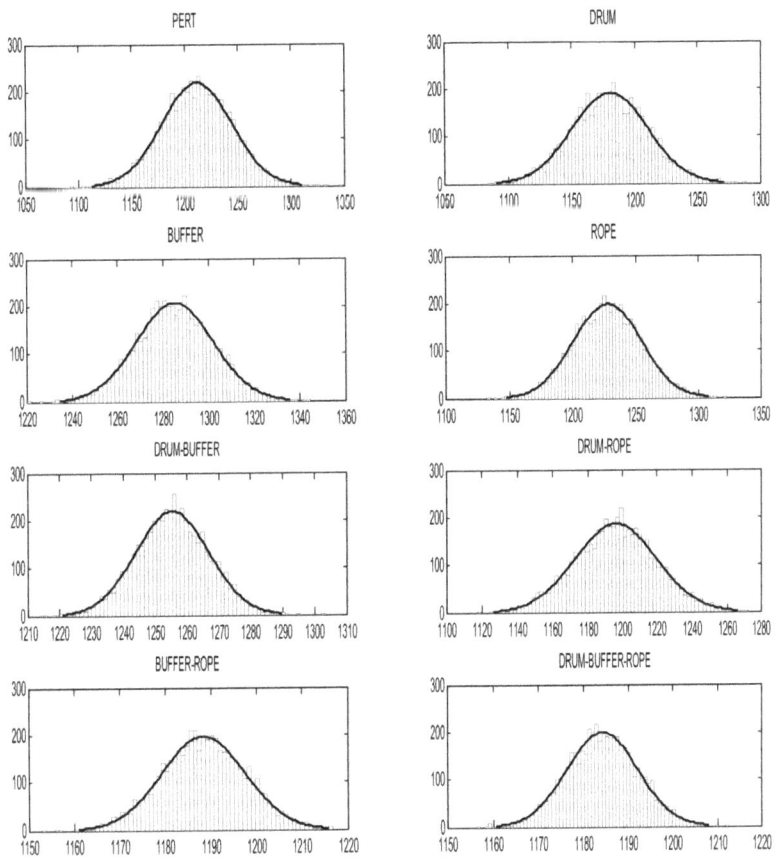

Figure 4. Simulation results under different combinations.

Table 2. Comparison table of simulation uncertainty of various combinations.

Number	Control Elements of Progress Schedule	Average Completion Period	Standard Deviation	The Most Optimistic Completion Time	The Most Pessimistic Completion Time	Uncertainty Reducing Compared with PERT
1	PERT	1211.63	33.56	1092.29	1348.92	-
2	drum	1180.04	29.76	1083.51	1288.12	3.8
3	buffer	1284.64	16.78	1220.30	1344.39	16.78
4	rope	1227.07	26.59	1132.12	1322.37	6.97
5	drum-buffer	1255.15	11.40	1213.72	1303.21	22.16
6	drum-rope	1196.19	23.11	1117.47	1271.83	10.45
7	buffer-rope	1188.46	9.18	1152.89	1216.75	24.38
8	drum-buffer-rope	1184.25	7.92	1156.38	1212.46	25.64

Observing the comparison chart of completion period distribution of various project schedule network simulation under different combinations (Figure 5), and comparison chart of completion probability under different combinations (Figure 6), through comparison the results in Figures 5 and 6, it is clearly that although the average completion time in traditional PERT project scheduling techniques are not the longest completion time, the degree of uncertainty of completion time calculated in this method is the highest. The higher degree of uncertainty of completion duration, the more likely there is additional operation idle time, and the more likely the allocation of operation resources fluctuates greater. These all result in the failure of management. However, under the conditions of independent application of drum elements, buffer elements and rope elements, the uncertainty of the progress plan can be reduced to some extent compared with PERT technology. In other words, in the case that these elements are independently applied, the results obtained are able to improve the uncertainty distribution of project completion time. However, in reality, the actual situation is that the case is also likely to result in a great degree of the extension of the project completion time. It is not the best approach to reduce the uncertainty of project progress. Similarly, in the case of the mutual application of drum elements, buffer elements and rope elements, the combination technology can reduce the uncertainty of completion duration of the plan to some extent, but may also cause increasing of the project completion time. In the joint application of drum elements, buffer elements and rope elements, although it is not the best choice to reduce uncertainty of the project network plan, it does not increase the project completion time. Under the constraints of the uncertainty of the project schedule and project completion time, the schedule control technique that drum elements, buffer elements and rope elements are jointly applied can get more appropriate effects on project progress scheduling.

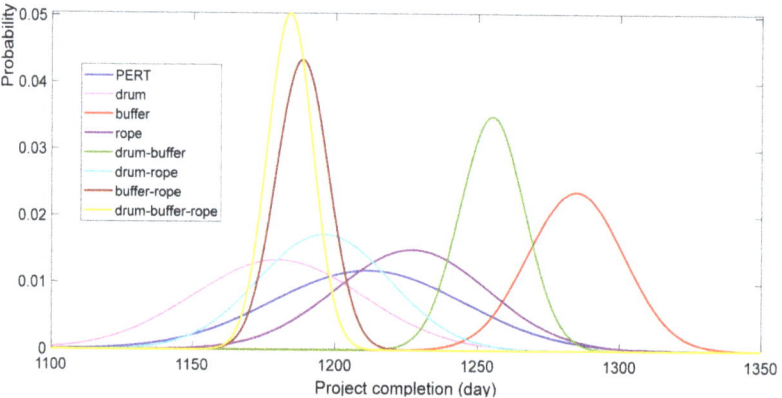

Figure 5. Simulation comparison of completion period distributions between different combinations.

As used herein, the management model of project progress proposed in this paper is adding each element of DBR scheduling and management technique in the TOC into the traditional PERT network schedule. The application of each element can effectively reduce uncertainty of completion period, and there is also a case study to demonstrate that the conjunctive use of three elements drum-buffer-rope can most effectively reduce the degree of uncertainty of the project construction period in this research.

This study also shows that when a progress schedule is made by using traditional PERT network plan, the application of drum elements can help project managers to decide the bottleneck process and arrange the plan of bottleneck process in order to achieve the purpose to reduce uncertainty of progress schedule. However, it still needs to be used in conjunction with the two elements of buffer and rope. Taking the buffer elements into account in progress scheduling process can help managers to remove redundant security protection time in each process, and make a centralized management as a mechanism to deal with the uncertain factors in the construction process of the project. Additionally, managers also are able to monitor the executed state of the progress under the help of buffer management technology, in order to reduce the uncertainty of scheduling. Traditional progress management techniques, such as CPM and PERT techniques, calculate the operation activities starting time by a method of forward reasoning. However, the construction process of a project is a propulsion system; a delay on the critical path will result in the completion period not meeting the schedule plan. In other words, the uncertainty of the construction schedule will increase. If the use of rope elements can help to control the start time of project bottlenecks, the impact from subsequent operation duration on the uncertainty of completion period will be reduced. If the project schedule makers can make the three elements of drum–buffer–rope be fully fit when they are developing the progress schedule, it is possible to form a good DBR schedule and control technology, which can be effectively applied to the actual schedule control of the project.

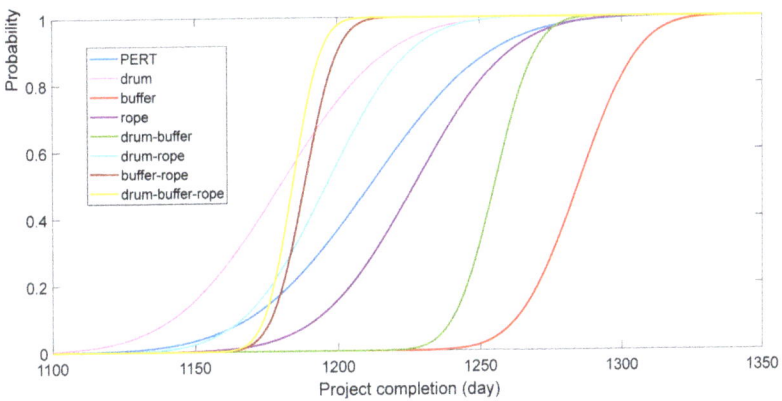

Figure 6. Simulation comparison of completion probability between different combinations.

5. Conclusions

The project schedule management model proposed in this study uses each element of the DBR schedule and management technology in the theory of constraints to join the formulation of the traditional PERT network schedule, and the application of each element can effectively reduce the uncertainty of the project completion period. At the same time, the calculation example also proves that the combined use of the three elements of Drum–Buffer–Rope can most effectively reduce the degree of uncertainty of the project completion period. The research in this study also shows that when using the traditional PERT network to prepare schedules, the use of Drum elements can help the project manager to determine the bottleneck process and arrange the schedule plan of the bottleneck process to reduce the inconsistency of the schedule and to obtain the degree of certainty, but this still requires the use of the two elements of Buffer and Rope; the Buffer element is taken into account in the formulation of the project schedule, which can remove the excess safety protection time in each process and concentrate it. Management is used as a mechanism to deal with uncertain elements in the construction process of a project. At the same time, it can monitor the execution status of the project schedule through the buffer management technology to achieve the purpose of reducing the uncertainty of the schedule; and the traditional project schedule management technology, such as CPM and PERT technology, the forward calculation method used when calculating the start time of the process activity, and the construction process of the process operation belongs to a propulsion system. Once the process on the key line is delayed, the project completion period will not meet the schedule plan. That is, the uncertainty of the construction schedule is increased. If the Rope element can be used to control the start time of the project bottleneck, the impact of the subsequent process operation duration on the uncertainty of the completion period will be reduced. The three elements of Drum–Buffer–Rope can be fully coordinated when the schedule is formulated to form a good DBR schedule and control technology, and can be effectively applied to the actual schedule control of the project, so that the benefits of DBR's engineering project schedule and control technology can be more effectively revealed.

Author Contributions: Conceptualization, X.L. and L.S.; methodology, X.L. and L.S.; formal analysis, X.L. and K.Z.; investigation, X.L. and K.Z.; data curation, X.L.; writing—original draft preparation, L.S.; writing—review and editing, X.L.; supervision, X.L. All authors have read and agreed to the published version of the manuscript.

Funding: Philosophy and Social Science Research in Colleges and Universities in Jiangsu Province (No. 2020SJA1394), Fundamental Research Funds for the Central Universities (No. 331711105), Jiangsu Provincial Construction System Science and Technology Project of Housing and Urban and Rural Development Department (No. 2017ZD074).

Institutional Review Board Statement: Ethical review and approval were waived for this study, due to this study not involving biological human experiment and patient data, which was not within the scope of review by the Institutional Review Board of Suzhou University of Science and Technology.

Informed Consent Statement: Informed consent was obtained from all subjects involved in the study.

Data Availability Statement: The data presented in this study are available on request from the corresponding author.

Acknowledgments: The authors would like to thank the reviewers for all helpful comments, and to thank the foundation of Philosophy and Social Science Research in Colleges and Universities in Jiangsu Province (No. 2020SJA1394), Fundamental Research Funds for the Central Universities (No. 331711105), Jiangsu Provincial Construction System Science and Technology Project of Housing and Urban and Rural Development Department (No. 2017ZD074), Jiangsu Province Joint Education Program High-Standard Example Project, for their support.

Conflicts of Interest: The authors declare no conflict of interest.

References

1. Wu, D.; Li, J.; Xia, T.; Bao, C.; Zhao, Y. A multi-objective optimization method considering process risk correlation for project risk response planning. *Inf. Sci.* **2018**, *467*, 282–295. [CrossRef]
2. Bao, C.; Wu, D.; Li, J. A Knowledge-Based Risk Measure from the Fuzzy Multicriteria Decision-Making Perspective. *IEEE Trans. Fuzzy Syst.* **2019**, *27*, 1126–1138. [CrossRef]
3. Zhang, Y.; Zuo, F. Selection of risk response actions considering risk dependency. *Kybernetes* **2016**, *45*, 1652–1667. [CrossRef]
4. Ahmad, Z.; Thaheem, M.J.; Maqsoom, A. Building information modeling as a risk transformer: An evolutionary insight into the project uncertainty. *Automat. Constr.* **2018**, *92*, 103–119. [CrossRef]
5. Li, G.L.; Abbasi, A.; Minchael, J.R. A simulation-based risk interdependency network model for project risk assessment. *Decis. Support. Syst.* **2021**, *148*, 113602.
6. Izmailova, A.; Kornevaa, D.; Kozhemiakinb, A. Project Management Using the Buffers of Time and Resources. *Procedia Soc. Behav. Sci.* **2016**, *235*, 189–197. [CrossRef]
7. Sackey, S.; Kim, B.S. Schedule Risk Analysis using a Proposed Modified Variance and Mean of the Original Program Evaluation and Review Technique Model. *KSCE J. Civ. Eng.* **2019**, *23*, 1484–1492. [CrossRef]
8. Hermans, B.; Leus, R. Scheduling Markovian PERT networks to maximize the net present value: New results. *Oper. Res. Lett.* **2018**, *46*, 240–244. [CrossRef]
9. Noemie, B.; Izack, C. A robust optimization approach for the multi-mode resource-constrained project scheduling problem. *Eur. J. Oper. Res.* **2021**, *291*, 457–470.
10. Jie, S.; Martens, A.; Mario, V. Using Schedule Risk Analysis with resource constraints for project control. *Eur. J. Oper. Res.* **2021**, *288*, 736–752.
11. Sin, T.; Wei, L.; Bing, H.; Hon, L. Debottlenecking cogeneration systems under process variations: Multi-dimensional bottleneck tree analysis with neural network ensemble. *Energy* **2021**, *215*, 119168.
12. Mukund, S.; Anders, S.; Azam, S.M.; Jon, B. A generic hierarchical clustering approach for detecting bottlenecks in manufacturing. *J. Manuf. Syst.* **2020**, *55*, 143–158.
13. Miklos, H.; Bokor, O. The Effects of Different Activity Distributions on Project Duration in PERT Networks. *Procedia Soc. Behav. Sci.* **2014**, *119*, 766–775.
14. Lee, D.E.; Arditi, D.; Son, C.B. The probability distribution of project completion times in simulation-based scheduling. *KSCE J. Civ. Eng.* **2013**, *17*, 638–645. [CrossRef]

15. Poshdar, M. A Probabilistic-Based Method to Determine Optimum Size of Project Buffer in Construction Schedules. *J. Constr. Eng. Manag.* **2016**, *142*. [CrossRef]
16. Kuklan, H. Project planning and control: An enhanced PERT network. *Int. J. Proj. Manag.* **1993**, *11*, 87–92. [CrossRef]
17. Ben-Haim, Y.; Laufer, A. Robust Reliability of Projects with Activity-Duration Uncertainty. *J. Constr. Eng. Manag.* **1998**, *124*, 125–132. [CrossRef]
18. Al-Momani, A.H. Construction delay: A quantitative analysis. *Int. J. Proj. Manag.* **2000**, *18*, 51–59. [CrossRef]
19. Cottrell, W.D. Simplified program evaluation and review technique (PERT). *J. Constr. Eng. Manag.* **1999**, *125*, 16–22. [CrossRef]
20. Leonardo, A.d.V.G.; Henry, L.V.; Ana, L.F.F. Playing chess or playing poker? Assessment of uncertainty propagation in open innovation projects. *Int. J. Proj. Manag.* **2021**, *39*, 154–169.
21. Lin, L.; Müller, R.; Zhu, F.; Liu, H. Choosing suitable project control modes to improve the knowledge integration under different uncertainties. *Int. J. Proj. Manag.* **2019**, *37*, 896–911. [CrossRef]
22. Tsai, W.H.; Lai, S.Y. Green Production Planning and Control Model with ABC under Industry 4.0 for the Paper Industry. *Sustainability* **2018**, *10*, 2932. [CrossRef]
23. Goldratt, E.M. *Critical Chain*; The North River Press Publishing Corporation: Great Barrington, MA, USA, 1997.
24. Cook, D.P. A simulation comparison of traditional JIT and TOC manufacturing systems in a flow shop with bottlenecks. *Prod. Invent. Manag. J.* **1994**, *35*, 73–78.
25. Blackstone, J.H.; Gardiner, L.R.; Gardiner, S.C. A framework for the systemic control of organizations. *Int. J. Prod. Res.* **1997**, *35*, 597–609. [CrossRef]
26. Gardiner, S.C.; Blackstone John, H., Jr.; Gardiner Lorraine, R. Drum-buffer-rope and buffer management: Impact on production management study and practices. *Int. J. Oper. Prod. Manag.* **1993**, *13*, 68–78. [CrossRef]
27. Spencer, M.S.; Cox, J.F. Optimum Production Technology (OPT) and the Theory of Constraints (TOC)—Analysis and Genealogy. *Int. J. Prod. Res.* **1995**, *33*, 1495–1504. [CrossRef]
28. Wu, S.Y.; Morris, J.S.; Gordon, T.M. A Simulation Analysis of the Effectiveness of Drum-Buffer-Rope Scheduling in Furniture Manufacturing. *Comput. Ind. Eng.* **1994**, *26*, 757–764. [CrossRef]
29. Steyn, H. Project management applications of the theory of constraints beyond critical chain scheduling. *Int. J. Proj. Manag.* **2002**, *20*, 75–80. [CrossRef]
30. Van de Vonder, S. The use of buffers in project management: The trade-off between stability and makespan. *Int. J. Prod. Econ.* **2005**, *97*, 227–240. [CrossRef]
31. Hu, X.; Cui, N.; Demeulemeester, E.; Bie, L. Incorporation of activity sensitivity measures into buffer management to manage project schedule risk. *Eur. J. Oper. Res.* **2016**, *249*, 717–727. [CrossRef]
32. Zhang, J.; Song, X.; Diaz, E. Project buffer sizing of a critical chain based on comprehensive resource tightness. *Eur. J. Oper. Res.* **2016**, *248*, 174–182. [CrossRef]
33. Bie, L.; Cui, N.; Zhang, X. Buffer sizing approach with dependence assumption between activities in critical chain scheduling. *Int. J. Prod. Res.* **2012**, *50*, 7343–7356. [CrossRef]
34. Goldratt, E.M.; Cox, J. *The Goal: A Process of Ongoing Improvement*; North River Press: Great Barrington, MA, USA, 1992.
35. Şimşit, Z.T.; Günay, N.S.; Vayvay, Ö. Theory of Constraints: A Literature Review. *Procedia Soc. Behav. Sci.* **2014**, *150*, 930–936. [CrossRef]
36. Hammad, M.W.; Abbasi, A.; Ryan, M.J. Developing a Novel Framework to Manage Schedule Contingency Using Theory of Constraints and Earned Schedule Method. *J. Constr. Eng. Manag.* **2018**, *144*, 225–236. [CrossRef]
37. Umble, M.M. Analyzing Manufacturing Problems Using V-A-T Analysis. *ProdInvent. Manag. J.* **1992**, *33*, 55–60.
38. Ye, T.; Han, W. Determination of buffer sizes for drum-buffer-rope (DBR)-controlled production systems. *Int. J. Prod. Res.* **2008**, *46*, 2827–2844. [CrossRef]
39. Thurer, M. Drum-buffer-rope and workload control in High-variety flow and job shops with bottlenecks: An assessment by simulation. *Int. J. Prod. Econ.* **2017**, *188*, 116–127. [CrossRef]
40. Izmailova, A.; Kornevaa, D.; Kozhemiakinb, A. Project Management with Theory of Constraints. *Procedia Soc. Behav. Sci.* **2016**, *229*, 96–103. [CrossRef]
41. Randm, G.K. Critical chain: The theory of constraints applied to project management. *Int. J. Proj. Manag.* **2000**, *18*, 173–177. [CrossRef]
42. Xie, X.-M.; Yang, G.; Lin, C. Software development projects IRSE buffer settings and simulation based on critical chain. *J. China Univ. Posts Telecommun.* **2010**, *17* (Suppl. 1), 100–106. [CrossRef]
43. Tukel, O.I.; Rom, W.O.; Eksioglu, S.D. An investigation of buffer sizing techniques in critical chain scheduling. *Eur. J. Oper. Res.* **2004**, *172*, 401–416. [CrossRef]
44. Demmy, W.S.; Demmy, B.S. Drum-buffer-rope scheduling and pictures for the yearbook. *ProdInvent. Manag. J.* **1994**, *35*, 45–47.
45. Slusarczyk, A. A comparison of buffer sizing techniques in the critical chain method case study. *J. Autom. Mob. Robot. Intell. Syst.* **2013**, *7*, 43–56.
46. Wei, C.-C.; Liub, P.-H.; Tsaic, Y.-C. Resource-constrained project management using enhanced theory of constraint. *Int. J. Proj. Manag.* **2002**, *20*, 561–567. [CrossRef]
47. Schragenheim, E.R.B. Buffer Management: A Diagnostic Tool for Production Control. *ProdInvent. Manag. J.* **1991**, *32*, 74–79.

48. Ronen, B.; Starr, M.K. Synchronized Manufacturing as in Opt—From Practice to Theory. *Comput. Ind. Eng.* **1990**, *18*, 585–600. [CrossRef]
49. Cohen, I.; Mandelbaum, A.; Shtub, A. Multi-Project Scheduling and Control: A Process-Based Comparative Study of the Critical Chain Methodology and Some Alternatives. *Proj. Manag. J.* **2004**, *35*, 39–50. [CrossRef]
50. Kock, A.; Schulz, B.; Kopmann, J.; Gemünden, H.G. Project portfolio management information systems' positive influence on performance—The importance of process maturity. *Int. J. Proj. Manag.* **2020**, *38*, 229–241. [CrossRef]
51. Papadonikolaki, E.; van Oel, C.; Kagioglou, M. Organising and Managing boundaries: A structurational view of collaboration with Building Information Modelling (BIM). *Int. J. Proj. Manag.* **2019**, *37*, 378–394. [CrossRef]
52. Pinedo, M.; Zacharias, C.; Zhu, N. Scheduling in the service industries: An overview. *J. Syst. Sci. Syst. Eng.* **2015**, *24*, 1–48. [CrossRef]
53. Drechsler, A.; Breth, S. How to go global: A transformative process model for the transition towards globally distributed software development projects. *Int. J. Proj. Manag.* **2019**, *37*, 941–955. [CrossRef]
54. Budayan, C.; Dikmen, I.; Birgonul, M.T.; Ghaziani, A. A Computerized Method for Delay Risk Assessment Based on Fuzzy Set Theory using MS Project™. *KSCE J. Civ. Eng.* **2018**, *22*, 2714–2725. [CrossRef]
55. Cobb, B.R.; Alan, W.J.; Rumi, R. Accurate lead time demand modeling and optimal inventory policies in continuous review systems. *Int. J. Prod. Econ.* **2015**, *163*, 124–136. [CrossRef]
56. Hazır, Ö. A review of analytical models, approaches and decision support tools in project monitoring and control. *Int. J. Proj. Manag.* **2015**, *33*, 808–815. [CrossRef]
57. Akbar, H.; Mandurah, S. Project-conceptualisation in technological innovations: A knowledge-based perspective. *Int. J. Proj. Manag.* **2014**, *32*, 759–772. [CrossRef]

Article

7-Score Function for Assessing the Strength of Association Rules Applied for Construction Risk Quantifying

Hubert Anysz *, Jerzy Rosłon and Andrzej Foremny

Department of Production Engineering and Construction Management, Faculty of Civil Engineering, Institute of Building Engineering, Warsaw University of Technology, 00-637 Warsaw, Poland; j.roslon@il.pw.edu.pl (J.R.); a.foremny@il.pw.edu.pl (A.F.)
* Correspondence: h.anysz@il.pw.edu.pl

Abstract: There are several factors influencing the time of construction project execution. The properties of the planned structure, the details of an order, and macroeconomic factors affect the project completion time. Every construction project is unique, but the data collected from previously completed projects help to plan the new one. The association analysis is a suitable tool for uncovering the rules—showing the influence of some factors appearing simultaneously. The input data to the association analysis must be preprocessed—every feature influencing the duration of the project must be divided into ranges. The number of features and the number of ranges (for each feature) create a very complicated combinatorial problem. The authors applied a metaheuristic tabu search algorithm to find the acceptable thresholds in the association analysis, increasing the strength of the rules found. The increase in the strength of the rules can help clients to avoid unfavorable sets of features, which in the past—with high confidence—significantly delayed projects. The new 7-score method can be used in various industries. This article shows its application to reduce the risk of a road construction contract delay. Importantly, the method is not based on expert opinions, but on historical data.

Keywords: association analysis; tabu search; delay; risk; construction project

Citation: Anysz, H.; Rosłon, J.; Foremny, A. 7-Score Function for Assessing the Strength of Association Rules Applied for Construction Risk Quantifying. *Appl. Sci.* **2022**, *12*, 844. https://doi.org/10.3390/app12020844

Academic Editor: Francesco Colangelo

Received: 5 December 2021
Accepted: 11 January 2022
Published: 14 January 2022

Publisher's Note: MDPI stays neutral with regard to jurisdictional claims in published maps and institutional affiliations.

Copyright: © 2022 by the authors. Licensee MDPI, Basel, Switzerland. This article is an open access article distributed under the terms and conditions of the Creative Commons Attribution (CC BY) license (https://creativecommons.org/licenses/by/4.0/).

1. Introduction

The early stage of a construction project planning process is characterized by a high level of uncertainty. Although every construction project is unique, the data collected from previously completed projects help to plan the new one. The problem of estimating the time necessary to complete a project becomes more complicated if "design & build" orders are applied. There are several factors influencing the completion time of construction projects, including the properties of the planned structure, the details of an order, macroeconomic factors, and prices of materials. The delayed completion date of the construction contract makes the contractor's costs much higher than expected [1–3]. The negative impact of such a delay also concerns the client and the community for whom the built object serves [4]. This is why identifying the most important causes of delays is crucial. Different methods are applied for identifying and validating their importance [4,5]. Lowering the possibility of delay occurrence can concentrate on either proper planning of work execution (planning the duration of the execution of work [1–6]), scheduling [7], or on avoiding unfavorable circumstances for project execution [8]. As the contractors base their decisions on their experience [9,10] (carrying out decisions about participation in a given tender procedure), completed projects can be analyzed to avoid circumstances that have resulted in a significant delay in the completion of projects in the past. The field of project management aimed at reducing the impact of threats of implementation not in accordance with the adopted plan is the interdisciplinary science of risk management. It is often reduced to the application of qualitative and quantitative risk analysis. The construction industry in general, as well as individual construction projects, deal with various risks [11,12]. Especially, infrastructure projects, as they are large in the volume of works and results, involve huge budgets. This means that failures may result in huge monetary losses, which are caused by the various risks linked with such projects [13]. This is why risk must be properly identified and mitigated [14,15].

The risk assessment process requires the introduction of several important assumptions regarding, inter alia, the distribution of the probabilities of occurrence and the occurrence costs of individual risk factors, as well as assumptions regarding the efficiency and costs related to the implementation of activities provided for in the schedule. At the preparatory stage of an investment project, a risk matrix is often created, which is a graphic representation of the risk analysis process (Table 1).

Table 1. Sample risk matrix of a construction project [2].

		The Likelihood of a Hazard Occurring		
		Small (0–33%)	Medium (34–66%)	Large (67–100%)
Consequences of the threat to the project	Small	Protests of environmentalists	Protests of the local population	Unfavorable contracts with contractors
	Medium	Changes in regulations Lack of renewable resources Lack of non-renewable resources	Construction equipment failure Interruptions in access to the media Availability of key employees	Low performance of work teams Late delivery of materials
	Large	No building permit Investor's financial problems	Bad weather conditions Design errors	Loss of financial liquidity Lack of funds Subcontractor errors

Over the course of years, many approaches to construction project risk management have been developed by various researchers. Wang et al. [16] developed an alien eyes' risk (AER) model, which uses hierarchical levels of risk and the mutual relationships between the risks and a qualitative risk mitigation framework. Schieg [17] proposed a risk management process in construction project management, which puts more emphasis on personal area risks. Choudhry and Iqbal [18] identified and prioritized common risks, management techniques to address them, the current status of the risk management systems implemented in organizations, and barriers for effective risk management in the construction industry. Taroun and Yang [19] introduced a combination of the Dempster–Shafer theory of evidence, a reasoning algorithm for structuring personal experience and professional judgment, and a classic spreadsheet-based decision support system. Serpella et al. [20] used a knowledge-based approach. The approach addresses project risks in the construction management industry based on a threefold arrangement and risk management function. Ebrat and Ghodsi [21] proposed the adaptive neuro-fuzzy inference system and stepwise regression model as a means of identifying and evaluating the risks in construction projects. Iqbal et al. [22] developed a risk management framework that allows for reporting the significance of different types of risks and the effectiveness of various risk management techniques commonly practiced in the construction industry. Vafadarnikjoo et al. [23] proposed the use of an intuitive fuzzy decision-making trial and evaluation laboratory (DEMATEL) to prioritize the risks associated with construction projects by using the risk breakdown structure (RBS). Kao et al. [24] suggested using an integrated fuzzy ANP (analytical network process)-based balanced scorecard system for the evaluation of relevant bilateral factors for the Taiwanese construction sector collaborating with local Chinese contractors. Ahmadi et al. [25] analyzed the criteria, prioritized potential risk events, and used the fuzzy AHP technique to quantify them. Li et al. [26] adopted text mining methods to identify safety risk factors and participants in urban rail projects. Chatterjee et al. [11] developed a hybrid D-ANP-MABAC model including the ANP methodology in the D numbers domain and extended multi-attributive border approximation area comparison (MABAC) method.

Anysz et al. [27] have found the set of unfavorable conditions usually accompanying the significant delays of construction projects with the use of association analysis. This tool is suitable for uncovering the rules in data, i.e., unusually frequent simultaneous appearance of factors or phenomena [28,29]. Although the speed of calculation is high, because of the use of dedicated software, the input data to the association analysis have to be preprocessed—every feature influencing, e.g., the duration of the project, has to be divided into ranges. The number of features and the number of ranges (for each feature) can create a very complicated combinatorial problem. The authors decided to use a metaheuristic algorithm to find the acceptable thresholds in association analysis, increasing the strength of the rules found. The sequence of the previous and current findings is presented in Figure 1.

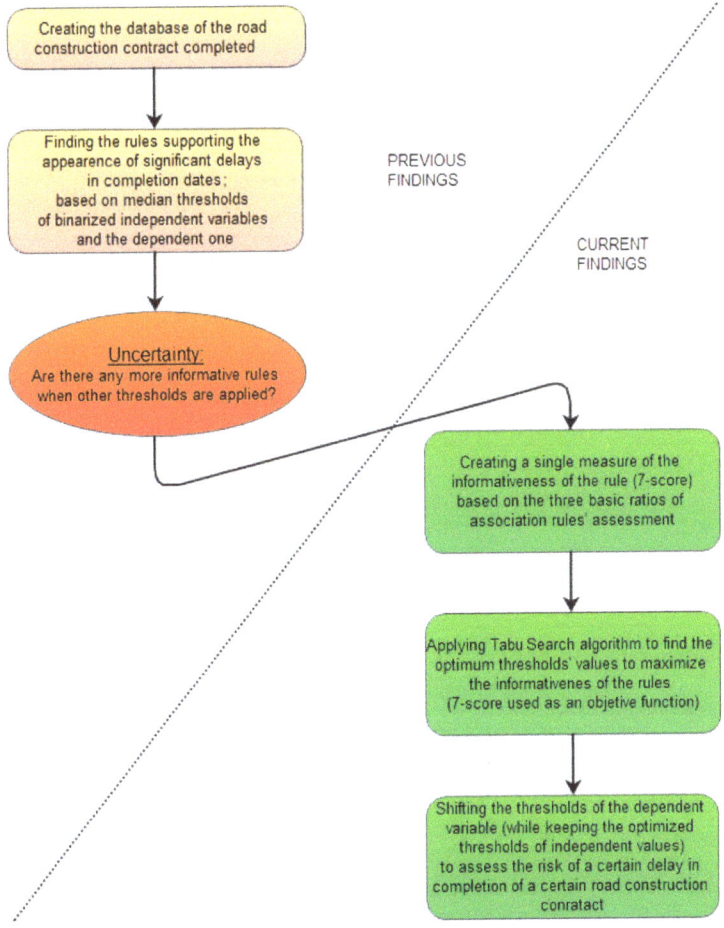

Figure 1. The sequence of the findings introduced and presented in the article.

The increase in the strength of the rules can help clients to avoid unfavorable sets of features, which in the past—with high confidence—significantly delayed projects. Data presented in the previous article [27] serve as a base to this work and concern the road construction projects (express roads and highways) completed between 2009 and 2013 in Poland. After presenting materials and methods, the invented 7-Score function is defined. It combines, in one formula, the typical ratios assessing the rules. The 7-Score assesses the strength of rules, so their importance can be ranked. Creating a 7-Score function is necessary to apply the tabu search algorithm that maximizes the objective function (it must be a single one). As a result, the most powerful and the most informative rules can be found. They are presented and discussed in the Section 4. Based on them, it is possible to assess the risk of delay in the completion of a road construction contract that meets the criteria applied in the analysis. This is to emphasize that the introduced innovative method of quantitative risk assessment is not based on the experts' opinions, but rather on evidence concerning the collected and completed construction contracts of the same kind.

2. Materials and Methods

2.1. Association Analysis

Association analysis was invented to increase sales in supermarkets. The contents of clients' trolleys were analyzed to find the rules for the appearance of specific goods in a trolley by a cash desk. Thus, the synonym for association analysis is market basket analysis [30]. Each rule found consists of a predecessor (body of the rule) and the consequent (head of the rule). The rule can be presented as $if\ body,\ then\ head$ or $body \rightarrow head$. Having a dataset comprising many cases consisting of their bodies and heads, it is possible to assess the meaning of the rule by three ratios called confidence ($conf$), support (sup), and lift (marked with its full name). They can be calculated as follows:

$$conf = \frac{n_{bh}}{n_b} \quad (1)$$

$$sup = \frac{n_{bh}}{N} \quad (2)$$

$$lift = \frac{conf}{P(h)} \quad (3)$$

where

n_{bh} is the number of cases where the criteria for body (predecessor) are met and simultaneously the criterium (or criteria) for head are also met;
n_b is the number of cases where the criteria for body are met;
N is the total number of cases in the database;
$P(h)$ is the probability of appearing head meeting the criteria set for head.

This probability can be calculated as follows:

$$P(h) = \frac{n_h}{N} \quad (4)$$

where n_h is the number of cases with heads meeting the criteria (set for heads). The rule with 100% confidence means that, every time a specific predecessor appears, then the specific consequent also always appears. This kind of rule is even more informative when there is a significant number of cases meeting the rule. Then, the support of the rule is relatively high (the total range of support values is (0, 1)). If the support is at a minimum, this means that there is only one case meeting the rule in the whole database of the cases. The lift has a secondary function. It protects against considering the rules (even of high confidence) for which the probability of a specific head is higher than the calculated confidence. If lift < 1, the rule is useless [28–31]. The importance of rules is further discussed in Section 3.1., where the total measure of the importance of rules is introduced. The body of the rule can be described by several features and conditions to be met, formulated with any logic expression (with OR, AND operators). That, and the simplicity of parameters describing each rule, allow association analysis to be used in a variety of applications. Nowadays, association analysis is still applied for the designed purpose ([32] as an example). However, smart applications can be found in several areas, e.g., for the following:

- precipitation prediction [33];
- insurance risk assessment [34];
- traffic safety analysis [35–37];
- assessment of construction project risk [27];
- assessment of risk in construction disputes [38];
- a variety of problems in biology [39–41];
- preferences' discovering in social sciences [42];
- collusion detection in tender procedures [43];
- quality management problem-solving in production [44].

The rule-finding processes have to be computer-aided as the number of rules is usually huge even if the database searched is not large. It is a common case where, within several thousand rules found, only several are meaningful.

2.2. The Analysed Case and Its Database

This paper is based on previous research that analyzed the studies on all projects of building express roads and highways completed in Poland between January 2009 and December 2013 [2,27]. Additional Polish and international literature research for possible reasons for delays in construction contracts was summarized in [2]. The result of the aforementioned research was the list of 142 possible reasons for delays. A huge number of them were reduced, mainly according to the fact that the moment of analysis took place before the choice of the contractor (by the client), before the start of building works. The final list is presented in Table 2 [27].

Table 2. Possible causes of delays and their values [28].

ID	The Cause of Delay	Values
A	Value of works	rational number (in PLN)
B	Length of the section built	rational number (in km)
C	Planned duration of the project	integer number (in days)
E	Project scope	binary: design & build = 1; build = 0
F	Project type	binary: build = 1; modernize etc. = 0
G	The total, average number of employees employed by contractor [1]	integer number (no. of persons)
H	Half of the year of works commencement	binary: first half = 0; second half = 1
I	The trend of unemployment rate in Poland [2]	binary: decreasing = 0; increasing = 1
J	The trend of price index in Polish construction industry [2]	binary: decreasing = 0; increasing = 1
K	The trend of total sales in Polish construction industry [2]	binary: decreasing = 0; increasing = 1
L	Number of partners in consortium (acting as contractor)	integer number
M	Summarized yearly total sales of consortium partners [1]	rational number (in PLN)

[1] Calculated for the year preceding the commencement of works. [2] Calculated year to year (the year preceding the commencement of works, to the year before).

Label D is left for marking a delay. Its integer value is calculated for each project based on the following formula:

$$D_i = \begin{cases} T_i^{(r)} \leq T_i^{(pl)} \to 0 \\ T_i^{(r)} > T_i^{(pl)} \to T_i^{(r)} - T_i^{(pl)} \end{cases} \quad (5)$$

where
1. $T_i^{(pl)}$ is a planned duration of the project given in days;
2. $T_i^{(r)}$ is an observed real duration of the project given in days;
3. i is an index of analyzed project.

The twelve factors listed in Table 2 that may influence the delay of the completion date of road construction projects can be categorized into three main groups by origin. That is, client-decision-dependent (B, C, E, H), contractor-dependent (A, G, L, M), or based on macroeconomic factors (I, J, K). Factor F arises from the technical matters and the standing of the national economy. The majority of data were provided by the Polish General Directorate for National Roads and Highways (GDDKiA) at the request of the Warsaw University of Technology. Macroeconomic factors were found in the Polish Central Statistical Office (GUS). For the real completion dates, approximately 500 websites were scraped. The data concerning the number of employees and the yearly sales of contractors were obtained commercially. The complete set of twelve feature values was completed for 139 projects, and only these were analyzed further in previous studies [27].

2.3. The Problem to Solve

As association analysis works well for dichotomous types of bodies and dichotomous types of head, the collected data (their types are presented in Table 2) as well as each type of body and head need to be divided into two subsets. In [27], the thresholds were assumed as median values. However, it is possible that, if other thresholds are set, the rules found can then be more informative. The problem is illustrated in Figure 2.

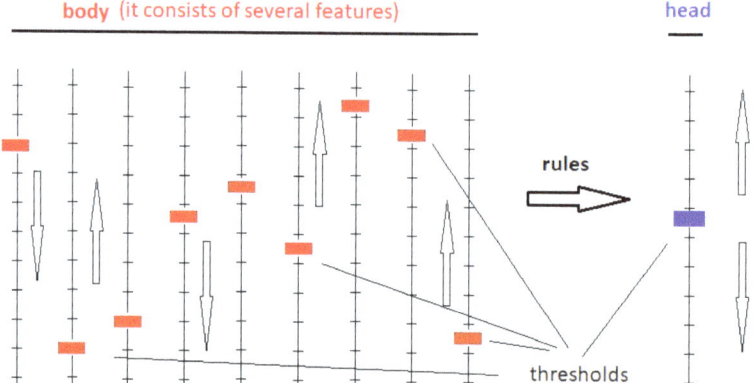

Figure 2. The problem: how to set the thresholds (red and blue) to allow finding the most powerful and most informative rules (if body then head) based on a specific database.

2.4. Tabu Search

Some practical problems in construction can be easily qualified as NP-hard (non-deterministic polynomial-time hard) problems. The time needed to solve these problems grows exponentially with the increase in the problem's size [45]. This is why mathematical methods do not allow for finding solutions for complicated construction problems in an acceptable time. For the same reasons, metaheuristic algorithms seem to be the most appropriate measures for scheduling and task sequencing. These algorithms do not guarantee finding the optimal solution to the given problem; however, they are very useful when it comes to solving NP-hard problems because they allow for finding suboptimal solutions in an acceptable time [46]. Finding the number of features and the number of ranges (for each feature) proved to be such a combinatorial problem.

It was decided to use the tabu search algorithm. Its advantages have been proven in many scientific publications [47–50]. Like many other IT solutions used in various industries, it can be adopted to construction problems [51]. The basic idea behind this algorithm is to search the solution space by a sequence of moves [50]. In this sequence, some moves are considered tabu moves—they are forbidden. The TS algorithm avoids getting stuck in local optima by storing the information about previously checked solutions in the form of tabu lists. The list grows as the algorithm proceeds. However, when it reaches its maximum capacity, the oldest entries of the tabu list are overwritten by the new ones. The simplified tabu search pseudocode in Table 3 presents its principles. It was decided to use the tabu search algorithm to find the thresholds in association analysis, which provides an increase in the strength of the rules found. It is a new approach and has never been applied before.

Table 3. Simplified tabu search pseudocode.

Line of Code	Code
1	sBest = s0
2	bestCandidate = s0
3	tabuList = []
4	tabuList.push(s0)
5	repeat (loop)
6	sNeighborhood ← getNeighbors(bestCandidate)
7	for (sCandidate in sNeighborhood)
8	if ((not tabuList.contains(sCandidate)) and (fitness(sCandidate) > fitness(bestCandidate)))
9	bestCandidate = sCandidate
10	end
11	end
12	if (fitness(bestCandidate) > fitness(sBest))
13	sBest = bestCandidate
14	end
15	tabuList.push(bestCandidate)
16	if (tabuList.size > maxTabuSize)
17	tabuList.removeFirst()
18	end
19	until stopping-criteria satisfied
20	return sBest

3. Results

3.1. Assessing the Strength of Association Rules with 7-Score

Considering the three basic ratios describing the rules, i.e., confidence, support, and lift, the most powerful is confidence. If a certain type of predecessor appears, a certain type of a consequent appears too every time. The confidence of this kind of rule is 100%. This kind of information gathered by a user of association analysis is very strong. The collected data provide the user with a high likelihood of a certain result if the same type of predecessor appears again. However, not every rule of 100% confidence gives the same level of certainty of appearing to be a specified consequent. The three examples of phenomena that can be described with 100% confidence are presented in Figure 3.

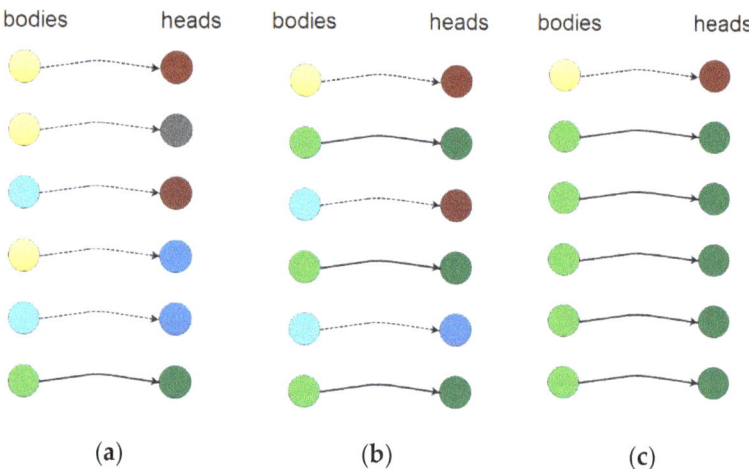

Figure 3. Three different exemplary datasets (**a**–**c**) with the rules of the same confidence of 100%. The rule: if light green, then dark green.

As presented in Figure 3b, predicting the effect—dark green—based on this dataset seems more powerful than in the case presented in Figure 3a. There, the rule is based on one case only. It is unknown if the case is caused by the nature of the analyzed phenomenon, or if it has happened by chance. The rule seems to be the most powerful in the case presented in Figure 3c. Support calculated for the rule, for cases (a), (b), and (c), is 1/6, 3/6, and 5/6, respectively. It can be concluded that, for the rules of the same confidence, the more powerful (meaningful) is the rule with higher support. Then, the following question can be asked: which rule is stronger of the following two: rule 1: conf = 100% and sup = 33.3%, rule 2: conf = 75% and sup = 66.7%? To answer this, the large database should be considered. Then, if the rule of 100% confidence is supported by 33.3% of cases, it still a large number of cases where the appearance of a light green body always makes the head dark green. For much smaller databases being analyzed, it seems sufficient if support is higher than its minimum value, i.e., $1/N$ (where N is a total value of cases in the database). Minimum support means that the rule is based on one case meeting the conditions of the rule. It can be stated that, for the rules with support higher than the minimum of one, confidence is more meaningful than support. The rules of $sup = 1/N$ should be excluded from the analysis.

The influence of lift on the strength of the rule should also be considered, as two rules of identical confidence and support can have different lifts (as presented in Figure 4).

Aiming at predicting a dark green head, based on a light green appearance, the rule for the dataset presented in Figure 4a seems to be a bit stronger, as the dark green head appears only if the light green body has appeared earlier. In case (b), dark green heads can also appear for bodies other than light green ones, but in case (a), the rule gives the full explanation for the appearance of the dark green bodies. For both cases, conf = 100% and sup = 33.3%; however, lift = 3 for (a) and lift = 1.2 for (b). It can be concluded that, for two rules of the same confidence and the same support, the stronger is the rule with the higher lift. When comparing the rules of different confidences and different supports, considering a lift seems unreasonable as—as discussed earlier—the meaning of confidence is higher than the meaning of support.

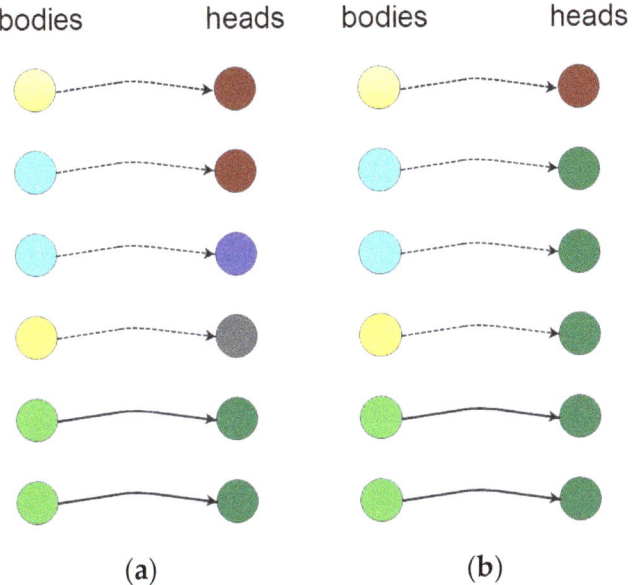

Figure 4. Two different exemplary datasets (**a**,**b**) with the rules of the same confidence of 100% and support of 33.3%. The rule: if light green, then dark green.

The next issue is assessing the rules of low confidence. Please observe the two examples illustrated in Figure 5.

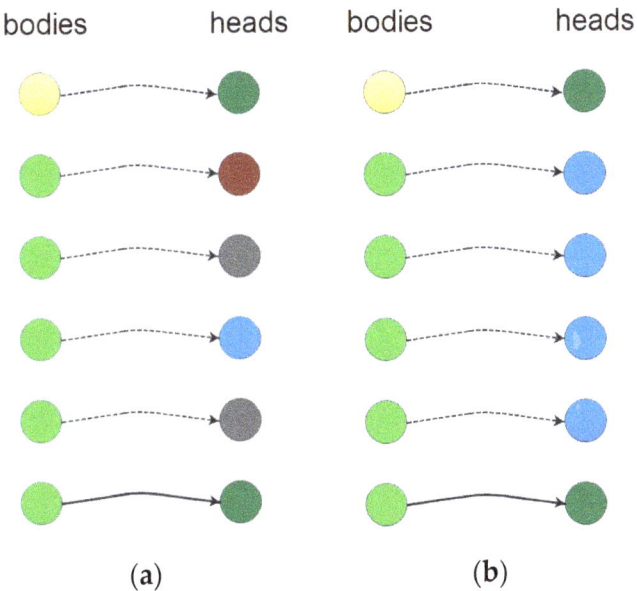

Figure 5. Two different exemplary datasets (**a**,**b**) with the rules of the same confidence of 20% and support of 16.7%, and lift of 1. The rule: if light green, then dark green.

In case (a), the heads are multi-colored and the rule—if light green, then dark green—seems meaningless. In case (b), where the head is dichotomous, it seems that finding the opposite rule (if light green, then blue) brings a better result (conf = 80%, sup = 66.7%, lift = 1). The same result will be achieved in case (a) if the rule will be stated as follows: if light green, then not dark green. It can be concluded that the rules of low confidence are meaningless. To assess the strength of rules, the following aim function is created:

$$\text{strength of rule} = \text{lift} + N^2 \times \text{sup} + N^2 \times \text{conf} \quad (6)$$

where N is for the total number of cases in a database. Equation (6) considers the following assumptions. Assumption 1:

$$I(\text{sup}) > I(\text{lift}) \quad (7)$$

where I is a function of the importance of the rule. Equation (7) is achieved by making the sum component of support equal or higher than lift, as follows:

$$\text{sup} \times N = n_{bh} \quad (8)$$

where n_{bh} is the number of cases meeting the rule and, as the maximum lift is N and the minimum support is $\frac{1}{N}$, the following Equation is met:

$$N^2 \times \text{sup} \geq \text{lift} \quad (9)$$

Meeting assumption 2 presented in Equation (10), using Equation (6),

$$I(conf) > I(sup) \quad (10)$$

is achieved by multiplying the confidence by the same number as the support, i.e., by N^2, as the confidence is greater than the support for each rule (as the number of bodies meeting the rule is always lower than N). Equation (6) for the strength of rule introduces possible cases where the joint impact of lift and support is greater than the impact of confidence on the strength of the rule. These kinds of cases are partially limited by excluding from the analysis the cases of low confidence (below 50%). To observe how the rules are assessed, the exemplary database is created of 10 bodies and 10 heads. The number of bodies meeting the rule n_b changes from 1 to 9, and the number of heads meeting the rule n_h also changes from 1 to 9. The number of cases meeting the rule n_{bh} changes from 1 to a number defined as $min(n_b, n_h)$. All possible combinations are assumed, and all rules are found in the created cases. Confidence, support, lift, and the strengths of the rules are calculated. From the full set of rules, regardless of the rules of confidence lower than 0.5, the cases with lift lower than 1 are excluded too. When the lift is lower than 1, this means that, when predicting the head, the better result can be achieved by applying the probability of appearance of a specific head, rather than basing it on a specific body appearance. The remaining data and results (scores) are presented in Appendix A Table A1. As it is difficult to present a 4-dimentional chart in a 2D figure, Figure 6 is prepared. Support and confidence are on the horizontal axes and 7-Score values are on the vertical axis.

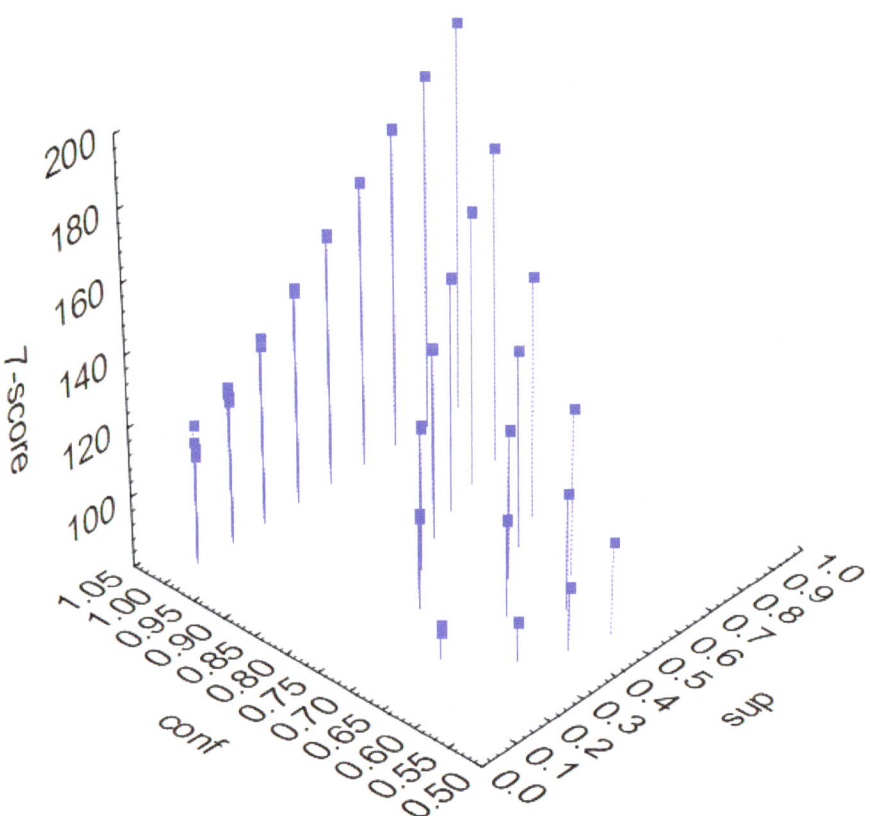

Figure 6. 7-Score values are presented for every combination of important rules presented in Table A1.

It can be observed that, for several pairs of identical conf and sup, there are several values of 7-Score. This is because of the influence of lift—which is also considered in 7-Score and in Figure 6. Lift differentiates 7-Score for the cases of the same support and confidence, as was assumed while the formula for 7-Score was created. Observing Figure 6 and, especially, Figure 7 i.e., the 2-dimentional scatter-plot for support and confidence, the shape of the 7 sign can be recognized—the basis of the name of the proposed method for scoring the strength of rules.

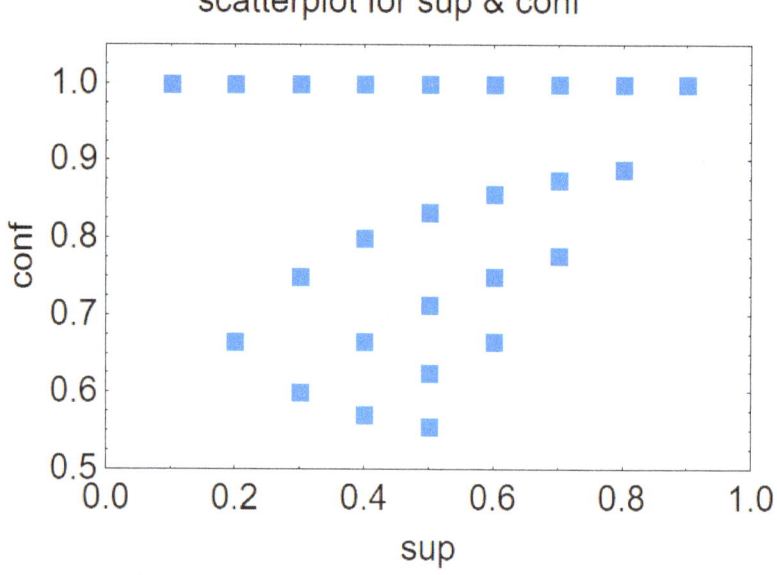

Figure 7. Two-dimensional scatterplot of sup and conf for all rules presented in Appendix A Table A1.

The database assumed to create the 7-Score is 10 × 10, considering that
- every combination of n_b, n_h, n_{bh} is assumed for creating the exemplary database and rule finding (presented in Table A1 and Figures 6 and 7);
- the values of sup and conf are always ≤ 1.

It can be stated that, for more numerous databases, the general shape of the scatter-plot will remain unchanged. It will be denser, especially between the points of very similar confidence (as the impact of confidence on the 7-Score is the highest). The plane presented in Figure 8 is an approximation of the 7-Score of the rules; however, it is presented to better explain the areas of the highest importance of the rules.

The aim of introducing the 7-Score measure is to compare the rules found based on a specific database (comprising bodies and heads) concerning a specific, analyzed phenomenon. For that reason, it can be used as is (not as a percentage of the highest 7-Score value). In order to compare the rules calculated for the databases of a different size, the relative 7-Score measure should be applied, as the values of 7-Score defined in (1) strongly depend on N, i.e., on the number of cases in a database.

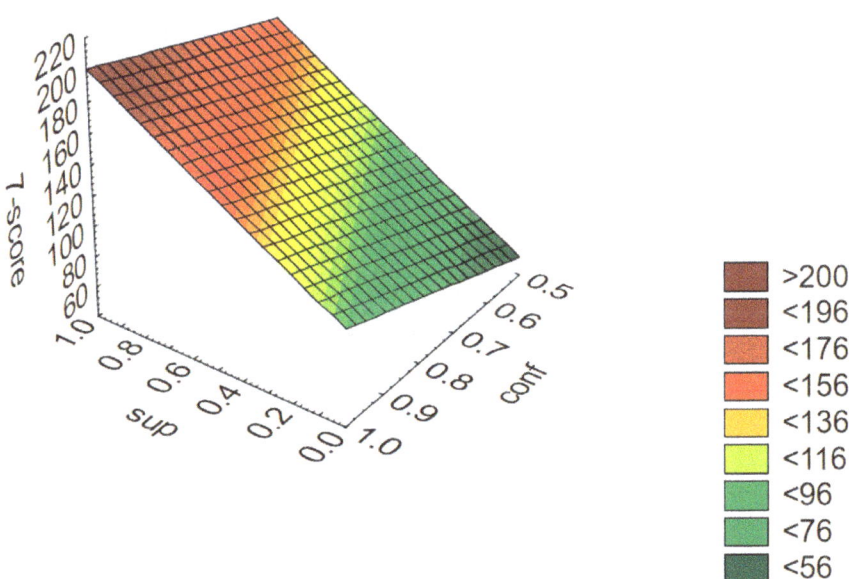

Figure 8. The approximate plane of 7-Score values.

3.2. Solving the Analysed Case

The previous results presented in [27] were very promising; however, only median values were used as bodies' thresholds. Testing different thresholds even for 139 projects proved to be a complex combinatorial problem, with up to 7.5×10^{31} potential variants. However, finding the right solution could improve the support and confidence parameters, thus providing better outcomes for the clients. This is why it was decided to use a metaheuristic algorithm. Such an approach proved to be very useful and might be used even for bigger databases.

Metaheuristic optimization of thresholds was done for three cases: two best sets of criteria established by [27] ($C_r - E - J - L$ and $A - E - K$), and for all 12 criteria from Table 2. The best results are presented below in Tables 4–6. The presented results were obtained with the use of commercial software OptQuest® Engine package, OptTek Systems, Inc., based on the tabu search algorithm. However, additional tests showed that similar results can be obtained by other applications of tabu search. The decision variables were the thresholds of criteria, and the objective function (SCORE) is as follows:

$$\text{Max}: \text{SCORE} = \text{lift} + N^2 \times \text{sup} + N^2 \times \text{conf} \qquad (11)$$

The results are presented in following Tables 4–6 together with the comparison to the results achieved in the previous study [27].

Table 4. Optimization results for criteria set Cr—E—J—L.

Support (%)	Confidence (%)	Lift	Score	Case
8.6	100	2.044	2,098,263	Median threshold
25.9	90	1.191	2,239,305	Metaheuristic

Table 5. Optimization results for criteria set A—E—K.

Support (%)	Confidence (%)	Lift	Score	Case
22.3	75.6	1.460	1,891,527	Median threshold
50.4	84.3	1.066	2,602,540	Metaheuristic

Table 6. Optimization results for all (presented in Table 2) criteria considered as the predecessor.

Support (%)	Confidence (%)	Lift	Score	Case
-	-	-	-	Median threshold
5.8	100.0	1.390	2,044,163	Metaheuristic

The rule wherein all features of the body are considered is excluded from further analysis according to its low support (even if this formula is found—as in the two other rules—with the use of metaheuristic). This makes its 7-Score much lower than the 7-Score of the two other rules. The maximum informativeness is found for the following rules:

- if (Cr and E and J and L), then D; that is, if (planned duration is lower than 1126 days and the contract is not "design & built" and price index in the construction industry is decreasing and the contractor has the form of consortium), then the contract is delayed;
- if (A and E and K), then D; that is, if (the contract value is over 5.77 million PLN and the contract is not "design & built" and the total sales in Polish construction industry is decreasing), then the contract is delayed.

4. Discussion

The most promising two rules for the appearance of delayed completion of construction were found in [27] with the use of association analysis. The bodies of these rules consist of several parameters, and it was decided to make their value dichotomous. The same is made with the head, i.e., the size of delay. Through the use of the tabu search algorithm, the settings of the thresholds (necessary to make the sets of values dichotomous) are found, making the two rules (if Cr-R-J-L, then D; if A-E-K, then D) the most informative. As can be seen in the tables presented (Tables 4 and 5), the determination of thresholds using the metaheuristic algorithm significantly improved the parameters describing the rules (in comparison with the median values used in [27]). There was a drastic improvement in the support for the rules in every case. Moreover, the scores for each case were significantly higher. The results obtained using the tabu search algorithm are significantly better than those obtained in the traditional way with the use of median values. The proposed innovative solution may be particularly useful when analyzing larger databases, where it is even more difficult to select the threshold levels. As already mentioned, metaheuristic algorithms are currently the best way to find solutions to particularly complex combinatorial problems. The results of the study only confirmed this thesis.

The assessment of the level of informativeness of the rules is possible because of the created measure named 7-Score. A significant improvement is achieved. For the rule with the body Cr-E-J-L, the confidence is lowered from 100% to 90%. However, support for these rules is increased from 8.5% to 25.9%. This means that there are three times more cases supporting the rule. Despite that the confidence and the lift are slightly lowered, owing to the significant support increase, 7-Score is approximately 10% higher than for median thresholds. For the most informative rule with A-E-K body, the increase is noted for both support (22.3% to 50.4%) and confidence (75.6% to 84.3%). Despite the lowered lift (1.460 to 1.046), 7-Score is more than 37% higher (up to 2,239,305). For these two very informative rules, the same threshold was found—zero. The head of this rule is defined as follows: the delay of a construction completion greater than the threshold. It has to be stated that there are several contracts (cases) in the database completed on time (not delayed, i.e., delay = 0). Considering the values of the thresholds found of 5,765,055.35 for A, 0 for E, and 0 for K, the rule if A-E-K, then D brings the following information based on the passed construction contract:

If:
- the contract value was above 5.77 million PLN,
- the contract scope was to build (design provided by a client), and
- the total sales of the Polish construction industry were decreasing (year to year),

then the completion of this type of contract was delayed with conf = 90%, sup = 25.9%, and lift = 1.191. This is to emphasize that such a calculation can be done before any new contract that is ordered and signed. Shifting the threshold for the head (the size of delay) from 0 to its maximum value, the set of results (conf, sup. lift, 7-Score) can be achieved for the rule if A-E-K, then the delay greater than the threshold value. This scenario is presented in Figure 9.

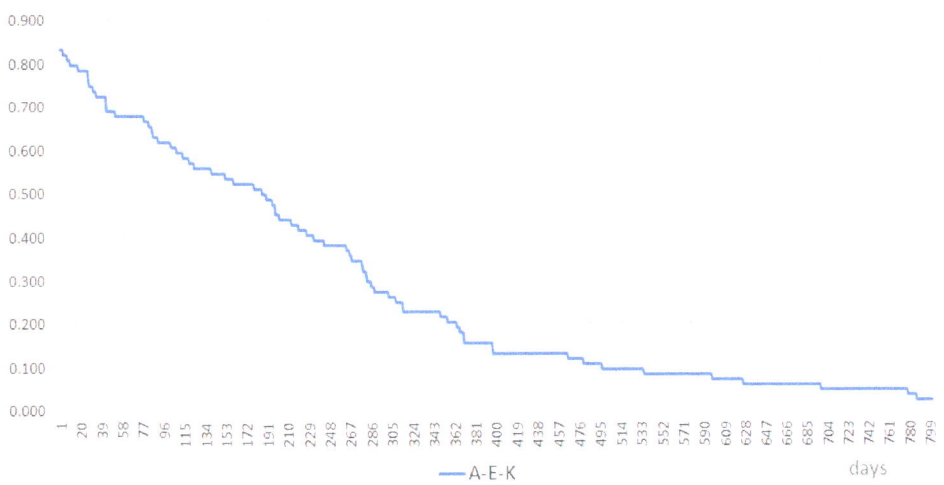

Figure 9. The confidence of the rule if A-E-K, then D for different number days as the threshold of the head (delay).

It can be observed that, the higher the threshold of the head, the lower the confidence in the delay appearance being greater than the threshold. It is to be noted that the thresholds of the body parameters (A, E, K) are left on the unchanged levels (as found for the highest 7-Score). As a natural result of shifting to the right, the thresholds of the head, supports, and 7-Scores lower, with the head threshold increasing. The full set of parameters of the rules (for the threshold D of the head being set from 0 to 800) is presented in Table 7 (and Table 8 for Cr-E-J-L body).

Table 7. Parameters of the rule with A-E-K body calculated for several thresholds of delay.

Support (%)	Confidence (%)	Lift	Score	D
50.4%	83.3%	1.034	26,205	0
37.4%	61.9%	1.195	19,467	100
27.3%	45.2%	1.338	14,226	200
16.5%	27.4%	1.312	8611	300
7.9%	13.1%	1.300	4119	400
5.8%	9.5%	1.203	2996	500
4.3%	7.1%	1.241	2247	600
2.9%	4.8%	1.655	1499	700
1.4%	2.4%	3.310	752	800

Table 8. Parameters of the rule with Cr-E-J-L body calculated for several thresholds of delay.

Support (%)	Confidence (%)	Lift	Score	D
26.6%	92.5%	1.148	23,016	0
20.1%	70.0%	1.351	17,418	100
15.1%	52.5%	1.553	13,064	200
7.9%	27.5%	1.318	6844	300
3.6%	12.5%	1.241	3111	400
2.2%	7.5%	0.948	1867	500
0.7%	2.5%	0.434	622	600
0.0%	0.0%	0	0	700
0.0%	0.0%	0	0	800

Let us analyze the opposite rule, i.e., if A-E-K, then delay is not greater than the threshold for the head. The number of bodies meeting the original rule n_b remains unchanged in the opposite rule. The parameters of the opposite rule are calculated just for the unchanged body. It can be written as follows:

$$conf\left(b \to h^{(-)}\right) = \frac{n_{bh^{(-)}}}{n_b} \qquad (12)$$

where
- $h^{(-)}$ is the opposite side of the dichotomous head;
- $n_{bh^{(-)}}$ is the number of cases meeting the opposite rule (where the head is inverted).

There are several (or even hundreds of) types of bodies, but only one type of body is analyzed. There are n_b bodies of this kind. From this subset, only n_{hb} bodies meet the rule, i.e., the number of heads is greater than the threshold. This means that the rest of the subset meets rule that the head values are not greater the threshold. Thus, the number of bodies meeting the inverted head can be calculated as follows:

$$n_{bh^{(-)}} = n_b - n_{bh} \qquad (13)$$

Considering Equation (12),

$$conf\left(b \to h^{(-)}\right) = \frac{n_b - n_{bh}}{n_b} = 1 - \frac{n_{bh}}{n_b} = 1 - conf(b \to h) \qquad (14)$$

The confidences of the rules found for the same body and upper and lower part of a dichotomous head are complementary, i.e., their sum equals 1. The confidence of the appearance of delay in completion of a construction contract can be read as a risk of the delay appearance being greater than the threshold (number of days). This kind of confidence has identical features to risk (risk as a probability of appearing unfavorable conditions or phenomena). Their values are 0 to 1. The probability of favorable conditions added to risk gives 1, and is identical for confidences for original and inverted heads.

Therefore, the risk values (of the delay appearance being greater than a certain number of days) can be read from Figure 9. It is consistent with common sense. The greater the delay, the lower possibility of its occurrence. However, it must be emphasized that the content of Figure 9 is created based on real data.

There is also another rule found based on the Cr-E-J-L body, and it has the same head. The confidences for these two rules are presented in Figure 10.

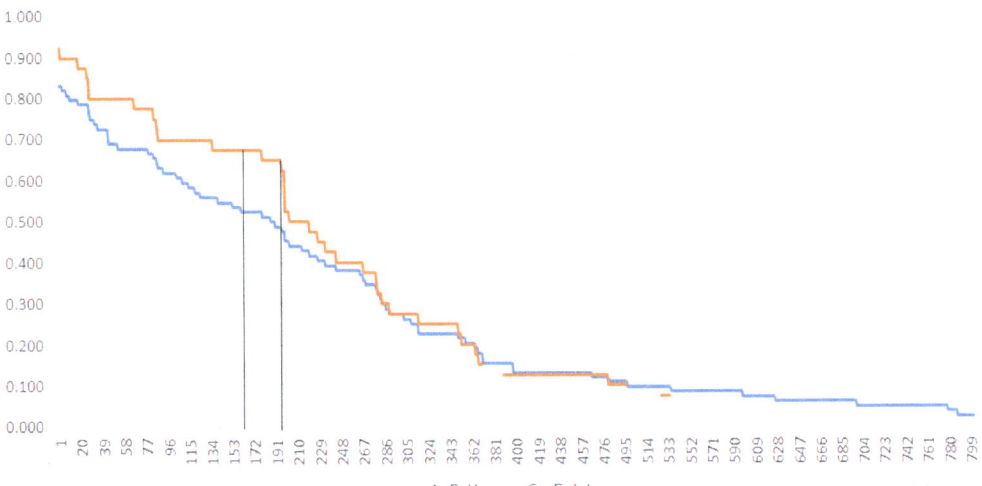

Figure 10. The confidences of the rules for A-E-K and CrEJL for different numbers of days as the threshold of the head (delay).

Confidence is a discrete function, as the nominator and denominator (defining confidence) are discrete by nature. However, confidence can be calculated for the continuous threshold (time), but is useless for the cases from the construction industry. Despite that, the lines in Figure 10 are presented as continuous. The blue line based on A-E-K body is continuous for the whole domain presented in Figures 9 and 10. The orange one (based on the Cr-E-J-L body) has two breaks (discontinuities). For days ranging from 370 to 386 and from 495 to 524, as the lift calculated for these rules is below 1, the rules are useless. There are no cases supporting this rule being delayed for more than 533 days, so the orange line ends there. In order to read the risk of delay greater than a certain threshold (given in days) and if there is more than one body for the rules found (as in Figure 10), it is recommended to use the confidence of a higher 7-Score. The calculated 7-Scores are higher for the rules based on the A-E-K body (blue line), except for the range from day 159 to day 196, as presented in Figure 11.

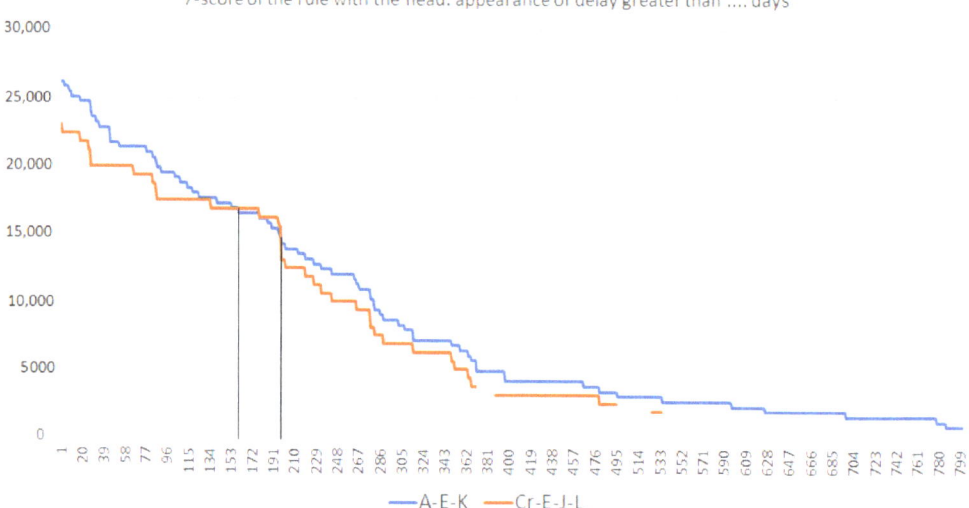

Figure 11. The 7-Score values of the rules for A-E-K and Cr-E-J-L for different numbers of days as the threshold of the head (delay).

This range is marked with black vertical lines in Figures 10 and 11. There, the rule with the other body (Cr-E-J-L) should be used (confidence read based on the orange line that has a higher 7-Score in this range).

The traditional approach to a construction contract risk estimation is based on statistics and on experts' opinions. It requires the experience of experts gained before a new assessment. The proposed method omits involving human's opinions. It is purely based on data. The experience—that is, past construction contracts completed—is necessary, but the risk is calculated based on formulas, algorithms, and a set of data collected. The higher the experience, i.e., the more cases serving as a source data, the more reliable the risk estimation. This statement points to the possible weakness of the proposed method. Analysis based on small databases can produce unreliable risk estimation. The other limitation of the invented method is the necessity of basing the risk estimation on the information gathered from the construction contracts of a similar scope of works. Assessing the risk of a road construction contract based on several completed apartment buildings is irrelevant and improper. Thus, the method can be applied by specialized contractors or clients (e.g., in the road construction, as in the analyzed case). Thirdly, the new, analyzed contract may not meet the criteria of the predecessors of the rules found to be the most informative. Then, the risk assessment is not possible. Considering the limitations of the invented method, it can be stated that the traditional approach to risk assessment (also based on experts' opinions) and the invented method should be used complementarily. If it is impossible to assess the risk with the invented method (owing to the limitation described above), the traditional method of risk assessment should be applied.

5. Conclusions

A typical software or a software package enables one to search for the rules in a database. The proposed method extends the scope of analysis by modifying the dataset. If values of any feature of a predecessor or a consequent are continuous or discrete, it is proposed to make them binary, and search—for a certain rule—for the set of thresholds dividing features' values into 0 and 1 (see Figure 2). The aim is to find the combination of these thresholds making the analyzed rule the most informative. As the three basic ratios (sup, conf, and lift) describe every rule, based on them, the measure is created and named as 7-Score. It was also necessary, owing to the need for applying the selected metaheuristic algorithm, to find the setup of thresholds maximizing the 7-Score for the analyzed rule. The results are superior when compared with the previous study. Moreover, the most informative rules are for the threshold of a construction project delay set to 0. As there are also projects in the database that were not delayed, it was decided to shift the threshold of the consequent up and observe the confidence (and other parameters) of the rule (or the set of the rules). It is concluded that the read-out is the construction risk of a delay in completion greater than the threshold (given in days). This risk decreases together with an increasing number of days. The 7-Score (the level of informativeness of the rule) decreases too. It is proved that, together with the threshold rising, the opposite rule, i.e., based on inverted consequent, is complementary to the basic rule. The sum of their confidences is 1. It can be read that the likelihood of completing a construction project (that meets the conditions of the predecessor) with the delay not greater than the threshold rises as the threshold increases. This innovative method of assessing the construction risk can be applied by clients and contractors. The results depend on the quality and size of the database being analyzed. The quality of data also refers to types of features creating the predecessor. They will be different for a contractor and for a client. Moreover, the consequent can describe a cost overrun, not exclusively delay. The invented method of risk assessment will be developed. The presented method of risk assessment is more accurate when more past cases are collected in the database. A given entity (a client or a contractor) with a rather short business history cannot expect precise quantitative risk estimations with the invented method. It is recommended to apply it to assess the risk of a contract for similar types of works. Despite that the type of contracted works can serve as an independent variable, the results will then be based on the limited number of cases. This lowers the accuracy of the method. However, the invented measure of the informativeness of association rules, i.e., 7-Score, can be broadly applied if the market basket analysis is applied.

Author Contributions: Conceptualization, H.A. and J.R.; methodology, H.A. and J.R.; software, J.R.; validation, H.A., J.R. and A.F.; formal analysis, A.F.; resources, H.A.; data curation, H.A., J.R. and A.F.; writing—original draft preparation, H.A. and J.R.; writing—review and editing, A.F. and J.R.; visualization H.A. and A.F.; supervision, A.F. and J.R.; project administration, A.F.; funding acquisition, H.A. All authors have read and agreed to the published version of the manuscript.

Funding: This research received no external funding.

Institutional Review Board Statement: Not applicable.

Informed Consent Statement: Not applicable.

Data Availability Statement: The database is published in [2]. As there is no electronic version of the Ph.D. thesis, data are available on request.

Conflicts of Interest: The authors declare no conflict of interest.

Appendix A

Table A1. Parameters and scores of the rules for the 10×10 database.

Lbl	n_b	n_h	n_{bh}	Sup	Conf	Lift	7-Score
1	9	9	9	0.9	1	1.111	191.1
2	8	8	8	0.8	1	1.25	181.3
3	8	9	8	0.8	1	1.111	181.1
4	7	7	7	0.7	1	1.429	171.4
5	7	8	7	0.7	1	1.25	171.3
6	7	9	7	0.7	1	1.111	171.1

Table A1. Cont.

Lbl	n_b	n_h	n_{bh}	Sup	Conf	Lift	7-Score
7	9	8	8	0.8	0.889	1.111	170
8	6	6	6	0.6	1	1.667	161.7
9	6	7	6	0.6	1	1.429	161.4
10	6	8	6	0.6	1	1.25	161.3
11	6	9	6	0.6	1	1.111	161.1
12	8	7	7	0.7	0.875	1.25	158.8
13	8	8	7	0.7	0.875	1.094	158.6
14	5	5	5	0.5	1	2	152
15	5	6	5	0.5	1	1.667	151.7
16	5	7	5	0.5	1	1.429	151.4
17	5	8	5	0.5	1	1.25	151.3
18	5	9	5	0.5	1	1.111	151.1
19	9	7	7	0.7	0.778	1.111	148.9
20	7	6	6	0.6	0.857	1.429	147.1
21	7	7	6	0.6	0.857	1.224	146.9
22	7	8	6	0.6	0.857	1.071	146.8
23	4	4	4	0.4	1	2.5	142.5
24	4	5	4	0.4	1	2	142
25	4	6	4	0.4	1	1.667	141.7
26	4	7	4	0.4	1	1.429	141.4
27	4	8	4	0.4	1	1.25	141.3
28	4	9	4	0.4	1	1.111	141.1
29	8	6	6	0.6	0.75	1.25	136.3
30	8	7	6	0.6	0.75	1.071	136.1
31	6	5	5	0.5	0.833	1.667	135
32	6	6	5	0.5	0.833	1.389	134.7
33	6	7	5	0.5	0.833	1.19	134.5
34	6	8	5	0.5	0.833	1.042	134.4
35	3	3	3	0.3	1	3.333	133.3
36	3	4	3	0.3	1	2.5	132.5
37	3	5	3	0.3	1	2	132
38	3	6	3	0.3	1	1.667	131.7
39	3	7	3	0.3	1	1.429	131.4
40	3	8	3	0.3	1	1.25	131.3
41	3	9	3	0.3	1	1.111	131.1
42	9	6	6	0.6	0.667	1.111	127.8
43	2	2	2	0.2	1	5	125
44	2	3	2	0.2	1	3.333	123.3
45	7	5	5	0.5	0.714	1.429	122.9
46	7	6	5	0.5	0.714	1.190	122.6
47	2	4	2	0.2	1	2.5	122.5
48	7	7	5	0.5	0.714	1.020	122.4
49	2	5	2	0.2	1	2	122
50	5	4	4	0.4	0.8	2	122
51	2	6	2	0.2	1	1.667	121.7
52	5	5	4	0.4	0.8	1.6	121.6
53	2	7	2	0.2	1	1.429	121.4
54	5	6	4	0.4	0.8	1.333	121.3
55	2	8	2	0.2	1	1.25	121.3
56	5	7	4	0.4	0.8	1.143	121.1
57	2	9	2	0.2	1	1.111	121.1
58	1	1	1	0.1	1	10	120
59	1	2	1	0.1	1	5	115
60	8	5	5	0.5	0.625	1.25	113.8
61	8	6	5	0.5	0.625	1.042	113.5
62	1	3	1	0.1	1	3.333	113.3
63	1	4	1	0.1	1	2.5	112.5

Table A1. Cont.

Lbl	n_b	n_h	n_{bh}	Sup	Conf	Lift	7-Score
64	1	5	1	0.1	1	2	112
65	1	6	1	0.1	1	1.667	111.7
66	1	7	1	0.1	1	1.429	111.4
67	1	8	1	0.1	1	1.25	111.3
68	1	9	1	0.1	1	1.111	111.1
69	6	4	4	0.4	0.667	1.667	108.3
70	6	5	4	0.4	0.667	1.333	108
71	6	6	4	0.4	0.667	1.111	107.8
72	4	3	3	0.3	0.75	2.5	107.5
73	4	4	3	0.3	0.75	1.875	106.9
74	9	5	5	0.5	0.556	1.111	106.7
75	4	5	3	0.3	0.75	1.5	106.5
76	4	6	3	0.3	0.75	1.25	106.3
77	4	7	3	0.3	0.75	1.071	106.1
78	7	4	4	0.4	0.571	1.429	98.6
79	7	5	4	0.4	0.571	1.143	98.3
80	5	3	3	0.3	0.6	2	92
81	5	4	3	0.3	0.6	1.5	91.5
82	5	5	3	0.3	0.6	1.2	91.2
83	3	2	2	0.2	0.667	3.333	90
84	3	3	2	0.2	0.667	2.222	88.9
85	3	4	2	0.2	0.667	1.667	88.3
86	3	5	2	0.2	0.667	1.333	88
87	3	6	2	0.2	0.667	1.111	87.8

References

1. Anysz, H. Managing Delays in Construction Projects Aiming at Cost Overrun Minimization. *IOP Conf. Ser. Mater. Sci. Eng.* **2019**, *603*, 032004. [CrossRef]
2. Anysz, H. Wykorzystanie Sztucznych Sieci Neuronowych Do Oceny Możliwości Wystąpienia Opóźnień w Realizacji Kontraktów Budowlanych. Ph.D. Thesis, Oficyna Wydawnicza PW, Warsaw, Poland, 2017. [CrossRef]
3. Kulejewski, J.; Ibadov, N.; Rosłon, J.; Zawistowski, J. Cash Flow Optimization for Renewable Energy Construction Projects with a New Approach to Critical Chain Scheduling. *Energies* **2021**, *14*, 5795. [CrossRef]
4. Gluszak, M.; Leśniak, A. Construction Delays in Clients Opinion–Multivariate Statistical Analysis. *Procedia Eng.* **2015**, *123*, 182–189. [CrossRef]
5. Ibadov, N. Determination of the Risk Factors Impact on the Construction Projects Implementation Using Fuzzy Sets Theory. *Acta Phys. Pol. A* **2016**, *130*, 107–111. [CrossRef]
6. Juszczyk, M. A concise review of methods of construction works duration assessment. *Tech. Trans.* **2014**, *2014*, 193–202.
7. Krzemiński, M. KASS v.2.2. Scheduling Software for Construction with Optimization Criteria Description. *Acta Phys. Pol. A* **2016**, *130*, 1439–1442. [CrossRef]
8. Ibadov, N. Selection of Construction Project Taking into Account Technological and Organizational Risk. *Acta Phys. Pol. A* **2017**, *132*, 974–977. [CrossRef]
9. Leśniak, A. Classification of the Bid/No Bid Criteria–Factor Analysis. *Arch. Civ. Eng.* **2015**, *61*, 79–90. [CrossRef]
10. Ibadov, N.; Kulejewski, J. The assessment of construction project risks with the use of fuzzy sets theory. *Czas. Tech.* **2014**, *2014*, 175–182.
11. Chatterjee, K.; Zavadskas, E.K.; Tamosaitiene, J.; Adhikary, K.; Kar, S. A Hybrid MCDM Technique for Risk Management in Construction Projects. *Symmetry* **2018**, *10*, 46. [CrossRef]
12. Kowalski, J.; Połoński, M.; Lendo-Siwicka, M.; Trach, R.; Wrzesiński, G. Method of Assessing the Risk of Implementing Railway Investments in Terms of the Cost of Their Implementation. *Sustainability* **2021**, *13*, 13085. [CrossRef]
13. Nawaz, A.; Waqar, A.; Shah, S.A.R.; Sajid, M.; Khalid, M.I. An innovative framework for risk management in construction projects in developing countries: Evidence from Pakistan. *Risks* **2019**, *7*, 24. [CrossRef]
14. PMI. *Guide to the Project Management Body of Knowledge (PMBoK Guide)*; Project Management Institute: Newtown Square, PA, USA, 2019.
15. Yaseen, Z.M.; Ali, Z.H.; Salih, S.Q.; Al-Ansari, N. Prediction of Risk Delay in Construction Projects Using a Hybrid Artificial Intelligence Model. *Sustainability* **2020**, *12*, 1514. [CrossRef]
16. Wang, S.Q.; Dulaimi, M.F.; Aguria, M.Y. Risk management framework for construction projects in developing countries. *Constr. Manag. Econ.* **2004**, *22*, 237–252. [CrossRef]

17. Schieg, M. Risk Management in Construction Project Management. *J. Bus. Econ. Manag.* **2006**, *7*, 77–83. [CrossRef]
18. Choudhry, R.M.; Iqbal, K. Identification of Risk Management System in Construction Industry in Pakistan. *J. Manag. Eng.* **2013**, *29*, 42–49. [CrossRef]
19. Taroun, A.; Yang, J.-B. A DST-based approach for construction project risk analysis. *J. Oper. Res. Soc.* **2013**, *64*, 1221–1230. [CrossRef]
20. Serpella, A.F.; Ferrada, X.; Howard, R.; Rubio, L. Risk management in construction projects: A knowledge-based approach. *Procedia Soc. Behav. Sci.* **2014**, *119*, 653–662. [CrossRef]
21. Ebrat, M.; Ghodsi, R. Construction project risk assessment by using adaptive-network-based fuzzy inference system: An empirical study. *KSCE J. Civ. Eng.* **2014**, *18*, 1213–1227. [CrossRef]
22. Iqbal, S.; Choudhry, R.M.; Holschemacher, K.; Ali, A.; Tamošaitienė, J. Risk management in construction projects. *Technol. Econ. Dev. Econ.* **2015**, *21*, 65–78. [CrossRef]
23. Vafadarnikjoo, A.; Mobin, M.; Firouzabadi, S.M.A.K. An intuitionistic fuzzy-based DEMATEL to rank risks of construction projects. In Proceedings of the 2016 International Conference on Industrial Engineering and Operations Management, Kuala Lumpur, Malaysia, 8–10 March 2016; pp. 23–25.
24. Kao, C.H.; Huang, C.H.; Hsu, M.S.C.; Tsai, I.H. Success factors for Taiwanese contractors collaborating with local Chinese contractors in construction projects. *J. Bus. Econ. Manag.* **2016**, *17*, 1007–1102. [CrossRef]
25. Ahmadi, M.; Behzadian, K.; Ardeshir, A.; Kapelan, Z. Comprehensive risk management using fuzzy FMEA and MCDA techniques in highway construction projects. *J. Civ. Eng. Manag.* **2016**, *23*, 300–310. [CrossRef]
26. Li, J.; Wang, J.; Xu, N.; Hu, Y.; Cui, C. Importance Degree Research of Safety Risk Management Processes of Urban Rail Transit Based on Text Mining Method. *Information* **2018**, *9*, 26. [CrossRef]
27. Anysz, H.; Buczkowski, B. The association analysis for risk evaluation of significant delay occurrence in the completion date of construction project. *Int. J. Environ. Sci. Technol.* **2018**, *16*, 5369–5374. [CrossRef]
28. Morzy, T. *Eksploracja Danych. Metody i Algorytmy*; Wydawnictwo Naukowe PWN: Warsaw, Poland, 2013.
29. Larose, D.T.; Larose, C.D. *Discovering Knowledge in Data*; John Wiley & Sons: Hoboken, NJ, USA, 2016; ISBN 978-81-265-5834-6.
30. Statsoft Electronic Statistics Textbook. Available online: https://www.statsoft.pl/textbook/stathome.html (accessed on 20 November 2021).
31. Hahsler, M.; Grün, B.; Hornik, K. Introduction to arules–Mining Association Rules and Frequent Item Sets. *SIGKDD Explor* **2007**, *4*, 1–28.
32. Ünvan, Y.A. Market basket analysis with association rules. *Commun. Stat.-Theory Methods* **2020**, *50*, 1615–1628. [CrossRef]
33. Ahmed, A.M.; Bakar, A.A.; Hamdan, A.R.; Abdullah, S.M.S.; Jaafar, O. Sequential Pattern Discovery Algorithm for Malaysia Rainfall Prediction. *Acta Phys. Pol. A* **2015**, *128*, B324–B326. [CrossRef]
34. Roodpishi, M.V.; Nashtaei, R.A. Market basket analysis in insurance industry. *Manag. Sci. Lett.* **2015**, *5*, 393–400. [CrossRef]
35. Geurts, K.; Wets, G.; Brijs, T.; Vanhoof, K. Profiling of High-Frequency Accident Locations by Use of Association Rules. *Transp. Res. Rec. J. Transp. Res. Board* **2003**, *1840*, 123–130. [CrossRef]
36. Xu, C.; Bao, J.; Wang, C.; Liu, P. Association rule analysis of factors contributing to extraordinarily severe traffic crashes in China. *J. Saf. Res.* **2018**, *67*, 65–75. [CrossRef]
37. Anysz, H.; Włodarek, P.; Olszewski, P.; Cafiso, S. Identifying factors and conditions contributing to cyclists' serious accidents with the use of association analysis. *Arch. Civ. Eng.* **2021**, *LXVII*, 197–211. [CrossRef]
38. Anysz, H.; Apollo, M.; Grzyl, B. Quantitative Risk Assessment in Construction Disputes Based on Machine Learning Tools. *Symmetry* **2021**, *13*, 744. [CrossRef]
39. Shi, A.; Mou, B.; Correll, J.C. Association analysis for oxalate concentration in spinach. *Euphytica* **2016**, *212*, 17–28. [CrossRef]
40. Klimanek, T.; Szymkowiak, M.; Józefowski, T. Analiza koszykowa w badaniu zjawiska niepełnosprawności biologicznej. *Pr. Nauk. Uniw. Ekon. Wrocławiu* **2018**, 95–105. [CrossRef]
41. Atluri, G.; Gupta, R.; Fang, G.; Pandey, G.; Steinbach, M.; Kumar, V. Association Analysis Techniques for Bioinformatics Problems. In *Bioinformatics and Computational Biology*; Rajasekaran, S., Ed.; Springer: Berlin/Heidelberg, Germany, 2009; Volume 5462, pp. 1–13. ISBN 978-3-642-00726-2.
42. Lasek, M.; Pęczkowski, M. Analiza Asocjacji I Reguły Asocjacyjne W Badaniu Wyborów Zajęć Dydaktycznych Dokonywanych Przez Studentów. Zastosowanie Algorytmu Apriori. *Ekon. J.* **2013**, *34*, 67–88.
43. Anysz, H.; Foremny, A.; Kulejewski, J. Comparison of ANN Classifier to the Neuro-Fuzzy System for Collusion Detection in the Tender Procedures of Road Construction Sector. *IOP Conf. Ser. Mater. Sci. Eng.* **2019**, *471*, 112064. [CrossRef]
44. Nicał, A.; Anysz, H. The quality management in precast concrete production and delivery processes supported by association analysis. *Int. J. Environ. Sci. Technol.* **2020**, *17*, 577–590. [CrossRef]
45. Rosłon, J. The multi-mode resource constrained project scheduling problem in construction. State of the art review and research challenges. *Tech. Trans.* **2017**, *5*, 67–74.
46. Rosłon, J.; Zawistowski, J. Construction Projects' Indicators Improvement Using Selected Metaheuristic Algorithms. *Procedia Eng.* **2016**, *153*, 595–598. [CrossRef]
47. Sroka, B.; Rosłon, J.; Podolski, M.; Bożejko, W.; Burduk, A.; Wodecki, M. Profit optimization for multi-mode repetitive construction project with cash flows using metaheuristics. *Arch. Civ. Mech. Eng.* **2021**, *21*, 1–17. [CrossRef]

48. Tang, F.; Zhou, H.; Wu, Q.; Qin, H.; Jia, J.; Guo, K. A Tabu Search Algorithm for the Power System Islanding Problem. *Energies* **2015**, *8*, 11315–11341. [CrossRef]
49. Choi, J.; Xuelei, J.; Jeong, W. Optimizing the Construction Job Site Vehicle Scheduling Problem. *Sustainability* **2018**, *10*, 1381. [CrossRef]
50. Fridgeirsson, T.V.; Rosłon, J. *Optimisation of Construction Processes*; Civil Engineering Faculty of Warsaw, University of Technology: Warsaw, Poland, 2017.
51. Böde, K.; Różycka, A.; Nowak, P. Development of a Pragmatic IT Concept for a Construction Company. *Sustainability* **2020**, *12*, 7142. [CrossRef]

Article

Randomized Estimation of the Net Present Value of a Residential Housing Development

Tadeusz Kasprowicz [1], Anna Starczyk-Kołbyk [1,*] and Robert Wójcik [2]

[1] Faculty of Civil Engineering and Geodesy, Military University of Technology, ul. gen. Sylwestra Kaliskiego 2, 00-908 Warsaw, Poland; tadeusz.kasprowicz@wat.edu.pl

[2] PFR Nieruchomości S.A., Polish Development Fund Group, ul. 6 Przeskok Str., 00-032 Warsaw, Poland; robertwojcik63@gmail.com

* Correspondence: anna.starczyk@wat.edu.pl

Abstract: Randomized estimation of the net present value of a housing development allows for the assessment of the efficiency of projects in random implementation conditions. The efficiency of a project is estimated on the basis of primary input data, usually used in project planning. For this purpose, random disturbances are identified that may randomly affect the course and results of the project. The probability and severity of disturbances are determined. The primary initial data is then randomized, and a randomized probabilistic index of the project's net present value is calculated, the value of which indicates whether the project is profitable or whether implementation should be stopped. Based on this data, the expected total revenue, the expected total cost, the expected gross profit, and the net present value of the randomized performance of the project are calculated. The values of these are estimated for expected, favorable, and unfavorable conditions of implementation. Finally, the risks for the total revenue and total cost of the project are calculated and plotted for comparative revenue values in the range [1, 0] and cost in the range [0, 1]. Their analysis makes it possible to make the right investment decisions before starting the investment at the preparation stage.

Keywords: efficiency; risk; randomization; construction project

1. Introduction

The term efficiency [1–5] in a general sense describes the achievement of objectives in an economic manner. It is a search for a good balance between the resources used (time, money, space, equipment, and materials) and the achievement of the objectives of the activity [6].

The term efficiency is a measurable concept that can be applied to the quantitative assessment of production or performance when a given number of resources (money, time, work, etc.) has been used. All well-known methods of assessing efficiency, although they consider the time value of money, actually use deterministic data. The results obtained in this way are also deterministic. In such cases, the efficiency is usually measured as the ratio of useful result ("product") to total outlay. It can be expressed with the mathematical formula E = P/C that is the ratio of the amount P of useful result (benefit) of project implementation to the amount (cost) C of consumed resources.

Generally, these indicators are reliable and well describe efficiency when the project would be implemented in stable and balanced economic and environmental conditions. Usually, inner and outer random events can strongly affect the project implementation. Such disturbances can increase costs and decrease revenues of the project [7,8].

In published books and papers relevant to the project lifecycle, problems predominate the deterministic approach, for example: Kasprowicz, T. (2015) [9] or Kasprowicz T. (2011) [10] and Bizon-Górecka, J. (2008) [11].

However, because of random conditions of project implementation, the better approach is probabilistic analysis when the efficiency of the project implementation can be characterized as a random variable.

The term of risk (probabilistic conditions) means that a situation in which the result of the project is unknown, uncertain, or there is a possibility that something will succeed or fail, and the probability of the random events appearance and project execution according to the plan is known or it can be estimated. These problems have been studied by Cao, J., Song, W. 2016 [12], Radło M. J. 2015 [13], Kalkhoran S. H. A., Liravi, G., Rezagholi F. 2014 [14], Stephen C Ward, S. C., and Chris B Chapman, Ch. B. 1995 [15], et al. Here, the risk conditions describe circumstances in which individual capabilities and benefits associated with each possible action within the project are known or can be predicted with some probability—the probability of possible conditions of the project implementation is known or can be reliably estimated. In the conditions of risk variability, different random disturbances can interfere with the process of project implementation. In the aftermath of these disturbances, the impact variability of technical, technological, social, economic, and environmental random factors can impede or preclude the project accomplishment according to the design documentation and technical specification of the works execution and acceptance [16].

These may randomly increase prices and consumption of labor and resources, expenditures of machinery and equipment and reduce the productivity of project executors.

So, in random conditions, the efficiency is a random variable and must be estimated, taking into account the impact of random disturbances on course and results of the project lifecycle implementation [17,18].

This is a process of creating random variables of costs and revenues as well as connected and derived quantities that reflect likely random conditions of the construction project's implementation. Based on the randomized data, the proposed method of estimating the net present value of the construction project's efficiency can be estimated. The method should be applied when random disturbances can disturb or disrupt the project implementation. That is, in a situation, during the project implementation there are likely disturbances that can randomly change the cost, time, and quality of projects.

The method of assessing the effectiveness of unstable construction projects in the conditions of the impact of significant disturbances on the course and results of implementation uses specific concepts and mathematical formulas in a different scope and at different stages of the investment.

The end result of the presented method is a reliable estimation of profitability and a realistic assessment of the cost-benefit balance of unstable construction projects already at the planning stage.

Often, a distinction is made between economic and technical efficiency. Economic efficiency means producing and distributing goods at the lowest possible cost. Technical efficiency means the maximum amount of production for a given input, or the maximum amount of an output that can be obtained from a certain input.

In many studies, the term "efficiency" is often used to evaluate economic activities (Blue Book, [19]). From this point of view, economic and financial efficiency are most often distinguished.

The randomized method for estimating the net present value of a construction project's efficiency has been worked out recently. The method has been theoretically finished and verified on simple examples of projects. However, the method is still being tested and verified in the process of estimating the net present value of the housing investment's efficiency.

So, the objective of the research was to verify the correct probabilistic method for the analysis and assessment of the net present value of the effectiveness of unstable construction projects for real building investments.

In the article, research results for real building investments by the randomized method of the estimation of construction projects' efficiency are presented.

Such building investments have been realized by a developer in Warsaw, the capital city of Poland. Here, the randomized method of efficiency estimating of the net present value of one such residential housing development's efficiency is presented.

The subject of the study is one of the stages of a five-stage construction of a residential housing development in Warsaw. The settlement was built by a Warsaw real estate developer. The development area takes up approximately 1.3 hectares of land. Inside this area, six multi-family buildings of different size, set on a single garage plate, have been built. Complete with the buildings, some concomitant facilities have been built, for example, new public roads, additional parking spaces, and usable premises. The residential housing development has been built in the system of a general contractor. Implementation cycle of the project consisted of four basic stages, namely: feasibility study (6 months), designing (12 months), construction (18 months), and operation and maintenance (60 months). However, the stage of operation and maintenance concerned only activities resulting from the general guarantee and warranty for defects. The efficiency analysis was carried out as part of the feasibility study. In the first stage of the analysis, the project life cycle was projected, i.e., the stages of implementation and tasks cost of these stages, as well as tranches of revenue payments. Moreover, in this stage, discount rates were established, and the types of random disturbances of the project implementation were determined. In the case of this project, the primary initial data were developed collectively by the team for the real estate purchases and the team for analysis and preparation of investments. The teams identified and estimated this data, taking into account the predicted conditions and projected prices in the country for the entire project lifecycle implementation. The discount rate was determined on the basis of historical data of the investments realized by the developer and accordingly to the current market forecasts. The investment in the initial phase, such as all investments of this company, was financed by an operating loan for the activities of the developer. The first and second tranches of payment were flown in with the start of sales of the first and second parts of the apartments, respectively. The third and fourth tranches of revenue payments began with the start of sales of concomitant facilities, such as parking lots and commercial premises. Detailed tally of the project cost and revenue was shown in spreadsheets later in the case study. In random implementation conditions, costs and revenues of the project can be disturbed by random disturbances. Such cost random disturbances have been defined in accordance with conditions and requirements of the project stages and tasks implementation, whereas revenue random disturbances have been projected in accordance with likely prices on domestic and foreign property markets, as well as conditions and quality of business environment. Unfortunately, until now, the impact of the disturbances on the course and results of the project was analyzed mostly qualitatively. In this way, both costs and revenues were estimated approximately only on the basis of experience and knowledge of the analysts. The company has not used yet any probabilistic method that would have allowed quantitative analysis and estimation of costs and revenues when random disturbances may have an impact on the project implementation. The randomized method of estimation of the net present value of a construction project's efficiency were developed in order to bridge this gap. It is assumed that such an analysis would be carried out in the "feasibility study" stage and could be used by teams that are also currently involved in the planning of construction projects.

2. Materials and Methods

Estimation of the primary initial data is the first phase of the randomized estimation of the net present value of the residential housing development's efficiency. These data are the basis for further analysis and data randomization for the estimation of efficiency. They directly describe the kinds and values of parameters and characteristics of the project implementation. In the primary analysis, based on an in-depth analysis of the facilities structure and the construction conditions, the residential housing development implementation cycle has to be projected. Such a lifecycle has been defined in a similar way by De Wilde, P. 2018 [20], Ding, L., Zhou, Y., Akinci, B. 2014 [21], Halpin, D. W.,

Woodhead R. W. 1998 [22], Ritz, G., 1994 [23], et al. In this meaning, the construction project lifecycle consisted of several successive stages $s = 1, 2, \ldots, t$. In each stage, the subset A^s of tasks a_i should be executed, $A^s = \{a_i : i = 1, 2, \ldots, i^s\}$. During the project lifecycle implementation, the set A of all tasks a_i of the project should be carried out, $A = \{A^s : s = 1, 2, \ldots, t\} = \{a_i : i = 1, 2, \ldots, i^a\}$. Consistently, for each subset A^s of tasks a_i, the subset C^s of the initial preliminary execution costs c_i should be estimated, $C^s = \{c_i : i = 1, 2, \ldots, i^s\}$. In this way, the subsets C^s, $s = 1, 2, \ldots t$, contain all costs c_i, and each stage s, of the project and form the set $C = \{C^s : s = 1, 2, \ldots, t\} = \{c_i : i = 1, 2, \ldots, i^t\}$. According to the stage of construction, the apartments and other facilities can be sold after some objects are finished, or they can be sold after the whole residential housing development' completion. Each sold part represents one tranche b_i of revenue $p_i, i = 1, 2, \ldots, i^b$. All paid tranches b_i of revenues p_i constitute the set $P = \{p_i : i = 1, 2, \ldots, i^b\}$. Because of money value changes in time, for both costs and revenues discount rates, ρ_i paid per year should be set.

The primary initial data of costs for the residential housing development implementation cycle can be determined as follows:

1. Stages s and subsets A^s of task a_i, and subsets C^s of cost c_i:

   feasibility study—s1, the subset A^1 of tasks a_i, $A^1 = \{a_i, i = \overline{1,5}\}$ and the subset $C^1 = \{c_i : i = \overline{1,5}\}$ of costs c_i:

   a_1—initial analysis and its cost c_1.
   a_2—land acquisition and its cost c_2.
   a_3—determination of the technical conditions and its cost c_3.
   a_4—coverage of overheads and its cost c_4.
   a_5—tracing of economic and financial conditions and its cost c_5;

   design documentation development—s2, the subset A^2 of tasks a_i, $A^2 = \{a_i, i = \overline{1,6}\}$ and the subset $C^2 = \{c_i : i = \overline{1,6}\}$ of costs c_i:

   a_1—initial concept and nets and their cost c_1.
   a_2—conceptual architectural design and its cost c_2.
   a_3—general conceptual design and its cost c_3.
   a_4—construction project and its cost c_4.
   a_5—detailed design (executive project) and its cost c_5.
   a_6—complementary design documentation and its cost c_6;

   construction—s3, the subset A^3 of tasks a_i, $A^3 = \{a_i, i = \overline{1,8}\}$ and the subset $C^3 = \{c_i : i = \overline{1,8}\}$ of costs c_i:

   a_1—construction site development and its cost c_1.
   a_2—structure of construction, state zero, and its cost c_2.
   a_3—plumbing and electrical wiring, state zero, and its cost c_3.
   a_4—structure of construction—superstructure and its cost c_4.
   a_5—plumbing and electrical wiring—superstructure and its cost c_5.
   a_6—nets and connections to main supply and their cost c_6.
   a_7—roads and land development and decorative green and their cost c_7.
   a_8—technological startup of construction and its cost c_8;

   operation and maintenance—s4, the subset A^4 of tasks a_i, $A^4 = \{a_i, i = \overline{1,3}\}$ and the subset $C^4 = \{c_i : i = \overline{1,3}\}$ of costs c_i:

   a_1—unpredicted additional completion activities and their cost c_1.
   a_2—additional activities under contract warranty and their cost c_2.
   a_3—maintenance of flats unsold according to the plan and its cost c_3:

2. The set A of subsets A^s, $A = \{A^s : s = \overline{1,4}\} = \{a_i : i = \overline{1,22}\}$.
3. Discount rates ρ_i paid per year τ during the period of incurring costs.
4. The set C of subsets C^s, $C = \{C^s : s = \overline{1,4}\} = \{c_i : i = \overline{1,22}\}$.

The basic primary initial data of revenue for the residential housing development implementation cycle can be determined as follows:

1. The set B of tranches b_i and the set P of the revenues p_i, $B = \{b_i : i = \overline{1,4}\}$, $P = \{p_i : i = \overline{1,4}\}$:

   b_1—sale of apartments—part 1 and revenue p_1.
   b_2—sale of apartments—part 2 and revenue p_2.
   b_3—sale of parking lots and revenue p_3.
   b_4—sale od usable premises and revenue p_4.

2. Discount rates ρ_i paid per year τ for revenues $p_i, i = \overline{1,4}$.

In the case of the residential housing development implementation, the primary initial data for costs have been listed in Table 1 and for revenues in the Table 2 in the chapter Results.

Table 1. Primary initial data for costs.

Stages s	$a_i \in A^s$	$a_i \in A$	c_i PLN	ρ_i %	Years for ρ_i	K_i PLN
s1	a_1	a_1	100,000	3	0.6	98,242
	a_2	a_2	12,300,000	3	0.6	12,083,779
	a_3	a_3	3,500,000	3	0.6	3,438,474
	a_4	a_4	6,400,000	3	0.6	6,287,495
	a_5	a_5	3,400,000	3	0.6	3,340,232
s2	a_1	a_6	200,000	4	1.5	1,885,732
	a_2	a_7	200,000	4	1.5	188,573
	a_3	a_8	200,000	4	1.5	188,573
	a_4	a_9	600,000	4	1.5	565,720
	a_5	a_{10}	600,000	4	1.5	565,720
	a_6	a_{11}	200,000	4	1.5	188,573
s3	a_1	a_{12}	200,000	4	3	177,799
	a_2	a_{13}	10,500,000	4	3	9,334,462
	a_3	a_{14}	3,500,000	4	3	3,111,487
	a_4	a_{15}	16,500,000	4	3	14,668,440
	a_5	a_{16}	3,500,000	4	3	3,111,487
	a_6	a_{17}	3,950,000	4	3	3,511,536
	a_7	a_{18}	1,000,000	4	3	888,996
	a_8	a_{19}	500,000	4	3	444,498
s4	a_1	a_{20}	100,000	4	3	88,900
	a_2	a_{21}	200,000	4	6	158,063
	a_3	a_{22}	150,000	4	5	123,289
Sum		Total primary cost	67,800,000		Discounted total primary cost	62,752,911

Table 2. Primary initial data of revenues.

Revenue	$b_i \in B$	p_i PLN	ρ_i %	Years for ρ_i	D_i PLN
Tranches	b_1	11,100,000	4	3	9,867,860
	b_2	68,900,000	4	5	56,630,778
	b_3	6,200,000	4	5	5,095,948
	b_4	4,100,000	4	4	3,504,697
Sum	Overall revenue	90,300,000	Discounted overall revenue		75,099,282

In Tables 1 and 2, project primary initial costs c_i and revenues p_i were discounted by the use of formulas $K_i = \frac{c_i}{(1+\rho_i)^\tau}$ and $D_i = \frac{p_i}{(1+\rho_i)^\tau}$, respectively. In this way, one can define the set $\mathcal{K} = \{K_i : i = 1, 2, \ldots, i^t\} = \{K_i : i = \overline{1,22}\}$ of discounted costs K_i of tasks a_i

execution and the set $\mathcal{D} = \left\{ D_i, i = 1, 2, \ldots, i^b \right\} = \left\{ D_i, i = \overline{1,4} \right\}$ of discounted revenues D_i paid in tranches b_i.

The primary initial data describe project implementation without taking into account likely influences of random events on the course and results of the project. In practice, it can cause incorrect estimation of the costs and revenues of the project. In order to take cognizance of such disturbances, the primary initial data must be randomized.

2.1. Randomization of the Primary Initial Data

2.1.1. Estimation of the Impact of Disturbances on the Task's Costs and the Revenues Tranches

Here, randomization means transforming the deterministic primary initial data into probabilistic computing data. Such modification should be undertaken when future random inner and outer disturbances can significantly impact and change the values of project costs and revenues. This is a process of creating random variables that reflect likely random conditions of the project implementation. In this area, the issues of costs that must be incurred on a project task's execution and questions of revenues that should be paid in tranches for the completed tasks have to be analyzed.

As part of the cost analysis, it is necessary to analyze probabilistic characteristics of cost random variables depending on internal and external random disturbances related to the cost of project tasks implementation. In the considered situation, the sets $A^s, s = \overline{1,t}$, of tasks a_i may be disturbed by relevant sets E^s of disturbances $e_{i,j}$, $E^s = \left\{ e_{i,j} : i = 1, 2, \ldots, i^s, j = 1, 2, \ldots, j^s \right\}$. In the result, costs K_i of tasks a_i may randomly change. In the analyzed case, the sets E^s as well as disturbances $e_{i,j}$ that belong to them are relatively independent of each other. Each set E^s of disturbances $e_{i,j}$ correspond with a particular stage s. This means that the execution of each task $a_i \in A^s$ may be randomly affected by each threat $e_{i,j} \in E^s$. All disturbances of costs must be identified for projected places, environment, and system surroundings of the project.

For the residential housing development sets $E^s = \left\{ e_{i,j} : i = \overline{1,i^s}, j = \overline{1,j^s} \right\}, s = \overline{1,4}$, of disturbances $e_{i,j}$ related to sets $A^s = \left\{ a_i : i = 1, i^s \right\}$ of tasks a_i have been identified as follows:

Stage of feasibility study—$s = 1$, $A^1 = \left\{ a_i : i = \overline{1,5} \right\}$. $E^1 = \left\{ e_{i,j} : i = \overline{1,5}, j = \overline{1,6} \right\}$:

1. Modification of the terms of financing operational analysis $j = 1$—costs of additional:
   - general re-analysis and complementary research;
   - land pre-purchasing analysis, soil property tests, checking soil pollution and possible protections, as well as the risk of land purchase costs and the final purchase price;
   - commissions for intermediaries and often remediation;
   - expertise and structural survey of existing buildings to be secured;
   - evaluation of the scale and size of the existing paid land rights, land servitude, and/or transit for gestors, etc.

2. Architectural survey—incorrect description of the scope of the reconstruction $j = 2$:
   - extra costs of additional analyzing and assessing of the type and scope of reconstruction of existing infrastructure.

3. Technical conditions of connecting utilities differ significantly from assumptions $j = 3$:
   - additional costs of re-analysis of technical conditions for connection to the system of utilities, e.g., water supply and sewerage, energy, communication technology, gas, etc., can significantly change costs of the project.

4. Reconstruction of the collision of the technical infrastructure in a much expanded, unpredicted range $j = 4$:
   - extra costs of re-analysis, redesign, and reconstruction of the technical infrastructure because of analytical errors or unpredicted changings of conditions or circumstances.

5. Analysis of the absorbency of terrain incompatible with possible to obtain of development $j = 5$:
   - additional costs of re-analysis of possible construction permit and redesign of terrain absorbency.
6. Changings of overall economic and business conditions $j = 6$:
   - possible direct changes of any costs of management, administrative staff, and all other employees.

Stage of design documentation development—$s = 2$, $A^2 = \{a_i : i = \overline{1,6}\}$, $E^2 = \{e_{i,j} : i = \overline{1,6}, j = \overline{1,6}\}$:

1. Collapse of the design office $j = 1$:
   - costs of searching and a contract negotiation with a new design office;
   - additional costs of development of new design documentations.
2. Modification of the rules of law or delivery system $j = 2$:
   - additional costs of modification of design details caused by changes in the rules of law, or technical or organizational requirements during design.
3. Collapse of the housing sales market $j = 3$:
   - it can be forced to completely redesign the investment and change the structure of apartments or even the standard of investment so that it is adapted to the current demand of the real estate market.
4. Changes of technical standards $j = 4$:
   - cost of implementation of a new technology;
   - additional costs of re-design and development of partly new design documentation.
5. Changes of overall economic and business conditions $j = 5$:
   - it is possible to directly change any costs of management, administrative staff, strictly design personnel, and all others, as well as the work of machines and auxiliary equipment.

Stage of construction—$s = 3$, $A^3 = \{a_i : i = \overline{1,8}\}$, $E^3 = \{e_{i,j} : i = \overline{1,8}, j = \overline{1,5}\}$:

1. Change in prices of goods and services:
   - Changes in the purchasing costs of goods and services.
2. Collapse of the services market of construction:
   - additional costs of likely delay of activities and obtaining contractors.
3. Collapse of the project's general contractor:
   - costs of searching for and a contract negotiation with a new contractor;
   - re-employment of the new contractor, beginning of works continuation.
4. Particularly unfavorable conditions for the implementation of works:
   - possible disturbances of individual outer works;
   - extra costs of likely delay of work implementation and other activities;
   - there are possible direct changes of any costs of employment of managerial staff, core workers, and auxiliary workers, as well as work of machines and auxiliary equipment.
5. Changings of economic and business conditions and deterioration of payment terms:

Stage of operation and maintenance—$s = 4$, $A^4 = \{a_i : i = \overline{1,3}\}$, $E^4 = \{e_{i,j} : i = \overline{1,3}, j = \overline{1,5}\}$:

1. Disclosure of hidden defects and removal of them within the product warranty or warranty for physical defects $j = 1$:
   - additional activities and costs related to removing hidden defects and faults (at the expense of the developer).

2. Maintenance of unsold homes (apartments, usable premises, parking lots, etc.) significantly exceeding budgetary assumptions $j = 2$:
   - extending time and additional cost of maintenance (at the expense of the developer).
3. Supplementary activities of construction completion—beyond the contract with the project's general contractor $j = 3$:
   - additional costs of the construction completion—beyond the contract with the project's general contractor (at the expense of the developer).
4. Changes of economic and business conditions and deterioration of payment terms $j = 4$:
   - there are possible direct changes of any costs of employment, core workers, auxiliary workers, the work of machines and auxiliary equipment, as well as goods and services.
5. Litigation and legal proceedings $j = 5$:
   - additional costs of litigation and legal proceedings;
   - additional costs related to the extension of maintenance of unsold homes and other higher costs.

As part of the revenue analysis, it is necessary to analyze probabilistic characteristics of revenue random variables depending on internal and external random disturbances related to the payment of revenues tranches. The random disturbances of revenues reflect the influence of likely random conditions that may exist in a given place, environment, or systemic situation. They also depend on financial market stability during the project implementation and revenue payment. Taking all mentioned terms, the impact of disturbances was projected in accordance with likely prices on domestic and foreign property market, as well as probable conditions and quality of business environment. In the considered situation, the set B of tranches b_i may be disturbed by relevant set E^b of disturbances $e_{i,j}$, $E^b = \left\{ e_{i,j} : i = 1, 2, \ldots, i^b, j = 1, 2, \ldots, j^b \right\}$. In the result, revenues D_i paid in tranches b_i may randomly change. In the analyzed case, the sets E^b as well as disturbances $e_{i,j}$ that belong to them are relatively independent of each other. Each set E^b of disturbances $e_{i,j}$ correspond with a particular tranche b_i. This means that the execution of each tranche $b_i \in B$ may be randomly affected by each threat $e_{i,j} \in E^b$.

For the residential housing development sets $E^b = \left\{ e_{i,j} : i = \overline{1,4}, j = \overline{1,5} \right\}$, of disturbances $e_{i,j}$ related to sets $B = \left\{ b_i : i = \overline{1,4} \right\}$ of tranches b_i, values have been identified as follows:

1. Price decline of apartments $j = 1$:
   - direct decrease of revenues.
2. Decline (slack) in sale of apartments $j = 2$:
   - increase of interest costs and rise of loan repayment;
   - possible disruptions of accounting liquidity and rise of various activities costs.
3. Slackening in sale of apartments $j = 3$:
   - additional cost of maintenance of unsold apartments (at the expense of the developer).
4. Tighter credit policies by banks (financial restriction of investment) $j = 4$:
   - increase of interest costs and rise of loan repayment;
   - possible disruptions of accounting liquidity and rise of various activities costs.
5. Changing of economic and business conditions and deterioration of payment terms $j = 5$:
   - possible direct changes of any costs of employment, core workers and auxiliary workers, work of machines, and auxiliary equipment, as well as goods and services.

The random disturbances of costs and revenues reflect the influence of likely conditions of tasks execution and tranches payment. In the various random situations, disturbances can arise with different probabilities, and they can have distinct severity on the project implementation (see Ding, L., Zhou, Y., Akinci, B., 2014 [21]). This means that random implementation conditions directly decide the probability of occurrence $r_{i,j}$ and severity $c_{i,j}$ of disturbances $e_{i,j}$. Disturbances of costs $e_{i,j}$, $i = 1, 2, \ldots, i^s$, $j = 1, 2, \ldots, j^s$ that belong to the subsets E^s, or disturbances $e_{i,j}$, $i = 1, 2, \ldots, i^a$, $j = 1, 2, \ldots, j^i$ that belong to the set E, may emerge with probability $r_{i,j} \in [0,1]$ and severity $c_{i,j}[0,1]$, and they may variously impact the execution of tasks $a_i \in A$. The disturbances of revenues $e_{i,j}$, $i = 1, 2, \ldots, i^b$, $j = 1, 2, \ldots, j^b$, or $j = 1, 2, \ldots, j^i$ that belong to the set E^b may emerge with probability $r_{i,j} \in [0,1]$ and severity $c_{i,j}[0,1]$ and may variously impact payment of tranches $b_i \in B$. The probability and severity of the disturbances for costs and revenues of the residential housing development should be estimated by construction and financial experts based on their knowledge, experience, and even intuition. Based on that evaluation, the influences of disturbances on project costs and revenues can be estimated. Depending on values of the disturbances' probability and severity, costs of task execution and revenues paid in tranches may change in varying degree. Degree of average random changes of costs and revenues can be estimated by the use of probabilistic coefficients of optimism and probabilistic coefficients of pessimism for costs and revenues, respectively. The probabilistic coefficients of cost optimism $\underline{p}_i$, $i = 1, 2, \ldots, i^a$ reflect likely decreased costs K_i of tasks $a_i \in A$ due to the probable improvement of implementation conditions. The probabilistic coefficients of costs pessimism $\overline{p}_i$, $i = 1, 2, \ldots, i^a$ reflect likely increased costs K_i of tasks $a_i \in A$ due to the probable degradation of implementation conditions. Average random impact of disturbances $e_{i,j}$, $i = 1, 2, \ldots, i^a$, $j = 1, 2, \ldots, j^i$ on costs K_i of tasks $a_i \in A$ can be calculated by using formulas:

1. Average probability r_i, and average severity c_i of disturbances $e_{i,j} \in E$, $i = 1, 2, \ldots, i^a$, $j = 1, 2 \ldots, j^i$, which may randomly change costs $K_i \in \mathcal{K}$ of the implementation of tasks $a_i \in A$:

$$r_i = \left(\sum_{j=1}^{j=j^i} r_{i,j}\right)\left(j^i\right)^{-1}; \quad c_i = \left(\sum_{j=1}^{j=j^i} c_{i,j}\right)\left(j^i\right)^{-1} \qquad (1)$$

2. Probabilistic coefficients of cost optimism $\underline{p}_i \in [0,1]$ and cost pessimism $\overline{p}_i \in [0,1]$, $i = 1, 2, \ldots, i^a$:

$$\underline{p}_i = 1 - r_i c_i; \quad \overline{p}_i = 1 - (1 - r_i)(1 - c_i) \qquad (2)$$

- coefficients $\underline{p}_i = 1$ indicate extremely favorable conditions for carrying out tasks $a_i \in A$ and possible maximum reduction of cost $K_i \in \mathcal{K}$ and $\underline{p}_i = 0$ points of an opposite case;
- coefficients $\overline{p}_i = 1$ indicate extremely difficult conditions for carrying out tasks $a_i \in A$ and possible maximum increase of costs $K_i \in \mathcal{K}$ and $\overline{p}_i = 0$ points of an opposite case.

Similarly, the probabilistic coefficients of revenue optimism $\overline{p}_i$ $i = 1, 2, \ldots, i^b$ reflect the likely increase of revenues D_i paid in tranches $b_i \in B$ due to the probable improvement of business and payment conditions. The probabilistic coefficients of revenue pessimism $\underline{p}_i$, $i = 1, 2, \ldots, i^b$ reflect the likely decrease of revenues D_i paid in tranches $b_i \in B$ due to the probable degradation of business and payment conditions. Average random impact of disturbances $e_{i,j}$, $i = 1, 2, \ldots, i^b$, $j = 1, 2, \ldots, j^i$ on revenues tranches D_i payment can be calculated by using formulas:

$$r_i = \left(\sum_{j=1}^{j=j^i} r_{i,j}\right)\left(j^i\right)^{-1}; \quad c_i = \left(\sum_{z=1}^{j=j^i} c_{i,j}\right)\left(j^i\right)^{-1} \qquad (3)$$

1. Probabilistic coefficients of revenue pessimism $\underline{p}_i = [0,1]$ and revenue optimism $\overline{p}_i = [0,1]$, $i = 1, 2, \ldots, i^b$:

$$\underline{p}_i = 1 - (1-r_i)(1-c_i); \quad \overline{p}_i = 1 - r_i c_i \qquad (4)$$

- the coefficient $\underline{p}_i = 1$ indicates extremely difficult conditions for carrying out tranches $b_i \in B$ payments and possible maximum decrease of revenues $D_i \in \mathcal{D}$ and $\underline{p}_i = 0$ points of an opposite case;
- the coefficient $\overline{p}_i = 1$ indicates extremely favorable conditions for carrying out tranches $b_i \in B$ payments and possible maximum increase of revenues and $\overline{p}_i = 0$ points of an opposite case.

Data depicted above allow comprehensively described conditions of the project implementation. In case of the residential housing development costs, the probability $r_{i,j}$ and severity $c_{i,j}$ as well as coefficients of optimism $\underline{p}_i$ and pessimism $\overline{p}_i$ have been estimated for all disturbances $e_{i,j}$ that may disrupt execution of tasks a_i that is for s = 1, 2, 3, 4, $i = \overline{1,22}$ and $j = \overline{1,6^1}$, $j = \overline{1,5^2}$, $j = \overline{1,5^3}$, $j = \overline{1,5^4}$. For the residential housing development revenues, the probability $r_{i,j}$ and severity $c_{i,j}$, as well as coefficients of optimism $\underline{p}_i$ and pessimism $\overline{p}_i$, have been estimated for all disturbances $e_{i,j}$ that may disrupt payment of tranches b_i that is for $i = \overline{1,4}$ and $j = \overline{1,5}$.

Random factors of optimism and pessimism of costs and revenues allow for comprehensive randomization of the primary initial data and thus take into account the impact of disturbances on the course and results of the project implementation.

2.1.2. Randomized Costs of Tasks and Revenues of Tranches

In reality, costs of task execution and revenue tranches payment are random variables of costs of carrying out tasks and random variables of revenue tranches payment. Such quantities (see, e.g., Benjamin, J. R. and Cornell, C. A. 2014 [24]) should be described by means of appropriate probability density functions or probability distribution functions. Unfortunately, these functions in an analyzed situation are unknown or are virtually impossible to directly elaborate. In that case, probabilistic characteristic of costs and revenues as random variables have been developed by the use of the data randomization method. With this end in view, values of primary initial costs $K_i, i = 1, 2, \ldots, i^a$ and primary initial revenues $D_i, i = 1, 2, \ldots, i^b$ after additional analysis have been verified as the most probable values of random costs $\hat{K}_i, i = 1, 2, \ldots, i^a$ of tasks a_i execution and the most probable random revenues $\hat{D}_i, i = 1, 2, \ldots, i^b$ paid in tranches b_i. Next, using coefficients of optimism $\underline{p}_i$ and pessimism $\overline{p}_i$, the probable bottom boundaries $\underline{K}_i$ and the probable upper boundaries $\overline{K}_i$ of costs K_i of tasks a_i have been calculated. Similarly, using coefficients of optimism $\underline{p}_i$ and pessimism $\overline{p}_i$, the probable bottom boundaries $\underline{D}_i$ and the probable upper boundaries $\overline{D}_i$ of revenues D_i paid in tranches b_i have been calculated. Based on these values (see e.g., Hajdu M., Bokor O. 2016 [25]), using PERT-beta probability distribution function and simplified formulas, the expected values $E[K_i]$, $i = 1, 2, \ldots, i^a$, of random variables of costs K_i and the expected values $E[D_i], i = 1, 2, \ldots, i^b$, of random variables of revenues D_i can be calculated reliably enough for practice. These values, for the residential housing development, have been estimated as follows:

1. The most probable costs $\hat{K}_i$, $i = \overline{1,22}$, of tasks a_i execution—primary initial costs verified by experts due to the predicted random implementation conditions.
2. Probable bottom boundaries of costs $\underline{K}_i = \hat{K}_i - \underline{p}_i \hat{K}_i, i = \overline{1,22}$ of tasks a_i execution.
3. Probable upper boundaries of costs $\overline{K}_i = \hat{K}_i + \overline{p}_i \hat{K}_i, i = \overline{1,22}$ of tasks a_i execution.
4. Expected costs $E[K_i], i = \overline{1,22}$ of tasks a_i execution:

$$E[K_i] = \frac{\underline{K}_i + 4\hat{K}_i + \overline{K}_i}{6} \qquad (5)$$

5. Variance of costs $\sigma^2[K_i]$, $i = \overline{1,22}$ of tasks a_i execution:

$$\sigma^2[K_i] = \left(\frac{\overline{K_i} - \underline{K_i}}{6}\right)^2 \qquad (6)$$

6. The most probable revenues $\hat{D}_i$, $i = 1, 2, 3, 4$, paid in tranches b_i—primary initial revenues verified by experts due to the predicted random implementation conditions.
7. Probable bottom boundaries of revenues $\underline{D}_i = \hat{D}_i - \underline{p}_i \hat{D}_i$, $i = 1, 2, 3, 4$, paid in tranches b_i.
8. Probable upper boundaries of revenues $\overline{D}_i = \hat{D}_i + \overline{p}_i \hat{D}_i$, $i = 1, 2, 3, 4$, paid in tranches b_i.
9. Expected revenues $E[D_i]$, $i = 1, 2, 3, 4$, paid in tranches b_i:

$$E[D_i] = \frac{\underline{D}_i + 4\hat{D}_i + \overline{D}_i}{6} \qquad (7)$$

10. Variance of revenues $\sigma^2[D_i]$, $i = 1, 2, 3, 4$, paid in tranches b_i:

$$\sigma^2[D_i] = \left(\frac{\overline{D}_i - \underline{D}_i}{6}\right)^2 \qquad (8)$$

Estimated in this way, values of the quantities were used for calculations of the randomized project total cost and overall revenue.

2.1.3. Randomized Total Cost and Overall Revenue of the Residential Housing Development

The total cost and overall revenue of the residential housing development are random variables. The random variable of project total cost K is equal to the sum of cost random variables K_i, $i = 1, 2, \ldots, i^a$ of tasks a_i execution. Alike, the random variable of project overall revenue D is equal the sum of revenue random variables D_i, $i = 1, 2, \ldots, i^b$ paid in tranches b_i. The values of random total cost, among others, depend on the dependence between the added random variables K_i. The values of random overall revenue, among others, depend on dependence between the analyzed random variables D_i. Taking into account rules and a way of tasks execution and tranches payment, one can confirm that individual tasks are executed and particular tranches are paid relatively independently of each other. They are physically relatively independent of each other. So, one can conclude that random variables K_i, $i = 1, 2, \ldots, i^a$ and random variables D_i, $i = 1, 2, \ldots, i^b$ are independent and uncorrelated. These variables also take nonzero values. In this case, the project total expected cost $E[K]$ is the sum of expected tasks costs $E[K_i]$, $i = 1, 2, \ldots, i^a$ incurred on an execution of all project tasks $a_i \in A$. Similarly, the project overall expected revenue $E[D]$ is the sum of expected revenues $E[D_i]$, $i = 1, 2, \ldots, i^b$ paid in all project tranches $b_i \in B$. For the residential housing development, these quantities can be estimated as follows:

1. Expected value of the project total cost $E[K]$:

$$E[K] = \sum_{i=1}^{i=22} E[K_i] \qquad (9)$$

2. Variance of the project total cost $\sigma^2[K]$:

$$E[K] = \sum_{i=1}^{i=22} E[K_i] \qquad (10)$$

3. Expected value of the project overall revenue $E[D]$:

$$E[D] = \sum_{i=1}^{4} E[D_i] \qquad (11)$$

4. Variance of the overall project revenue $\sigma^2[D]$:

$$[D] = \sum_{i=1}^{4} \sigma^2[D_i] \qquad (12)$$

Randomized data of costs and revenues of the residential housing development implementation have been tallied in Tables 3 and 4 in Results section, respectively.

Table 3. Randomized data of costs.

s	$a_i \in A^s$	$a_i \in A$	$E[K_i]$	$\underline{K_i}$	$\hat{K}_i$	$\overline{K_i}$	$\underline{p_i}$	$\overline{p_i}$
s1	a_1	a_1	98,379	24,888	98,242	172,415	0.75	0.76
	a_2	a_2	11,966,298	2,676,893	12,083,779	20,785,778	0.78	0.72
	a_3	a_3	3,371,615	670,741	3,438,474	5,805,051	0.80	0.69
	a_4	a_4	6,252,564	1,467,082	6,287,495	10,898,324	0.77	0.73
	a_5	a_5	3,312,396	753,640	3,340,232	5,759,812	0.77	0.72
		Sum	25,001,252					
s2	a_1	a_6	184,487	35,565	188,573	317,067	0.81	0.68
	a_2	a_7	185,745	38,808	188,573	321,366	0.79	0.70
	a_3	a_8	185,745	38,469	188,573	321,706	0.80	0.71
	a_4	a_9	554,405	109,297	565,720	954,256	0.81	0.69
	a_5	a_{10}	557,234	115,973	565,720	964,552	0.80	0.71
	a_6	a_{11}	185,745	38,469	188,573	321,706	0.80	0.71
		Sum	1,853,360					
s3	a_1	a_{12}	174,540	34,671	177,799	301,370	0.81	0.70
	a_2	a_{13}	9,210,002	1,941,568	9,334,462	15,980,599	0.79	0.71
	a_3	a_{14}	3,054,443	609,852	3,111,487	5,270,859	0.80	0.69
	a_4	a_{15}	14,350,624	2,745,932	14,668,440	24,684,051	0.81	0.68
	a_5	a_{16}	3,070,001	653,412	3,111,487	5,320,643	0.79	0.71
	a_6	a_{17}	3,470,568	752,171	3,511,536	6,025,093	0.79	0.72
	a_7	a_{18}	872,698	174,954	888,996	1,505,249	0.80	0.69
	a_8	a_{19}	436,349	87,477	444,498	752,624	0.80	0.69
		Sum	34,639,225					
s4	a_1	a_{20}	85,936	14,188	88,900	145,831	0.84	0.64
	a_2	a_{21}	152,267	23,978	158,063	257,374	0.85	0.63
	a_3	a_{22}	120,207	22,266	123,289	205,819	0.82	0.67
		SumE[K]	358,410 61,852,242					

Table 4. Randomized data of revenues.

Revenue	$b_i \in B$	$E[D_i]$	$\underline{D_i}$	$\hat{D}_i$	$\overline{D_i}$	$\underline{p_i}$	$\overline{p_i}$
Tranches	b_1	8,943,077	799,297	9,867,860	13,387,725	0.92	0.36
	b_2	51,734,858	5,436,555	56,630,778	78,449,484	0.90	0.39
	b_3	4,611,152	302,699	5,095,948	6,980,419	0.94	0.37
	b_4	3,138,317	208,179	3,504,697	4,602,936	0.94	0.31
	SumE[D]	68,427,404 68,427,404					

Calculated in the above way, the expected total cost $E[K]$ and the expected overall revenue $E[D]$ represent only a particular possible case. This means that values $E[K]$ and $E[D]$ describe only this one case of the project implementation. The range of probable changings of these values is interesting. These variations were measured by using coefficients of the risk of total cost $p(k)$ and the risk of overall revenue $p(d)$. Values of these quantities were calculated assuming student's t-distribution functions for the random variable $E[K]$ of total cost and for the random variable $E[D]$ of overall revenue. In accordance with the results of comprehensive analysis, the risk measure of total cost is the probability $p(k)$ that the actual total cost $\mathcal{K} = E(K)$ of the project is lesser than k. The risk measure of overall revenue is the probability $p(d)$ that the actual overall revenue $\mathcal{D} = E(D)$ of the project is greater than d. Values of $p(k)$ and $p(d)$, according to the comparative values k and d, can be calculated as follows:

1. Cost risk (contingency) of the project $p(k)$:

$$p(k) = P[E(K) \leq k] = Z\left[\frac{k - E(K)}{\sqrt{\sigma^2(K)}}\right]; \quad p(k) \in [0,1] \quad (13)$$

$p(k)$—means probability that the real total cost $\mathcal{K} = E(K)$ should be less than k.

2. Revenue risk (contingency) of the project $p(d)$:

$$p(d) = P[E(D) \geq d] = 1 - P[E(D) \leq d] = Z\left[\frac{d - E(D)}{\sqrt{\sigma^2(D)}}\right]; \quad p(d) \in [1,0] \quad (14)$$

$p(d)$—means probability that the real overall revenue $\mathcal{D} = E(D)$ should be greater than d.

Changes of the cost risk $p(k)$ depending on comparative costs k and changes of the revenue risk $p(d)$ depending on comparative revenues d are presented in Figure 1 in the Results section.

Classical and well-known estimation of the net present value was presented in many publications, e.g., Gorlewski, B. 2015 [17], JASPERS. 2008 [19], and Skov N. W. 1994 [26]. Unfortunately, classical approaches do not sufficiently allow the consideration of random conditions of the project implementation. This is possible in the proposed method. The net present value of the residential housing development efficiency has been estimated as a random variable equal to a quotient of the random variable of project overall revenue and random variable of project total cost. The values of such quotient function one can estimate using an appropriate derived probability density function of quotient of these two random variables. Unfortunately, such function is unknown and, in practice, impossible to define. Fortunately, the random variable of expected value of the project overall revenue and the random variable of expected value of the project total cost are independent, uncorrelated, and take positive and nonzero values. When abovementioned conditions are fulfilled, an expected value of a quotient of two random variables, as it has been proved by Frishman, F. in 1971 [27], is equal to a quotient of expected values of these random variables. Therefore, based on the randomized data, the net present value of the residential housing development efficiency has been estimated as follows [28]:

1. Net present value of the residential housing development efficiency (NPE):

$$NPE = E\left[\frac{D}{K}\right] = \frac{E[D]}{E[K]}, \text{ provided that } E(K) \neq 0 \quad (15)$$

2. Variance of efficiency:

$$Var(NPE) = Var\left[\frac{D}{K}\right] = \frac{Var(D) * [E(K)]^2 - Var(K) * [E(D)]^2}{[E(K)]^2 * \{Var(K) + [E(K)]^2\}} \quad (16)$$

3. Standard deviation of efficiency:

$$\sigma(NPE) = \sqrt{\sigma^2(RPE)} \qquad (17)$$

4. Expected gross profit $E[Z]$:

$$E[Z] = E[D - K] = E[D] - E[K] \qquad (18)$$

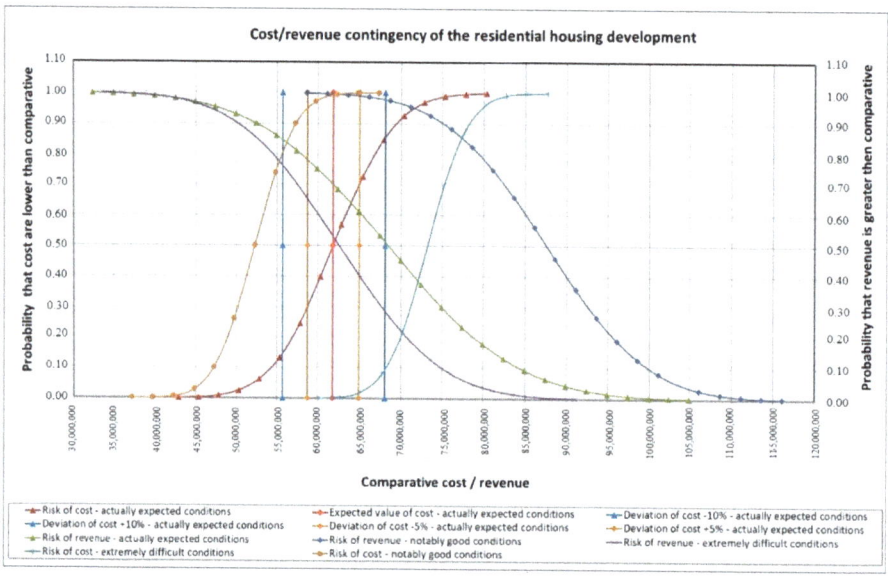

Figure 1. Comparative risk characterization of the costs and revenues of the residential housing development 1.

3. Results

The primary initial data of the residential housing development implementation 1 for costs have been listed in Table 1 and for revenues in Table 2.

Randomized data of costs and revenues of the residential housing development 1 implementation have been tallied in Tables 3 and 4, respectively.

Changes of the cost risk $p(k)$, depending on comparative costs k, and changes of the revenue risk $p(d)$, depending on comparative revenues d, are presented in Figure 1.

In the charts, the risk of total cost $p(k)$ and of the risk of overall revenue $p(d)$ have been presented for notably good, actually expected, and extremely difficult project implementation conditions. Coefficients of the risk of total cost $p(k)$ vary in the range of $[0, 1]$. The coefficients of risk $p(k) = 0$ are indicated by minimum costs $\underline{k}$, which amounts to 39,794,092 PLN for notably good conditions, 45,352,247 PLN for actually expected conditions, and 62,711,729 PLN for extremely difficult conditions. This means that the actual cost $\mathcal{K}$ of the project implementation should not be less than $\underline{k}$. Coefficients of risk $p(k) = 1$ are indicated by maximum costs $\bar{k}$, which amount to 64,794,092 PLN for notably good conditions, 77,852,247 PLN for actually expected conditions, and 85,211,729 PLN for extremely difficult conditions. This means that the actual cost $\mathcal{K}$ of the project implementation should be less than $\bar{k}$.

Coefficients of the risk of overall revenue $p(d)$ vary in the range of $[0, 1]$. Coefficients of risk $p(k) = 1$ are indicated by minimum revenue $\underline{d}$, which amount to 61,115,830 PLN for notably good conditions, 34,927,404 PLN for actually expected conditions, and 36,082,735 PLN for extremely difficult conditions. This means that the actual revenue

$\mathcal{D}$ of the project implementation should be greater than $\underline{d}$. Coefficients of risk $p(d) = 0$ are indicated by maximum revenue $\overline{d}$, which amount to 113,615,830 PLN for notably good conditions, 102,427,404 PLN for actually expected conditions, and 88,582,735 PLN for extremely difficult conditions. This means that the actual revenue $\mathcal{D}$ of the implementation should not be greater than $\underline{d}$.

Final results of the net present value of the residential housing development efficiency have been tallied in Table 5.

Table 5. Final results for conditions initially planned, actually expected, and really occurred for investment 1.

	Primary Initial Data	Prediction for Risk Conditions	Real Conditions ex Post
	Notably good conditions of property sale	Complicated conditions of property sale	Complicated conditions of property sale
	Notably good conditions at labor and construction products market	Extremely difficult conditions at labor and construction products market	Extremely difficult conditions at labor and construction products market
Specification	Primary value	Calculated value	Realized value
		Expected conditions (0,xx)	
Revenue	90,300,000	68,534,790	66,990,000
Cost	67,800,000	61,743,054	60,939,000
Efficiency	1.33	1.11	1.10
Gross profit	22,500,000	6,791,736	6,051,000
		Notably good conditions (0,00)	
Revenue		87,853,537	
Cost		52,293,772	
Efficiency		1.68	
Gross profit		35,559,765	
		Extremely difficult conditions (1,00)	
Revenue		62,229,589	
Cost		73,211,281	
Efficiency		0.85	
Gross profit		−10,981,692	

On the basis of the final results, the residential housing development efficiency was comprehensively assessed.

In order to present the practical application of the discussed method, the final results for two subsequent construction projects are presented below.

3.1. Investment 2

Changes of the cost risk $p(k)$ depending on comparative costs k and changes of the revenue risk $p(d)$, depending on comparative revenues d, are presented in the Figure 2.

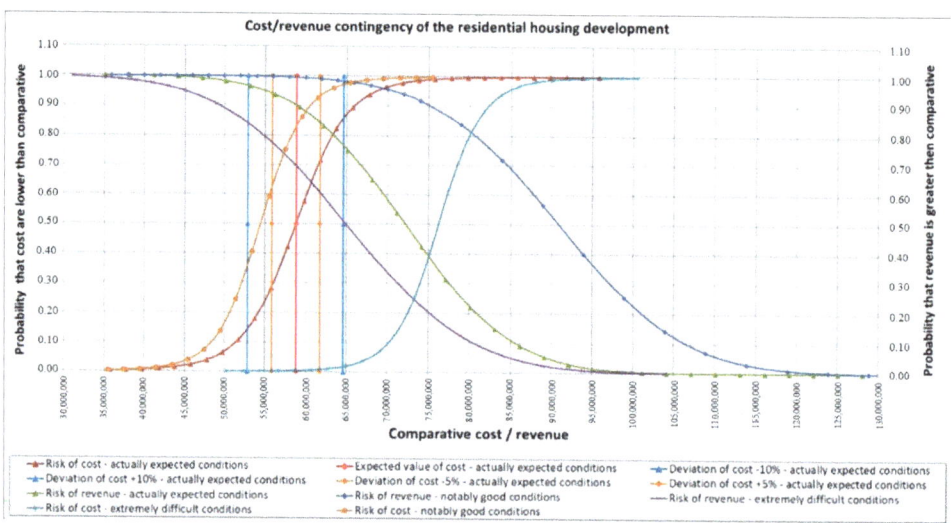

Figure 2. Comparative risk characterization of the costs and revenues of the residential housing development 2.

Final results of the net present value of the residential housing development efficiency have been tallied in the Table 6.

Table 6. Final results for conditions initially planned, actually expected, and really occurred for investment 2.

	Primary Initial Data	Prediction for Risk Conditions	Real Conditions ex Post
	Notably good conditions of property sale	Complicated conditions of property sale	Complicated conditions of property sale
	Notably good conditions at labor and construction products market	Extremely difficult conditions at labor and construction products market	Extremely difficult conditions at labor and construction products market
Specification	Primary value	Calculated value	Realized value
Expected conditions (0,xx)			
Revenue	87,700,000	72,190,231	69,638,000
Cost	70,530,000	58,782,084	56,847,000
Efficiency	1.24	1.23	1.23
Gross profit	17,170,000	13,408,147	12,791,000
Notably good conditions (0,00)			
Revenue		90,959,144	
Cost		54,454,291	
Efficiency		1.67	
Gross profit		36,504,853	
Extremely difficult conditions (1,00)			
Revenue		64,970,817	
Cost		76,236,008	
Efficiency		0.85	
Gross profit		−11,265,190	

On the basis of the final results, the residential housing development efficiency has been comprehensively assessed.

3.2. Investment 3

Changes of the cost risk $p(k)$ depending on comparative costs k and changes of the revenue risk $p(d)$, depending on comparative revenues d, are presented in the Figure 3.

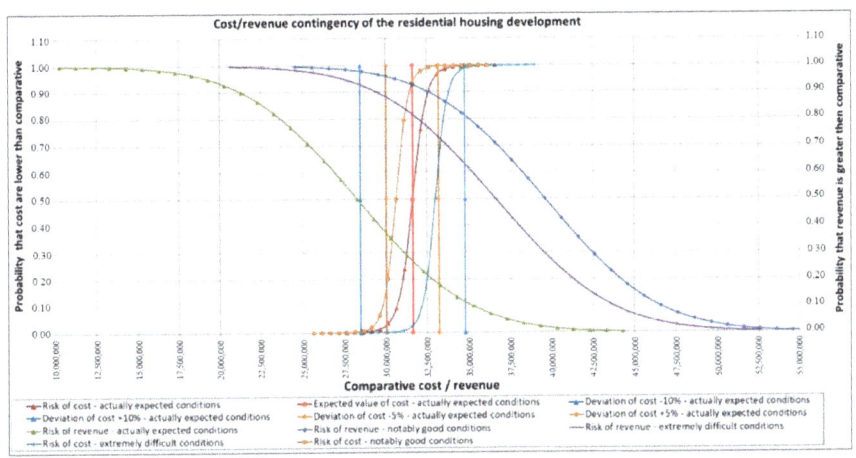

Figure 3. Comparative risk characterization of the costs and revenues of the residential housing development 3.

Final results of the net present value of the residential housing development efficiency have been tallied in Table 7.

Table 7. Final results for conditions initially planned, actually expected, and really occurred for investment 3.

	Primary Initial Data	Prediction for Risk Conditions	Real Conditions ex Post
	Notably good conditions of property sale	Complicated conditions of property sale	Complicated conditions of property sale
	Notably good conditions at labor and construction products market	Extremely difficult conditions at labor and construction products market	Extremely difficult conditions at labor and construction products market
Specification	Primary value	Calculated value	Realized value
	Expected conditions (0,xx)		
Revenue	38,190,000	36,631,885	40,585,000
Cost	32,500,000	31,641,704	34,591,000
Efficiency	1.18	1.16	1.17
Gross profit	5,690,000	4,990,181	5,994,000
Notably good conditions (0,00)			
Revenue		39,609,233	
Cost		30,632,977	
Efficiency		1.29	
Gross profit		8,976,255	
Extremely difficult conditions (1,00)			
Revenue		28,292,309	
Cost		33,016,968	
Efficiency		0.86	
Gross profit		−4,724,659	

On the basis of the final results, the residential housing development efficiency has been comprehensively assessed.

The table with the final results for each of the analyzed investments summarizes and compares the final expected values, such as revenue, cost, efficiency, or gross profit, calculated using the ex-ante method. The table also lists the theoretically possible extreme values between which the expected values are. Particularly good conditions, i.e., maximum income at minimum costs, and extremely difficult conditions, i.e., minimum income at maximum costs.

A graphic image of the above values is presented in the graph. The curves of the expected costs and revenues are located between the extreme, minimum, and maximum costs and revenue curves.

The risk chart of the total cost and total revenue from the implementation of residential investments is constructed in the following structure: the abscissa is the values for costs and revenues, respectively. The ordinate axis in the left-hand system is the probability values that the corresponding costs will be lower than the comparative costs, and the right-hand side is that the probability values of revenues will be higher than the comparative ones. This layout allows you to summarize all cost and revenue curves on one chart.

The expected values of costs and revenues correspond to the ordinate of 0.5.

Above and below the ordinate 0.5, we obtain the reading of the probability occurrence value and the corresponding costs and revenues. In such a system, the slope of the curves makes it possible to assess the speed of changes in the probability of threats, depending on possible changes in costs and revenues.

4. Discussion

In analyzed cases, the net present value of the residential housing development efficiency has been estimated on the basis of randomized primary initial data and is equal to: for investment 1: 1.11; for investment 2: 1.23; for investment 3: 1.16. The efficiency calculated on the basis of the primary initial data is equal to: for investment 1: 1.33; for investment 2: 1.24; for investment 3: 1.18. The efficiency calculated on the basis of realized data is equal to: for investment 1: 1.10; for investment 2: 1.23; for investment 3: 1.17.

It is easy to see that, in analyzed conditions of the investment 1 implementation, the randomized method of efficiency estimating allows for the determination of the project efficiency close to the actual value that has been calculated on the basis realized data. Such accuracy is especially important and reasonable when disturbances strongly interfere with the project implementation.

On the other hand, the investment 2 is a construction undertaking similar to the investment 1 in terms of size, location, and standard of investment execution. One can risk the statement that these are twin investments that were implemented under extremely different conditions.

The investment 2 was carried out in a period of very good economic conditions on the housing construction market, and particularly good conditions were adopted. The investment was completed successfully, in accordance with all the assumed parameters.

All presented above estimations are considered also for exceptionally possible project implementation conditions. Such an approach allows for the determination of a bottom boundary and top boundary of the net present value of the residential housing development efficiency. The bottom boundary of the net present value of randomized efficiency of the residential housing development is estimated for the notably favorable implementation condition. For the described example, it is equal to, for investment 1: 1.68; for investment 2: 1.67; for investment 3: 1.29. The top boundary value of the efficiency is estimated for extremely difficult implementation conditions. For the described example, it is equal to: for investment 1: 0.85; for investment 2: 0.85; for investment 3: 0.86. Such information will certainly be useful in making operational decisions.

5. Conclusions

The method presented in the article complements the publications and studies to date in terms of the probabilistic approach to the issue of assessing the effectiveness of a construction project [29–31]. Three construction projects were subjected to detailed tests: the first—stable and achieved parameters as expected; the second—burdened with a significant loss of stability in the multithreaded aspect; the third—an investment that did not show over-expected deviations from the adopted assumptions. The method proposed in the article, assuming appropriate parameters, has confirmed the effectiveness and correctness of the predicted current net value of probabilistic efficiency (REP). All analyzed investments are real construction projects of the Warsaw developing market. The adopted initial parameters and the obtained final results were verified by the final settlements of the investment. The authors' recommendation is to apply the method to various types and scopes of completed investments, which will enable the development of a broad database for future analyses of new construction projects.

The final result of the method used will be a reliable database that can significantly improve the efficiency, quality, and sustainability requirements of future investments.

Author Contributions: Conceptualization, T.K. and R.W.; methodology, T.K. and A.S.-K.; formal analysis, A.S.-K.; resources, R.W.; data curation, R.W.; writing—original draft preparation, A.S.-K.; writing—review and editing, T.K.; supervision, T.K.; project administration, R.W. All authors have read and agreed to the published version of the manuscript.

Funding: This research received no external funding.

Institutional Review Board Statement: Not applicable.

Informed Consent Statement: Not applicable.

Data Availability Statement: Not applicable.

Conflicts of Interest: The authors declare no conflict of interest. The funders had no role in the design of the study; in the collection, analyses, or interpretation of data; in the writing of the manuscript, or in the decision to publish the results.

References

1. Biznes, T. *Słownik Pojęć Ekonomicznych*; Wydawnictwo Naukowe PWN S.A.: Warszawa, Poland, 2007; ISBN 978-83-01-15279-6.
2. Black, J. *Słownik Ekonomii*; Wydawnictwo Naukowe PWN: Warszawa, Poland, 2008; ISBN 978-83-01-15079-2.
3. Pawłowski, J. *Metodyka Oceny Efektywności Finansowej Przedsięwzięć Gospodarczych*; Wydawnictwo Uniwersytetu Łódzkiego: Łódź, Poland, 2004; ISBN 83-7171-785-7.
4. Ziarkowski, R. *Opcje Rzeczowe Oraz ich Zastosowanie w Formułowaniu i Ocenie Projektów Inwestycyjnych*; Wydawnictwo Akademii Ekonomicznej: Katowice, Poland, 2004; ISBN 8372463018 (20+5).
5. Pyszka, A. *Istota Efektywności. Definicje i Wymiary*; Zeszyty Naukowe Uniwersytetu Ekonomicznego w Katowicach; Uniwersytet Ekonomiczny w Katowicach, Wydział Zarządzania, Katedra Zarządzania Zasobami Ludzkimi, Studia Ekonomiczne: Katowicach, Poland, 2015; ISSN 2083-8611.
6. Pačaiová, H.; Andrejiová, M.; Balažiková, M.; Tomašková, M.; Gazda, T.; Chomová, K.; Hijj, J.; Salaj, L. Methodology for Complex Efficiency Evaluation of Machinery Safety Measures in a Production Organization. *Appl. Sci.* **2021**, *11*, 453. [CrossRef]
7. Starczyk-Kołbyk, A.; Kruszka, L. The influence of construction works disturbances on the EVM analysis outcomes—Case study, Archives of Civil Engineering. *Arch. Civ. Eng.* **2020**, *66*, 161–177. [CrossRef]
8. Starczyk-Kołbyk, A.; Kruszka, L. Use of the EVM method for analysis of extending the construction project duration as a result of realization disturbances—Case study. *Arch. Civ. Eng.* **2021**, *67*, 373–393. [CrossRef]
9. Kasprowicz, T. *Inżynieria Przedsięwzięć Budowlanych*; Kasprowicz, T., Ed.; Inżynieria Przedsięwzięć Budowlanych, Rekomendowane Metody i Techniki, wyd.; PAN KIWiL, Sekcja IPB: Warszawa, Poland, 2015; pp. 10–20.
10. Kasprowicz, T. *Analiza Ryzyka Przedsięwzięć Budowlanych, Budownictwo i Inżynieria Środowiska, Zeszyt 58, nr 3/2011/III*; Oficyna Wydawnicza Politechniki Rzeszowskiej: Rzeszów, Poland, 2011; pp. 233–240.
11. Bizon-Górecka, J. O zarządzaniu projektami inwestycyjno-budowlanymi z uwzględnieniem czynników ryzyka. *Przegląd Bud.* **2008**, *79*, 42–46.
12. Cao, J.; Song, W. Risk assessment of co-creating value with customers: A rough group analytic network process approach. *Expert Syst. Appl.* **2016**, *55*, 145–156. [CrossRef]
13. Radło, M.J. *Risk Management in Integrated Management Systems*; Warsaw School of Economics: Warsaw, Poland, 2015.

14. Kalkhoran, S.H.A.; Liravi, G.; Rezagholi, F. Risk Management in Construction Projects. *Int. J. Eng. Trends Technol.* **2014**, *10*, 133–138. [CrossRef]
15. Ward, S.C.; Chapman, C.B. Risk-management perspective on the project lifecycle. *Int. J. Proj. Manag.* **1995**, *13/3*, 145–149. [CrossRef]
16. Korytárová, J.; Hromádka, V. Risk Assessment of Large-Scale Infrastructure Projects—Assumptions and Context. *Appl. Sci.* **2021**, *11*, 109. [CrossRef]
17. Gorlewski, B. *Project Effectiveness Evaluation*; Warsaw School of Economics: Warsaw, Poland, 2015.
18. Pham, T.Q.D.; Le-Hong, T.; Tran, X.V. Efficient estimation and optimization of building costs using machine learning. *Int. J. Constr. Manag.* **2021**. [CrossRef]
19. JASPERS. *Blue Book. Road Infrastructure. Joint Assistance to Support Projects in European Regions*; Ministry of Infrastructure: Warsaw, Poland, 2008.
20. De Wilde, P. *Building Performance Analysis*; John Wiley and Sons Ltd.: Hoboken, NJ, USA, 2018.
21. Ding, L.; Zhou, Y.; Akinci, B. Building Information Modeling (BIM) application framework: The process of expanding from 3D to computable nD. *Autom. Constr.* **2014**, *46*, 82. [CrossRef]
22. Halpin, D.W.; Woodhead, R.W. *Construction Management*, 2nd ed.; John Wiley & Sons Inc.: New York, NY, USA, 1998.
23. Ritz, G. *Total Construction Project Management*; McGraw Hill Professional: New York, NY, USA, 1994.
24. Benjamin, J.R.; Cornell, C.A. *Probability, Statistics, and Decision for Civil Engineers*; Manufactured in the United States by Courier Corporation: New York, NY, USA, 2014.
25. Hajdu, M.; Bokor, O. Sensitivity analysis in PERT networks: Does activity duration distribution matter? *Autom. Constr.* **2016**, *65*, 1–8. [CrossRef]
26. Skov, N.W. *Finance & Management*; The American Model Applied to Polish Private Enterprise; PRET: Warsaw, Poland, 1994.
27. Frishman, F. *On the Arithmetic Means and Variances of Products and Ratios of Random Variables*; Army Research Office: Durham, UK, 1971. Available online: https://www.semanticscholar.org/paper/On-the-Arithmetic-Means-and-Variances-of-Products-Frishman/5116dc6b2987ee3eb26eab0a514c8a4a2e2b953c (accessed on 20 December 2021). [CrossRef]
28. Parmenter, D. *Key Performance Indicators: Developing, Implementing, and Using Winning KPIs*; Copyright by David Parmenter; John Wiley & Sons, Inc.: Hoboken, NJ, USA, 2015.
29. Braganca, L.; Kokkari, H.; Veljkovic, M.; Borg, R.P. *Sustainable Construction. A Life Cycle Approach in Engineering*; International Training School: Hal Far, Malta, 2010.
30. Sicotte, H.; Delerue, H. Project planning, top management support and communication: A trident in search of an explanation. *J. Eng. Technol. Manag.* **2021**, *60*, 101626. [CrossRef]
31. Uher, T.E.; Lawson, W. Sustainable development in construction. In Proceedings of the CIB World Building Congress, Gaevle, Sweden, 7–12 June 1998; pp. 1–8.

Article

Project Risk in the Context of Construction Schedules—Combined Monte Carlo Simulation and Time at Risk (TaR) Approach: Insights from the Fort Bema Housing Estate Complex

Janusz Sobieraj [1] and Dominik Metelski [2,*]

1 Department of Building Engineering, Warsaw University of Technology, 00-637 Warsaw, Poland; jsob@il.pw.edu.pl
2 Faculty of Economics and Management Sciences, University of Granada, 18071 Granada, Spain
* Correspondence: dominik@correo.ugr.es

Abstract: In this article, we present our own construction process model consisting of 16 stages and eight phases, which is particularly applicable to large investment projects. In the context of each project phase, we examine how the appropriate way of scheduling construction processes affects the problem of the risk of prolonging individual phases and the whole project, as well as of not meeting deadlines (which is one of the main problems faced by management practitioners in the construction industry). There are many methods for assessing risk in this context, but they tend to be overly complex and rarely used by construction practitioners. On the other hand, the risks associated with potential schedule delays can be considered holistically. One tool that can serve this purpose is the combined Monte Carlo simulation and Time-at-Risk (TaR) approach, which originates from the world of finance. We show how the implementation of the process model (individual phases) and the whole project can be considered in the context of the covariance matrix between all its phases and how changes in the arrangement of these phases can affect the risk of time extension of the whole project. Our study is based on simulation data for a large development project (Fort Bema/Parkowo-Leśne housing estate complex) in Bemowo, a district of Warsaw, carried out between 1999 and 2012. The entire investment project involved the construction of almost 120,000 m^2 of floor space.

Keywords: time schedules; project risk; construction project management; Time-at-Risk (TaR); investment-construction process model; Monte Carlo simulation

1. Introduction

Time uncertainty is ubiquitous in all disciplines of project planning and scheduling, yet in the construction industry it is of particular importance [1–6]. In particular, the uncertainty and risk associated with scheduling is an important aspect of large and complex construction projects [7–10]. In fact, successful project completion is mainly the result of a variety of multidirectional and comprehensive activities in the area of the creation of companies' intellectual capital, preparation of flexible plans and schedules, and their posterior implementation [3]. The analysis of completion times for individual activity tasks as well as entire phases/stages of the construction process should help project managers understand the factors that influence the final estimate of project duration [1,10].

Among the methods used for project scheduling, the best known (and most commonly used) are the critical path method (CPM) [11,12] and the programme evaluation and review technique (PERT) [10]. These methods are based on the determination of minimum times assigned to individual tasks/activities of the project, without which its realisation is not possible. More precisely, these tasks are critical activities that have a direct impact on the overall duration of the project completion. In other words, any change in the duration of the critical activities affects the project deadline. However, schedules developed with the use of these classical deterministic methods represent completely unrealistic expected project deadlines that are very likely to be exceeded [8]. Methods like CPM and PERT were criticised by a number of researchers for being not true models of the construction process. These methods lack the adequacy to model complex logic and resource constraints in a construction process [13–19]. This form of scheduling seems to be particularly unsatisfactory for construction projects, where exceeding completion dates carries heavy penalties [9]. In any project there is also a whole range of activities whose possible disruption does not always affect the project's target completion date (in this sense they are non-critical activities); in such cases there are some flows or slacks, providing a certain degree of execution flexibility in terms of time buffers. In other words, a float, or slack is the amount of time by which a project or activity can be delayed without extending the completion time of the entire project; while the total slack is the amount of time by which the entire project can be delayed without postponing the ending date of the project. To some extent the shortcomings of the deterministic methods can be solved with the Critical Chain Project Management (CCPM) method, which addresses resource alignment (and resource constraints), requiring some flexibility in resource allocation and individual stages' starting dates [20]. In other words, an appropriate use of resources plays a key role in the CCPM method. The results confirm that CCPM can perform better than the traditional CPM method, which translates into significantly shorter construction times combined with accurate resource levelling at the early stages [20]. The application of CCPM leads to a reduction in indirect costs due to shorter project implementation times and a reduction in direct costs as a result of better allocation of resources and ensuring their timely availability.

Also, many problems related to project implementation and project management can be solved with the use of game theory, which allows for the shifting (and appropriate allocation) of resources between specific activities (and phases/stages) of a project. This requires some trade-off between the execution times and costs of each activity [21]. That is, it is possible to reduce the time for a given task by allocating more resources to its execution; this depletion of resources is later compensated at the expense of the execution times of some non-critical tasks. It is obviously a good idea to conduct an appropriate cost-benefit analysis in this context.

In addition to the aforementioned CPM, PERT or CCPM scheduling methods, there are many others that also find practical application, such as Line of Balance (LOB) [22–26], Q Scheduling [27,28], Resource Oriented Scheduling [29,30], Last Planner System [31], or a widespread schedule visualisation method such as Gantt Charts [32,33]. There are also numerous studies that facilitate accurate analysis and prediction of project delay risks. These include Bayesian networks [34,35], decision trees and Bayesian classification algorithms [36], Naïve Bayesian (machine learning models) [36], Bayesian Belief Network [37], logistic regression [38], Markov Dynamic PERT Model [39] and Correlated Schedule Risk Analysis (CSRA) [20]. It should be noted here that understanding schedules is an important thing that often determines project success [2,3,40], therefore the more tools and research findings to help understanding in the area of project scheduling, the better.

Thus, in the context of some studies mentioned above, it is possible to determine/quantify/model the probability of project schedule delays using various research methods [36–38]. On the other hand, when analysing the literature on the subject, it is hard not to notice the lack of appropriate methods that would allow a comprehensive (from the risk control perspective) assessment of the risk of delays of a construction project, taking particular account of individual stages of the project. It would be good if such a method accounted for dependencies between individual stages of the project, and even allowed for conducting appropriate stress-tests in the context of potential results of changes in the schedules. A method of this kind exists and works very well in the financial world. It is known as the Value at Risk (VaR) measure. VaR is a statistic that quantifies the extent of possible financial losses in a firm, portfolio or position over a specified time period. It is a measure that allows to assess the probability of losing a specific value of an investment portfolio and to simulate the impact of changes in the external environment on the value of the entire portfolio. In addition, since many interesting solutions are created by transferring already existing solutions from other fields, it is very likely that applying a solution similar to VaR would work well with construction projects. To this end, individual stages of the project can be perceived as a set/collection of stages/activities and can be treated similarly to an investment portfolio. Instead of VaR in this study we will use Time-at-Risk (TaR), which is constructed similarly to VaR, although its interpretation is slightly different. TaR represents a certain quantile for a given probability distribution, so it is similar to VaR; however, TaR measures the magnitude of risk as time (the time until an adverse event occurs) rather than value (loss amount). This measure has already been analysed in other research papers, but in a slightly different context [41,42].

TaR can potentially allow construction practitioners not only to understand the impact of postponing activities/stages/phases, but also to visualise how frequently such changes are likely to occur. More importantly, the Time-at-Risk approach applied in construction projects can potentially give construction firms and project managers a high degree of confidence that they are adequately equipped to withstand schedule changes with minimal losses.

In this study, we present a proprietary model of the investment process and show how the implementation of its individual stages/phases, as well as the entire project, can be evaluated in the context of the covariance matrix between time schedules of all its stages and how their appropriate schedule timing can affect the risk of extending the implementation time of the entire project. Our study is based on both real and simulation data for a large development project, Fort Bema in Warsaw (Parkowo-Leśne housing estate complex), which was implemented in 1999–2012 and involved the construction of almost 120,000 m^2 of floor usable space area.

We also review the literature and point out the most important aspects of project scheduling, describe the research methodology and finally present the results, discussion and conclusions. The rest of the paper is organised as follows. In the following part, we discuss some theoretical aspects related to the topic of our study. In the empirical part that follows, we focus on the analytical aspects, presentation of data and models, and methodology to better understand the intricacies of scheduling in construction and the aspects of the Time-at-Risk concept. The paper ends with the discussion and final conclusions.

2. Literature Review

One of the most important elements of every construction project management is the preparation of an appropriate schedule [10,43]. It should be detailed and realistic so that it reflects all the assumed (planned) activities, stages and whole phases of the project and facilitates their proper coordination. It is good when the schedule itself is understandable for all stakeholders involved in the project [32]. Research has clearly shown that Gantt Charts are the best tool to increase schedule comprehensibility [32]. A well-developed schedule should indicate important milestones against a timeline, and these can be considered as starting and ending dates for individual stages and phases. However, no matter what the project schedule is, it is usually difficult to implement and manage, and there are many reasons for this, but in general they can all be grouped under the category of uncertainty around the duration of activities and whole project phases [18]. To begin with, each construction project has very complex characteristics and consists of thousands of different activities [10], each of which is different. Even when assuming that these activities/stages are similar to the ones known from previously executed construction projects it is still difficult to accurately reproduce them in each subsequent project. In other words, it is difficult to standardise certain activities (in this respect every activity is project-specific). Therefore, it is difficult to estimate precisely the completion times of most activities, stages, not to mention the whole projects [18]. In general, there are many reasons why the majority of construction projects fail to achieve their objectives, exceeding the assumed completion times [44].

There is a group of researchers and scientists who attribute the highest proportion of the total uncertainty associated with a construction project to its early stages [10,45–48]. Adequate risk control related to project scheduling can help to mitigate the uncertainty that accompanies many construction projects. Therefore, risk management is becoming more and more important in projects.

The common practise in analysing the uncertainty of times for individual tasks is to divide the project into several construction processes and to capture the uncertainty of individual components (elements) probabilistically [1,16].

It is also worth remembering that risk is defined as any change (also positive) with respect to the assumed objectives [49]. Risk is always seen in the context of both the aforementioned deviations from expectations (both positive and negative) and is quantified by its probability of occurrence. In any case, in order to manage risks appropriately, they must first be adequately recognised and quantified by means of an appropriate model [50]. In the past, network diagram techniques were used for this purpose. Conventional deterministic methods such as CPM do not address the problem of project uncertainty. The opposite is true, i.e., these methods imply certainty about the successful completion of each of the activities under consideration. They indicate certain rigid values for the duration of each activity and do not take into account different scenarios. As it turns out, however, the effectiveness of the CPM method is very low, and the probability that it correctly indicates the duration of any activity does not even exceed 25% [16]. The situation does not look better when it comes to the PERT method [17–19], which typically underestimates schedule times in majority of cases [17,19].

According to Kong et al. [16], in order to develop a more realistic schedule, a formal identification and assessment of project uncertainty is essential. They list a number of different methods used in project risk analysis (e.g., sensitivity analysis, decision tree analysis and the Delphi method), accentuating the superiority of Monte Carlo simulations, which, in their opinion, offer the greatest potential [51,52].

Lindkvist and Soderlund [53] point out the importance of meeting deadlines in the form of a schedule as a key component of any project. In the same vein, the importance of the schedule in project development is further underpinned by Dille and Söderlund [54], who attribute a fundamental importance to the time component in the context of integrating all stakeholders in the project. In this sense, the Gantt method enables what Dille and Söderlund [54] call isochronism by addressing the temporal positioning of certain tasks/activities and thus emphasising the importance of timely project completion.

An interesting perspective seems to be that of Yakura [55], who perceives the Gantt method as a boundary object in the context of time, which leads the parties involved to perceive the project in concrete (rather than abstract) terms, subject to a common understanding that is appropriately communicated between project participants. According to Geraldi and Lechter [32], while supporting the Gantt method in projects is good, it is important that it does not override management efforts, as an excessive focus on time can neglect other aspects of project management such as project quality, value creation, maintaining appropriate relationships with other project stakeholders, etc. In this context, Maylor [33] warns in his work that project management should not be reduced to time management, especially since time is not necessarily always the decisive criterion for success. There are many examples where excessive time pressure to complete a project on time has led to a failure of the project [10,56,57]. The following are examples of such projects [10]: in Germany, the construction of a railway tunnel in Cologne, the construction of a bypass around Munich, the destruction caused by underground works that led to the collapse of the municipal archives building, etc.; in China, the construction of a railway line from Qinghai province to the Tibetan capital; in the USA, the World Trade Center reconstruction project; in Brazil, the construction of the Belo Monte dam; in the UK, a delay of more than 2 years in the construction of Terminal 5 of London Heathrow Airport; in Poland, delays in the construction of stadiums and road investments, such as the famous extension of the bridge in Mszana by more than 4.5 years, as well as the construction of a housing estate in Warsaw's Wilanów district, which has been dragging on for years, and many others.

Jugdev and Müller [58] point out that increasing pressure for sustainable development, as well as the customisation of certain solutions (customers expect products and services tailored to their needs), ensure that the focus on the time component of a project and performance is not the only important project criteria. According to Geraldi and Lechter [32], wherever non-time-related project performance criteria are more important than those in which time plays a key role—placing great emphasis on the use of Gantt charts will not be justified; the authors point to many projects that are perceived as very successful, although they did not meet time-related criteria.

According to Flyvbjerg et al. [59,60] and Beckers et al. [61], any negative changes in project scheduling are among the most important factors undermining successful completion of construction investments. Therefore, time schedule constraints are a key issue for every construction company. Sobieraj [10] argues that time overruns are so common these days that they need to be reflected in project implementation procedures [9,10]. Flyvbjerg et al. [60] see the reasons for this in the (systematically reproduced) overly optimistic assumptions (i.e., time, cost and benefit forecasts for such projects) made in the planning phase. In this respect, the observations of Flyvbjerg et al. [59,60] are in line with the findings of the studies presented by Bliński [45] and Obolewicz [46], namely that the conceptual and planning phases are the most important in terms of possible delays and costs overruns for the entire project. Sobieraj [48] notes that most of the project management methodologies applied in construction focus on implementation phases, while neglecting the phases and stages determining the success and quality of the project, i.e., preparation and adoption of the feasibility study and directions of the local spatial development plan in cooperation with urban planners, architects and investors [10,48].

Various types of risks with construction project time schedules were analysed with the use of such methods as naïve Bayesian classifiers [36], Bayesian belief networks [37], the logit and probit models [38], robustness [62–65], Monte Carlo simulation [66], etc.

Table 1 shows different methods addressing time schedule management in construction projects.

Different schedule management techniques viewed from construction practitioners' perspective are presented in Table A2 in Appendix A.

Moreover, Gondia [36] emphasises that many schedule management problems can be solved thanks to machine learning (a relatively young but rapidly developing method), which offers an ideal set of techniques that can deal with such complex systems. However, it must be stressed that the implementation of such techniques is not an easy task, especially in a sector such as construction; development in this area is still at an early stage. Nevertheless, Gondia [36] has developed two such models using decision tree and naïve Bayesian classification algorithms. These algorithms were identified and trained using a dataset, predicting the extent of project delays. However, it is important to note that the development of such complex models requires not only the identification of relevant sources and risk factors of a delay, but also the use of a multidimensional dataset of past project time performances and risk sources leading to the delays [36]. Risk factors are active, interdependent, and dynamic, but naïve Bayesian models leverage machine learning capabilities to facilitate evidence-based decision making to enable proactive project risk management strategies [36].

Table 1. Different methods addressing time schedule management in construction projects.

Method	Authors	Contribution
Naïve Bayesian; machine learning models; decision tree and naive Bayesian classification algorithms	Gondia [36]	Machine learning models facilitate accurate analysis and prediction of project delay risks (time overruns). In the case of Gondia's [36] study, the evaluation results show that the naïve Bayesian model provides better predictive performance for the data studied.
Bayesian networks	Khodakarami et al. [34]; Khodakarami and Abdi [35]	Khodakarami et al. [34] used Bayesian network modelling for project scheduling. They reflect the causal relationship between these sources of uncertainty and project parameters. This method has the advantage (over other methods) of considering both uncertainty and causality. Khodakarami and Abdi [35] studied project-related uncertainties by modelling different factors affecting project performance. In their study, they quantified different types of uncertainty by relying on a modelling method for complex project dependencies such as common causal factors, formal use of expert opinion, and learning from data to update prior beliefs and probabilities.
Bayesian Belief Network (BBNs)	Kim et al. [37]	Kim et al. [37] used BBNs as a tool for predicting the probability of schedule delays. It is a method that offers great flexibility in accepting inputs and providing outputs. In addition, BBNs have the ability to treat the value of a variable as a known input or to evaluate its probability as an output of the system. It is a very useful technique for calculating the probability of events before and after the entry of evidence and for making predictions with the use of expert opinions (BBNs do not necessarily require historical data). Kim et al. [37] quantified the probability of delays in construction projects using Bayesian belief networks. The top main causes of changes in time schedule construction projects turned out to be the owner's financial difficulties, inadequate experience and financial difficulties of contractors, shortage of materials, slow site handover, inappropriate construction methods, defective works and reworks and a lack of management capacity by owners/project managers.
Logistic regression	Anastasopoulos et al. [38]	The probability of a project having a time delay can be modelled as a binary outcome variable (1 if there was a time delay and 0 otherwise). Statistical approaches for such a model include the standard probit and logit models. To investigate the probability of a project delay, Anastasopoulos et al. [38] used a binary logit model with random parameters. The results of the model estimation show that the probability and duration of project delays are significantly influenced by factors such as project cost (bid amount), project type, planned project duration and the probability of bad weather.

Table 1. Cont.

Method	Authors	Contribution
Combination of Bayesian networks, support vector machines and Monte Carlo simulation to simulate project outcomes	Fitzsimmons et al. [67]	To improve the performance of critical path (pre-project) activity scheduling, Fitzsimmons et al. [67] proposed a method that integrates Bayesian networks to estimate the conditional delay probability of an activity based on its predecessor. It yields much better results than the traditional Monte Carlo simulations (by 52%). The Fitzsimmons et al. [67] method relies on data that originates from 302 completed infrastructure construction projects; Fitzsimmons et al.'s [67] model was appropriately trained/calibrated and validated on a large infrastructure road construction project. It works well in predicting project delays.
Monte Carlo simulation and Bayesian network	Namazian et al. [66]	Namazian et al. [66] brought together the Bayesian network and Monte Carlo simulation methods and presented the timing of a construction project in the context of a framework for assessing the overall impact of risks in such a project.
Confidence based scheduling procedure (CBSP)	Poh and Lam [1]	Poh and Lam [1] propose a method to determine the probability distribution of project completion times by estimating the duration of individual tasks/activities using confidence-based estimation.
Robustness	Bertsimas and Sim [62]; Al-Fawzan and Haouari [64]; Van de Vonder et al. [65]; Jaśkowski [63]	As one of the methods, Van de Vonder et al. [65] presented robustness, and more precisely, proactive heuristic methods for robust project scheduling. According to Al-Fawzan and Haouari [64], robustness of a schedule means that its validity is maintained despite small changes in the duration of processes (activities) and these changes are due to risks. The development of robust methods is described by Bertsimas and Sim [62]. Predictive scheduling with a proactive approach is about creating schedules that are robust to disturbances (hence the name—robust schedules).
Markov Dynamic PERT Model	Azaron and Ghomi [39]	Refinement of project time duration uncertainty bounds; this approach estimates the influence of factors such as war, strikes and inflation that make activity durations non-static over time. This is an untested model, but it combines externalities, deterministic CPM, PERT and correlation in an interesting way.
Correlated Schedule Risk Analysis Model (CSRAM)	Ökmen and Öztaş [68]	CSRAM accounts for covariance and correlation effects with PERT and CPM; it uses simple subjective inputs for a number of project risks, including weather factors, soil conditions, labour productivity and material/resource availability. However, CSRAM does not address the key problem of subjectivity and opinion-based analysis, a factor commonly associated with contract disputes [69,70]. Furthermore, there is a lack of empirical evidence that this technique is scalable and works across different types of projects.

A number of different factors need to be taken into account when developing construction projects schedules, including weather [71,72] and macroeconomic factors/conditions (external factors are unpredictable and cannot be controlled) [73]. Bragadin and Kähkönen [74] emphasise the importance of a quality schedule and attribute to it an important role in reducing delays in project implementation and achieving agreed outcomes and benefits (deliverables) [75]. A high-quality schedule is interpreted by the above authors as an optimal and feasible plan of activities. The quality of a schedule can be improved by contrasting expected results with resource and spatial constraints and technical knowledge, avoiding subjectivity and uncertainty by optimising all activities (groups of activities) to produce an achievable schedule/action plan [74].

Good scheduling technique aims to minimise the possibility of project delays [10]. This includes taking time risk into account when planning construction projects and factoring in adequate time reserves (i.e., slacks) for activities or whole groups of activities that carry relatively high risks of delays/disturbances.

Also, Ortiz-González et al. [76] highlight the gap between construction theorists and practitioners and the low willingness of construction practitioners to utilise theoretical solutions/methods, resulting in a high degree of subjectivity in the assessment of time risk in construction projects [76].

The fact is that the vast majority of errors in investment projects can be avoided, or at least their impact on the whole process reduced, if appropriate tools are used to create, visualise and manage schedules. Appropriate mathematical (statistical) models can be helpful in this regard. One such approach is robustness [62,64,65] and predictive scheduling with a proactive approach [63].

According to Mubarak [77], the uncertainty surrounding a construction project makes scheduling an extremely difficult task. In order to better understand the problems associated with the time element in the preparation of schedules, it is necessary to understand detailed historical records and make appropriate calculations.

The paper focuses on schedules, however, in the empirical part (case study analysis) there are many references to individual stages and phases of the investment process, therefore this issue cannot remain unaddressed also in the theoretical part. A review of the literature on this topic reveals significant differences in the approach to defining and representing construction investment processes. The discrepancies concern the nomenclature itself, the individual stages and phases of an investment project, and the list of specific activities and measures that make up each stage/phase [46]. It should also be remembered that the discrepancies are due to the fact that every investment project is different. This is why practitioners and scholars often argue about how to present individual stages and phases/activities of a construction investment process. For example, Biliński [45] divides each investment-construction process into three stages, i.e., preparation of the investment project (stage A), activities preceding the commencement of works (stage B) and construction works, followed by maintenance of the building facility (stage C). Grzywiński [78] describes four stages of the construction process and, within each of them, outlines the activities to be performed from the investor's perspective. The stages of the investment process that he distinguishes pursuant to the construction legal framework are determination of the legal status and purpose of the property (stage 1), elaboration of the development conditions and preparation of the construction works (stage 2), obtainment of the construction permit and execution of construction works (stage 3) as well as obtainment of the operation permit and commencement of operation/exploitation of the building facility (stage 4). Zabielski [79] distinguishes four project phases, namely determination of the location conditions for a given investment, preparation of appropriate documentation, execution of construction works and exploitation of construction facilities. In each of these phases he identifies appropriate actions of legal, administrative and factual character. In turn, Dzierżewicz and Dylewski [80] describe an investment process through the prism of legal norms and provide a detailed description of its four phases and stages, i.e., stage 1, in which the conditions for the development and land use are established, stage 2, which consists in preparing the investment for its realisation, stage 3, which involves the implementation of the investment itself, and stage 4, which relates to the maintenance of the completed construction facility [80]. Strzelecka et al. [81] introduce five phases of the construction investment process, i.e., formulation of the project (pre-investment phase), analytical and research activities (divided into two phases), execution of construction works (execution phase) and the use (exploitation phase). Połoński [82] describes the construction investment process as consisting of three stages, i.e., investment preparation for realisation (phase 1), investment execution (phase 2) and operation of completed construction facilities (phase 3, otherwise known as the exploitation phase) [80]. He defines the investment process as a sequence of coordinated activities and actions of legal, organisational, technical, technological and financial nature, which as a whole lead to the realisation and subsequent use (exploitation) of the planned construction investment within a specified period of time and within certain limited financial resources. Baryłka and Baryłka's [83] approach differs from those described above in that it clearly separates the part of the construction process associated with the preparation and implementation of the investment process from the use and decommissioning part. In total, they identify four stages of the investment process. Obolewicz [46] points to a wide variety of interpretations of the elements of the investment process in construction and makes an effort to systematise them [46]. As a result, each construction object (investment) is created based on a different set of activities that make up the whole investment process. This is due to the fact that many different factors determine the final outcome of the investment completion. For investors and practitioners in the construction industry, however, it is not so much the process itself that is important as the ratio of implementation costs to the original investor cost estimates. To understand the structure of such costs, it is worth reaching for the Dutch experience described by Biliński [45]. Namely, on its basis we know that the very conceptual phase of a construction investment has the greatest impact on its overall cost (up to 200–300%) [45]. It is followed by the preliminary design phase, the influence of which on the total cost of the project is estimated to be between 40% and 80%. On the other hand, the influences of the executive project and the implementation phase are much smaller and amount to 15–30% and 5–10%, respectively. Although these studies are not new, the sense of these experiences—as Biliński aptly notes—is meaningful [45]. Put another way, it is the conceptual phase that should receive the most attention, as its impact on the subsequent total cost of the project is the greatest. Unfortunately, as Obolewicz [46] notices, in the Polish reality, investors are particularly keen on reducing the investment implementation time as much as possible, which in practice results in limiting the time that is devoted specifically for preparation of an investment [46]. In an attempt to find a recipe for keeping the investment budget in check and completing it according to the original timeline, Obolewicz [46] points to the need for careful execution of the investment preparation stage, and for shortening as much as possible the execution stage of the investment itself (as this is when the accumulation of resources takes place) and maintaining integration and coherence of all stages of the investment process. In the latter respect, the institution of the substitute investor is of key importance, as it makes it possible to coordinate all activities and actions that are part of the construction investment process (e.g., civil-legal, administrative-legal, design, consultation, financial, and those related to construction works or operation of the completed building facility itself).

The use of an appropriate investment process model is essential in this study, as scheduling involves potential shifts between individual stages of such a process. In this sense, it constitutes a certain framework with which it is possible to describe and encapsulate every investment project. At the same time, an appropriate timing of individual stages of the investment process and any changes with regards to their synchronisation can be viewed as schedule management.

3. Materials and Methods
3.1. Fort Bema Housing Estate Complex (Parkowo-Leśne Housing Estate)

The project included the design and construction of a large residential complex in Warsaw-Bemowo, the modernisation of the Bemowo Sports and Recreation Centre, the relocation of the burdensome WZL-4 military production unit, the rehabilitation of 140 hectares of neglected military land acquired from the municipality and located in the immediate vicinity of Fort Bemowo, the replacement of the worn-out, highly inefficient technical infrastructure with new infrastructure with parameters that allowed the use of the site, the elimination of the particularly dangerous railway crossing at Księcia Bolesława street and the construction of a viaduct in this place, a functional connection of the residential areas of the Bemowo district with the recreational and sports area.

More precisely, the whole project was implemented in the period 1999–2012 and included eight parts, namely, the construction of (1) housing complexes at Osmańczyka Street, (2) infrastructure (networks, roads), (3) buildings of Acciona Real Estate, (4) buildings of Dom Development, (5) buildings of SBM "Idealne Mieszkanie" housing cooperative, (6) buildings of PBM Południe Development, (7) a multi-storey car park and (8) accompanying facilities (flyovers, sports facilities, communal infrastructure, etc.). Figure 1 illustrates the construction site of the Parkowo-Leśne housing estate complex (the picture of the site was taken in October of 2004).

Figure 1. Construction of the Parkowo-Leśne housing estate complex—view from the side of Osmańczyk 10 street (construction site, October 2004).

3.2. Construction Process Model and Time Schedules

The literature on the subject lacks a comprehensive investment–construction process management model encompassing many specific stages of the whole investment process, i.e., from the idea of the project itself, through its forecasting, development of a spatial development study, adoption of local spatial plans (with the participation of architects and developers) [48], planning, scheduling, execution and commissioning for operation, closing with the removal of all defects and imperfections after 3 years of exploitation and assessing as to whether the assumptions concerning the parameters of use have been achieved after this period. The lack of such a model often results in numerous implementation problems, leading to longer investment execution times and, ultimately, higher than expected construction costs (sometimes even a few times higher than anticipated).

Based on the literature on the subject, as well as by analysing a case study of the implementation of a large complex of housing estates in the Warsaw agglomeration (comprising eight smaller project parts executed in the period 1999–2012), 16 stages of the investment process have been identified. The model consists of 16 stages, grouped into eight phases of the investment process: conceptual phase (project initiation), planning phase (design), preparation phase (design), execution phase (construction), exploitation phase (operation), two project completion phases (investment efficiency evaluation)—evaluation just after project completion and second evaluation after initial operation period.

In conclusion, the paper uses a proprietary investment process model, which addresses the deficiencies of the construction process model outlined in the Literature Review section. Different stages and phases of the investment process are illustrated in Figure 2 and their detailed description is presented in Table A1 in Appendix A.

Individual stages of the entire investment process (that was used for the realisation of Fort Bema housing estate complex) and the execution times of its all stages/phases are shown in Table 2.

Table 2. Different stages of the Fort Bema investment process grouped in eight phases.

Activities/Phases/Stages	Abbreviation	Execution Time in Months
Opportunity study	OS	2
Participation in local spatial development plan approvals	PLSDPA	11
Conceptual stage	CONCS	6
Pre-feasibility study	PFS	1
Implementation planning stage—phase I	IPSP1	3
Implementation planning stage—phase II	IPSP2	5
Arrangements stage and execution of construction documentation	ASECD	81
Developing a detailed investment management map	DDIMM	6
The stage of obtaining and securing financing (feasibility study)	SOSF	6
The stage of executing executive documentation	SEED	69
Stage of selecting general contractor(s) and verification of executive documentation	SSGCVED	69
Project implementation stage	PIS	93
Commissioning stage	COMMS	16
Evaluation stage of obtaining results and effects of the project (1st phase of project closing)	ESOREP	1
Phase of drawing up proposals for future implementation after Project Closure Phase I	PDPFIPCP1	1
Initial operation stage (usually 3 years of warranty and guarantee)	IOS	102
The stage of final evaluation of the results and effects of the project (II stage of project closing)	SFEREP2	1
Phase of drawing up proposals for future implementation after Phase II of the project closure	PCPFIPCP2 II	1

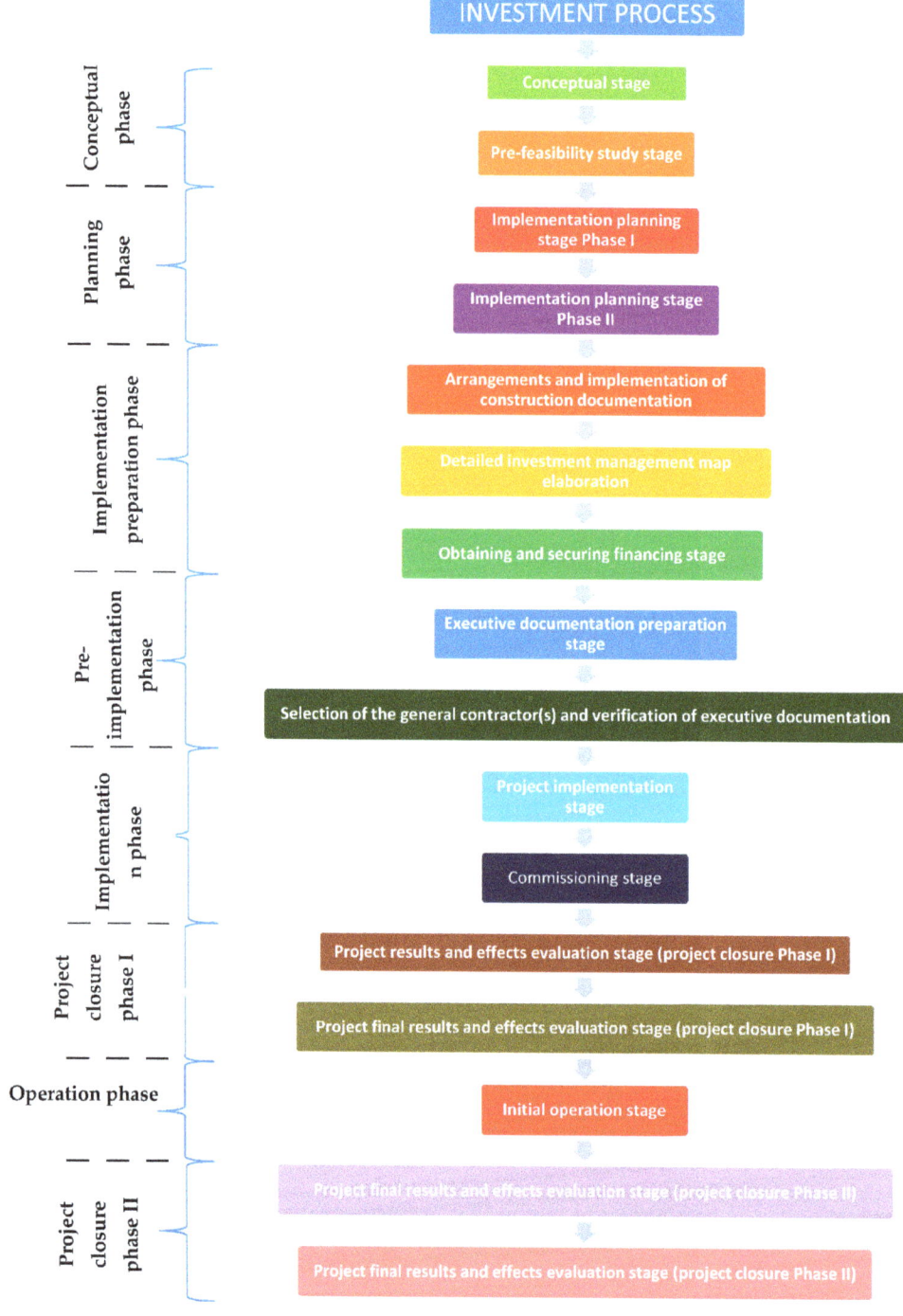

Figure 2. Investment process model (16 stages grouped in eight phases).

To better illustrate the durations of the individual construction stages in the context of a timeline, a typical Gantt chart used to reflect time schedules in various types of projects is presented below (see Figure 3).

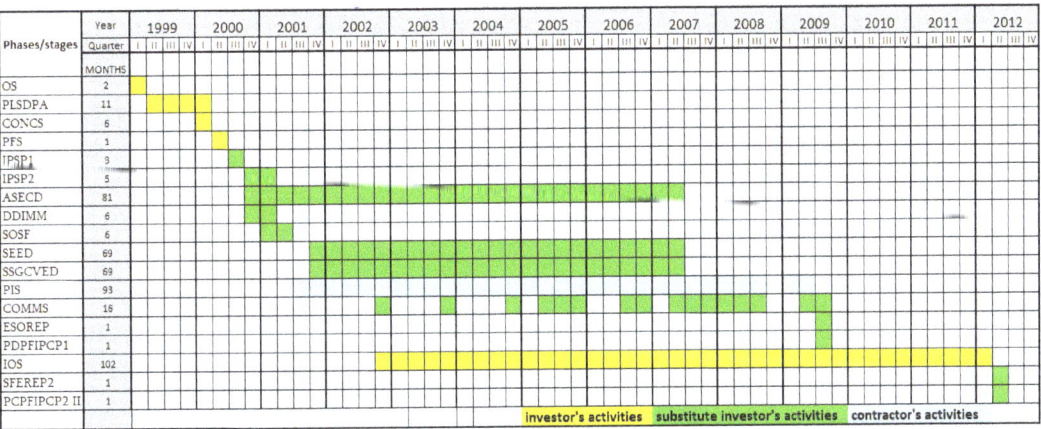

Figure 3. Gantt chart—Parkowo Leśne Housing Estate Project.

In the following Section 3.3 we will also show how to visualise the layout of individual stages of the project, which is best illustrated by Gantt charts widely used in investment projects, and activity-on-node (AON) graphs.

3.3. Research Method

In order to explain the research method used in this paper, in addition to the investment process model described above, we need to characterise the interdependencies between individual stages/phases of the project. These will later form the basis for determining the covariance matrix. Finally, we discuss the Time-at-Risk (TaR) method and the PERT distributions used in the Monte Carlo simulation, which were employed to simulate schedule changes in the context of the most likely values.

3.3.1. Relationships between Individual Stages of the Project

To better illustrate the dependencies between individual project stages, we use the method described in the paper by Gonçalves-Dosantos et al. [21]. There are four types of relationships between the activities undertaken in a project (the different stages are subject to the same principle), namely finish–start (FS), start–start (SS), finish–finish (FF), and start–finish (SF):

FS type: If an activity $i \in N$ precedes FS type to $j \in N$, then j cannot start until activity i has finished.

SS type: If an activity $i \in N$ precedes SS type to $j \in N$, then j cannot start until activity i has started.

FF type: Finish to finish (FF). If an activity $i \in N$ precedes FF type to $j \in N$, then j cannot finish until activity i has finished.

SF type: Start to finish (SF). If an activity $i \in N$ precedes SF type to $j \in N$, then j cannot finish until activity i has started.

In principle, not all types of relationships need to find their representation in every project, but usually at least the first two types of precedence (i.e., FS and SS) exist in all of them. The following is a schematic of the precedences occurring between the individual stages of Parkowo-Leśne Housing Estate Complex Project (see Table 3).

Table 3. Case of "Parkowo-Leśne" estate complex project: Types of precedence between individual stages of the project.

N	1	2	3	4	5	6	7	8	9	10	11	12	13	14	15	16
Finish–start (FS) type of precedence	NA	1	NA	2, 3	4	5	NA	NA	NA	9	NA	9	NA	12	NA	NA
Start–start (SS) type of precedence	NA	NA	2	NA	NA	NA	6	6,7	8	NA	10	NA	12	NA	14	13
Finish–finish (FF) type of precedence	NA	NA	NA	NA	NA	NA	NA	NA	NA	NA	NA	NA	NA	NA	NA	15
Start–finish (SF) type of precedence	NA	NA	NA	NA	NA	NA	NA	NA	NA	NA	NA	NA	NA	NA	NA	NA
Durations	2	11	6	1	3	5	81	6	6	69	69	93	16	1	1	102

The relationships between different stages of the construction process can be better illustrated with the use of the activity-on-node (AON) graph (see Figure 4). Individual activities carried out in a project can be represented in a similar way.

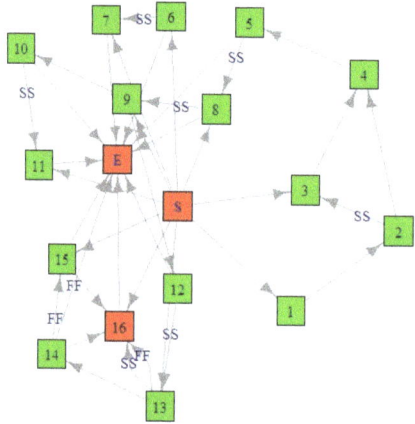

Figure 4. AON graph of the project. In an AON graph the activities are embodied in the nodes (squares) and the precedences of the various types, FS, SS, FF, SF, in the arcs (arrows). Nodes in red indicate critical activities.

3.3.2. Correlation Matrix

The relationships between different project stages are important and should be adequately reflected in the risk analytic tools, as they affect the accuracy of duration estimates for the whole investment process. However, these correlations are rarely studied and there are theoretical and practical obstacles when modelling them. For the sake of this study, a correlation matrix was created taking into account the individual phases/stages and their expected completion times over the entire project duration (see Figure 5).

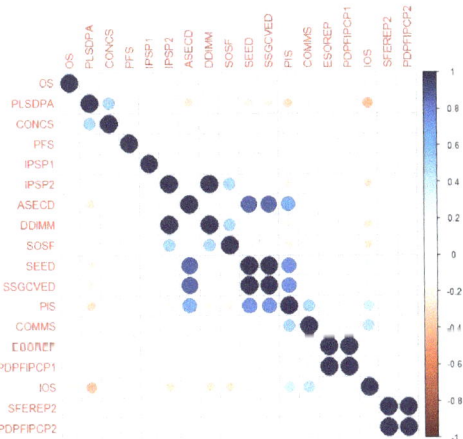

Figure 5. Correlation matrix reflecting dependencies between different phases and stages of the Fort Bema (Parkowo-Leśne Housing Estate) project.

A correlation heatmap is yet another graphical representation of the correlation matrix (see Figure 6).

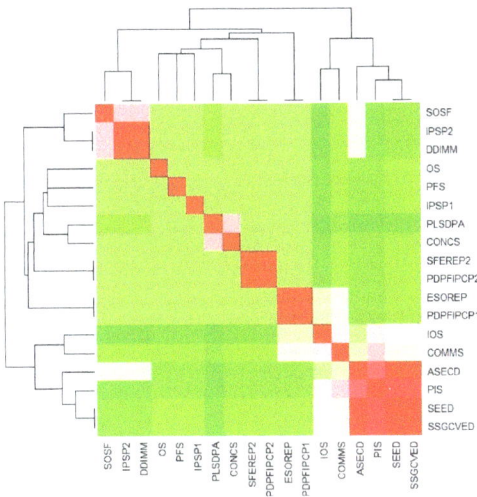

Figure 6. Heatmap reflecting dependencies between different phases and stages of the Fort Bema project.

3.3.3. Time-at-Risk (TaR) and Monte Carlo Simulation

As for the Time-at-Risk, Kovalenko and Sornette [42] proposed the concept of time@risk in the context of building a particular diagnostic system that provides continuous updates of possible scenarios and their probabilistic weights reflecting the diagnosis of hazardous regimes to target different types of instabilities. Such a system, based on the time@risk approach, could be used in many areas, e.g., to signal the possible occurrence of a crisis, provide insights for the adoption of appropriate policy measures and allow the assessment of future scenarios according to the chosen policy.

The Time-at-Risk (TaR) concept proposed by Bolgorian and Raei [41] differs from the concept proposed by Kovalenko and Sornette [42]. It is a quantile-based approach that refers to the Value-at-Risk (VaR) concept known from the financial world and is based on a probability distribution function (PDF) of the return times of peaks above the threshold value.

The method on which the study is based uses a function that simulates random variables with a PERT distribution (where each project stage has an assumed duration given in the original project documentation). The appropriateness of the use of PERT distributions, which are a special case of beta distributions, has already been addressed in other studies [16,84,85]. Naturally, researchers polemicize on this issue. Some justify the superiority of Weibull distributions [67,86]. Without going into special details, we choose PERT-distributions since they are easier to implement in the R package and they give the possibility to individually define the extremes and the most likely value (i.e., mode), which has already been defined for the Fort Bema project (see Tables 2 and 3).

PERT belongs to the group of continuous probability distributions that are characterised by its minimum (*a*), most likely (*b*) and maximum (*c*) values that a variable can take. PERT is derived from the transformation of a four-parameter beta distribution for which the expected value is: $E[X] = \frac{a+4b+c}{6} = \mu$. The PERT distribution has the characteristic property that its mean (of the distribution) is weighted by the (minimum and maximum) extrema values and the most likely value (mode), with the latter weighted (four times) more heavily than the former [87]. The original rationale for applying this distribution was to address the impact of the uncertainty surrounding the duration of tasks on the performance of a project schedule evaluated using the Programme Evaluation and Review Technique (PERT), one of the most popular methods for scheduling [87].

Unlike the triangular distribution, the PERT distribution uses the minimum, maximum and most likely parameters to create a smooth curve (Figure 7).

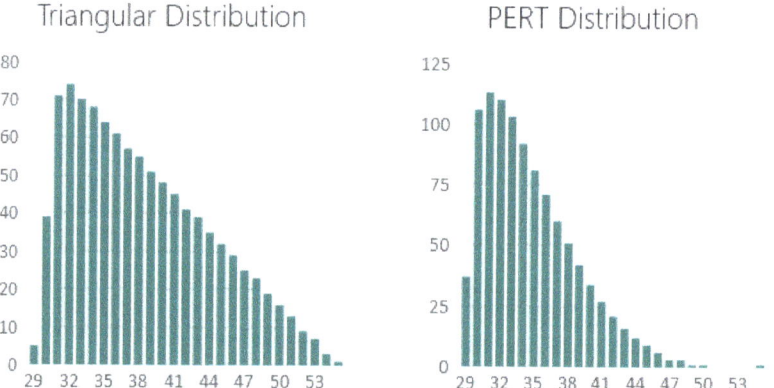

Figure 7. Comparison of triangular and PERT distributions for the assumed "Project implementation stage" assumptions (i.e., minimum, maximum and mode values).

From a mathematical perspective, the standard deviation of PERT distribution is equal to about 1/6 of a range [87,88]. It is well suited for any risk assessment that seeks to address (reflect) uncertainties in the value of some measure, specifically when one relies on some subjective estimates to quantify such a measure. In this regard, the minimum, maximum and most likely (mode) values are somehow intuitive, and they enable definition of the distribution at the researcher's discretion. To simulate the PERT distributions in the study, we use R packages (such distributional and ggdist) [89], and the simulation itself is performed with the rpert function.

The PERT distribution is also often compared with the triangular distribution (since it relies on the same parameters), but in contrast to the latter, the former is characterised by a smoother curve shape. In the case of a triangular distribution, the most likely value is not weighted in any particular way. Hence its mean μ can be expressed as: $\mu = \frac{a+b+c}{3}$.

The triangular distribution features with an angular shape, while the PERT distribution with a smoother shape (the difference is characteristic for subjective knowledge-type distributions). However, if the PERT distribution is strongly skewed, the extreme values may be underestimated [90]. One solution for controlling the risk specifically for construction schedules can be obtained with the use a modified PERT distribution [91], whereby the there is more control over tail values. Such a modified PERT distribution has an additional parameter by which the weight of the most likely value can be arbitrarily determined (the mean of this distribution can be expressed as: $\mu = \frac{a+\gamma b+c}{\gamma+2}$).

As for the Monte Carlo simulation method, there are numerous studies based on this approach. Barraza [92], for example, used Monte Carlo simulation to estimate the time unpredictability of project activities. Tokdemir et al. [23] proposed a method to assess the risk of delay in projects planned according to the line-of-balance method (LOB) and used Monte Carlo simulation to quantify the risk. Kirytopoulos et al. [93] investigated the importance of using historical information and the correct choice of distribution when estimating activity duration for the Monte Carlo simulation process. Vanhoucke [94] used Monte Carlo simulation and empirical project data from schedule risk analysis and earned value management process and proposed measures of project control efficiency. Additionally, Kong et al. [16] show explicitly the benefits of using Monte Carlo simulation to assess risk in practical scheduling. These are just a few examples, of which there are many more. A very thorough overview of many construction-related studies based on Monte Carlo simulations can be found in the paper by Koulinas et al. [95].

The method that we use yielded the deviations from the expected (most likely) completion times for each stage of the construction process. For this purpose, the Monte Carlo method was used (with 10,000 repetitions). In the next step, a covariance matrix was obtained taking into account the dependencies between schedules of individual project stages. Subsequently, the TaR of the entire project was calculated, i.e., for the entire "portfolio of stages", assuming that the weights of all project phases add up to 100%. The weights were calculated based on the durations and investment levels of each stage of the entire project.

Given a confidence level of $p \in (0,1)$ and assuming that the number of random samples s (of deviations from the estimated schedules) with α equal to a certain quantile of the distribution, we would like to determine the change in $\Delta T(\alpha)$ over time for the whole project (schedules).

Let $G_\alpha(x)$ be the cumulative distribution function (CDF) of $\Delta T(\alpha)$. Since the time change is $\Delta T(\alpha) \geq 0$, then we can define the TaR as a quantile of the CDF for a given p as $p = \mathbb{P}[\Delta T(\alpha) \geq \text{TaR}] = G_\alpha(\text{TaR})$.

Assuming that the project developer expects to complete all project phases in a period shorter than TaR, with a given α and probability p, which means that time, T, changes $\Delta T(\alpha) \leq 0$, the probability p with respect to TaR can be expressed as follows:

$$p = \mathbb{P}[\Delta T(\alpha) \leq \text{TaR}] = 1 - \mathbb{P}[\Delta T(\alpha) \geq \text{TaR}] = 1 - G_\alpha(\text{TaR})$$

Consequently the $p - quantile$ of $G_\alpha(x)$ for the CDF of $G_\alpha(x)$ and a given confidence level of $p \in (0,1)$ is $\text{TaR}_p = x_p = inf\{x | G_\alpha(x) \geq p\}$, where inf denotes the lowest real number. Therefore, the tail behaviour of the CDF of G_α or its quantile is a condition necessary for approaching TaR calculation.

In summary, the combined Monte Carlo simulation and TaR calculation is based on certain assumptions, some of which may change depending on project expectations:

1. Time deviations from the expected project stages' completion times, simulated with different type of a distribution, e.g., PERT, triangular or Weibull.
2. Covariance matrix reflecting the relationship between different project stages.
3. Monte Carlo simulation with a certain number of repetitions (e.g., 10,000).
4. A given confidence level of $p \in (0,1)$, and quantile of the distribution, for which the TaR value is determined.

4. Results

The method described in Section 3.3 allows the calculation of the Time-at-Risk for the entire project. Table 4 shows the TaR results for the 95 and 99 quantiles. Scenario A is the baseline scenario, assuming no changes to the schedule. The simulation was performed using the Monte Carlo technique with 10,000 repetitions. A PERT distribution of times for each stage of the investment process was assumed, with the most likely duration defined in the project documents and shown in Tables 2 and 3. The results show that project managers have to allow for a time delay of up to 4.47 and 6.35 quarters, resulting from the 95 and 99 quantiles, respectively. Of course, these are extreme values, which in fact represent the fifth and first percentiles of the least favourable trajectories of the investment process (when arranged in order from 1 to 10,000, i.e., from least favourable to most favourable). Such a view of the project and its associated schedule is the essence of the Time-at-Risk method. While managers do not have to expect such an extremely unfavourable development during the implementation of the entire process as indicated by the Time-at-Risk times shown in Table 4, they cannot rule out the worst-case scenario. Knowing these times allows them to be aware of what to expect if many elements of the project implementation fail.

Table 4. Different scenarios reflected in time schedules.

Time-at-Risk (TaR) (in Quantiles)	Scenario A	Scenario B [1]	Scenario C [2]
99% quantile	6.35866	6.77638	6.15269
95% quantile	4.47752	5.12026	4.40843

[1] Shift in stages: ASECD/seven stage/(three quarters) and SSGCVED/11 stage/(four quarters). [2] Each of the following phases: SEED/10 stage/, SSGCVED/11 stage/, COMMS/13 stage/, and IOS/16 stage/started one quarter earlier and ended earlier.

All calculations were done in R-Studio using the packages Distributional and PerformanceAnalytics. In the case of Scenarios B and C, some changes in the timing of some stages of the investment project were assumed. In the case of Scenario B, Stages 7 and 11 were postponed by 3 months (in line with previous slack assumptions). Scenario C results in a slightly lower TaR as it assumes a one month earlier start of four phases (10th, 11th, 13th and 16th), which slightly reduces the risk of project extension.

The use of PerformanceAnalytics package allows for reflecting the contribution of each stage to the overall TaR measure. For example, one of the stages with the largest contribution to TaR is the project implementation stage (PIS) whose deviations from assumed time schedules are shown in Figure 8. The histograms and Q–Q plot shows that PERT distribution was used to simulate changes from the assumed time schedules.

More importantly, different project stages can also be visualised in the context of their time completion distribution across the timeline of the whole project (see Figure 9). The individual distributions of changes in time schedules are based on the PERT distribution (more outliers in its right tail).

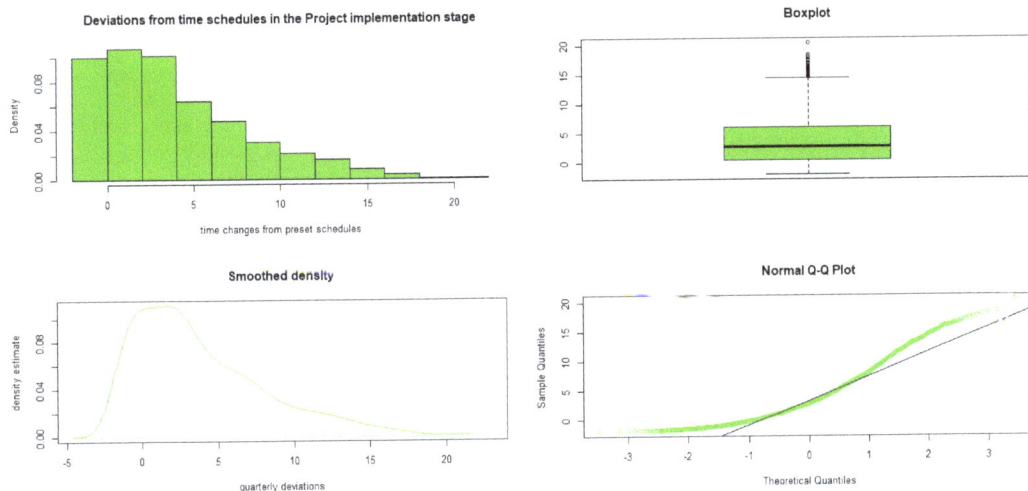

Figure 8. Deviations in time schedules for the project implementation stage (PIS).

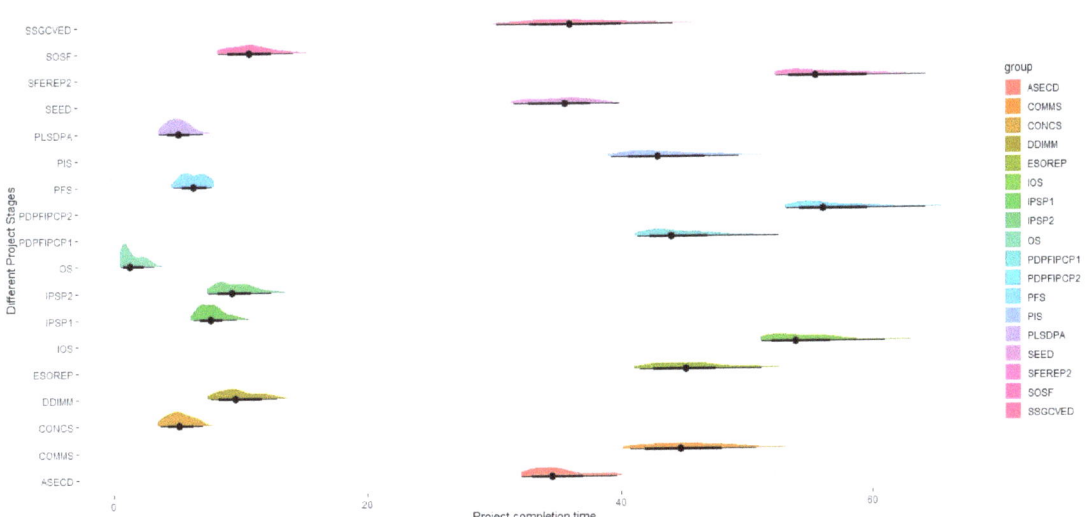

Figure 9. Individual project stages can be visualised in terms of the distribution of changes in the assumed time schedule on the timeline of the entire project.

5. Discussion

In a construction project where time really means money, schedule management is crucial. Therefore, predicting the probability of time overruns plays a key role in project success or failure. It is also much easier to manage what can be quantified. Therefore, project managers should know the probability of time overruns so that they can take the necessary preventive or corrective actions.

Researchers have developed various quantitative methods for assessing time schedules in construction projects. The aim is to examine the reliability of the project schedule taking into account available probabilistic information accounting for potential disruptions that may occur during project execution. We use the Time-at-Risk (TaR) metric, which is a static measure of risk. More specifically, it is a quantile that reflects a particular feature of the probability distribution of the underlying value, i.e., time changes of the whole project and the contributions of its individual stages. In this respect, a quantile represents a certain part of the data set (time duration of all project stages) and indicates how many values in its distribution are above a certain threshold value. This paper shows a practical application of the TaR approach, which provides a realistic view of the feasibility of project schedules within anticipated time windows. In short, we used this approach to assess the reliability of performing a project according to the anticipated schedule time span. The analysed case was the construction of a residential complex (Fort Bema, situated in the Warsaw-Bemowo district), realised on 148 hectares of land and comprising almost 120,000 m^2 of usable floor area of residential units. The construction of such a large residential complex took a total of 128 months. In particular, TaR allows visualising the risk of time changes related to the project as a whole, as well as to each individual stage. TaR enables the assessment of the uncertainty surrounding each project. It is a tool that helps managers create a realistic project plan that can be used as a guide to manage individual project stages/phases or even specific tasks, as well as to monitor the entire project and its expected completion time. This method, in combination with Monte Carlo simulation, where time lags (i.e., deviations in the execution times of specific stages or even tasks/activities themselves) are sampled, allows for arbitrary assumptions tailored to the specifics of each stage/task. The MC method allows for certain assumptions about the distribution of time offsets in specific stages of the project. For example, triangular distributions can be assumed instead of PERT distributions for short phases such as the realisation of the opportunity study (OS) or for the conceptual phase (CONC), while Weibull distributions can be used for the other longer implementation stages (such as the arrangement stage and execution of construction documentation [ASECD], or the project implementation stage [PIS]), as they are inherently characterised by fat tails on the right (and thin tails on the left) side of their distributions [67,86]. According to Abdelkader [86], the variation in the duration of activities/processes is best described by a Weibull distribution. However, we use PERT-distributions since they are easier to implement in the R package and they give the possibility to individually define the extremes and the most likely value, which has already been defined for the project under study.

Viewing a construction project through the prism of a series of sequential or parallel stages of the investment process makes it possible to improve the quality of implementation of the entire project. From the perspective of the risks associated with the project (assessed a priori), postponing the implementation of some stages, e.g., stage 7 and 11, by three quarters does not necessarily imply the risk of not meeting the final completion date of the project, provided, of course, that such postponements occur within a given timeframe of the project. In the case of the Parkowo-Leśne housing estate in Warsaw, this time span totalled 128 months (or 54 quarters). More precisely, if postponements of certain phases (in accordance with the original assumptions of the project) take place within a given time horizon, the quantified risk of delays in implementation is negligible. In the case of Scenario B (Table 4), it increased merely 0.42 of a quarter (99-quantile). However, if the individual stages were implemented one after the other, the risks in realising the entire project would be much higher. In this context, much depends on the synchronisation of the realisation of the individual project phases and the covariance matrix, which reflects the dependencies between the different phases and stages of the project. It is advisable to carry out these stages/activities in parallel, of course only if this is possible. For this purpose, it is advisable to develop a suitable implementation strategy, e.g., the formation of different teams that work in parallel and independently of each other. At the same time, the work of all teams should be coordinated by one person who has supervision over all teams. A similar method is used by some Silicon Valley technology companies that are leaders in bringing new technological solutions to market. Their aim is to significantly shorten their project schedules when working on new innovative technologies. One such technology company has established for this purpose three independent project teams that work independently in parallel [57,96]. The overlapping work schedules of all the teams is supposed to ensure that the final product is released to the market three times faster than it would be if a single team had worked alone. The difference, however, is that in the case of these technology giants, it is important to compete with the biggest competitors to see which one of them manages to deliver the new technology first (because the first usually skims the cream of the market) [97]. For construction companies, the situation is somewhat different. Nevertheless, the very idea of treating individual project phases (and the project as a whole) as a portfolio of certain stages or activities, while paying attention to their diversification wherever possible, seems to be reasonable. To understand this better, it is useful to use an analogy to the financial world, and more specifically, to the measure of value-at-risk (VaR) used by financial institutions for stress-testing and risk control. Stress tests are risk management tools widely used by both institutional investment managers and regulatory authorities. Value-at-risk calculates the worst case loss over a given time period that won't be exceeded with a given level of confidence.

The TaR we describe in this article plays a similar role to VaR's in evaluating a portfolio of assets for some investment fund or other financial institutions. In the case of a construction project, the portfolio comprises the phases/stages of the project, which, like the assets in an investment portfolio, may be subject to a certain diversification (thanks to time buffers), while the equivalent of value-at-risk in this case is Time-at-Risk. Looking at a project in this way can also allow the efficiency of entire projects to be improved. Similarly to VaRs in the investment portfolio of an investment fund, TaRs for construction companies with large investment projects can be used to perform stress tests in the context of the final completion times of such projects. Stress tests can be used to test changes in individual stages of a project to see how this affects the overall schedule. This type of method seems to be a good tool for controlling risks in a project.

It should also be noted that as new technologies are developed, the perception of the investment process is changing. There is an increasingly noticeable trend towards structuring the entire investment process [98]. It consists in the standardisation of all processes, the development of rules to ensure that each phase is carried out in an identical or similar way for each type of project, and the identification of sequences of repetitive activities aimed at standardising management processes with regard to planning and execution activities.

The standardisation and unification of all procedures, which has been taking place for some time, aims to minimise the impact of random events that cannot be excluded in the forward planning and scheduling of an investment. This manifests itself in the implementation of clearly defined regulations and specific rules, but also in the dissemination of knowledge in the field of project management, e.g., through management methodologies such as PMBoK or PRINCE2 [99], and more effective methods of achieving objectives, such as the control of construction progress or the application of standard contract terms, e.g., FIDIC [9,10,100].

It should also be emphasised that in the course of preparation and implementation of any investment project, disruptions will always occur. They are caused by the accumulation of various processes and factors and their interaction with each other. Project managers should also pay close attention to how logistics management activities are affected. Last but not least, all phases and activities of an investment project should be addressed and formulated in a comprehensive manner. They should be appropriately arranged and the relationships between them and the order in which they should be carried out should be defined. For this purpose, it is recommended to use Gantt charts, table of precedences and AON graphs such as the ones presented in this paper. In other words, the realisation of the investment project requires an appropriate arrangement of all stages/phases of the investment process and its individual activities. Against this background, the authors have developed their own model of the investment process, which is presented in this study (consisting of 16 stages grouped into eight phases).

Finally, it is worth highlighting one important limitation of one particular case study. It is rather problematic and impractical to make generalisations on the basis of one case study. Obviously, the best solution would be to collect relevant data for hundreds or even thousands of projects (preferably similar ones) and use historical distributions. It would then be possible to arrive at some generalisations. However, this is extremely difficult to achieve. Besides not all investors (and substitute investors) would have a motive to share such information (for some, possible disclosure of data on delays could have a very negative impact on their reputation), even if they were willing to provide such information (subject to anonymity), it would still be an extremely big logistical challenge to collect such data. For all intents and purposes, the case study and risk analysis that we present is aimed at providing a certain way of looking at the project and the risks that accompany the schedules. This was inspired by seeking analogies to the methods used in the capital markets and the VaRs. Drawing other analogies from the financial world, construction stakeholders wishing to adopt some more practical risk management models may seek to identify some of the largest realised historical outliers. To put it another way, they can create databases of the largest delays in other similar projects and these outliers (the most severe historical delays in schedules) can be taken as a reference point (i.e., the worst plausible scenario). To use an example from the capital markets, risk managers often use models in which the worst possible scenario is the worst loss amount that has materialised in the past (over a certain period of time).

Also, the validity of the approach used in this study is somehow strengthened by the Monte Carlo simulation, which, according to some researchers, offers the greatest potential in comparison to other methods, such as, e.g., sensitivity analysis, decision tree analysis or the Delphi method [16,51,84].

6. Conclusions

Scheduling of construction projects has attracted much attention in academic research, leading to numerous publications in the field. Authors have proposed various solutions to create effective schedules for construction projects. However, many of the current solutions are purely theoretical and often impractical for existing construction projects.

The paper presents a practical application of the combined Monte Carlo and Time-at-Risk (TaR) methods, which, under certain assumptions, can be used to estimate the risks associated with scheduling a construction project conducted in conditions of uncertainty. TaR is defined here as the maximum time deviations expected to be experienced within a given time window, at a pre-defined confidence level. For example, if the 95-quantile TaR is 4.477 quarters (as was the case for the Fort Bema project analysed in the study), it means that there is 5% confidence that over the course of the entire project there is going to be a delay of that magnitude. In other words, TaR is the time corresponding to a certain percentile of the most extreme deviations from the assumed duration of the entire project (from the right tail of the distribution, which is the product of a Monte Carlo simulation for 10,000 trials; the MC procedure simulates the distributions of deviations from the assumed durations of all the stages within the overall investment process). As some of the stages are carried out in a sequential order and some in parallel (they are associated with so-called slacks), the total duration of the project is not the sum of the durations of its individual stages. The final result also depends on the covariance matrix between the schedules of individual stages of the entire project, which can be well illustrated by a Gantt chart. In turn, the type of relationship between individual project stages can be best illustrated with an activity-on-node (AON) graph and a correlation (or covariance) matrix. To put things in another way, the final project completion times, i.e., the deviations of the project duration from the time assumed in the documentation, simulated with the use of the Monte Carlo method, are not linear combinations of sampled deviations of the duration of individual project stages.

The final conclusion is that the method presented makes it possible to quantify the risk involved in carrying out a construction project. To do this, one needs a good model of the investment process (which we present in this study) and certain assumptions with regards to the duration of individual stages of the investment process. In addition, certain assumptions must be made about possible deviations from the assumed durations of the individual stages (which can be described with a suitable distribution, e.g., a PERT or triangular distribution or even a Weibull distribution). A Monte Carlo simulation is then performed, and a new distribution is created from which the TaR can be determined.

On the other hand, the risk of time lags associated with any project can be considered from a qualitative perspective by running parallel activities wherever possible. It is also important to model the investment process itself, where great importance should be attached to both the conceptual and planning phases, as this is where the seeds are sown for future chains of events that both extend project duration and lead to excessive costs. In this regard it is obviously a good idea to conduct an appropriate cost–benefit analysis. Such a conclusion is also in line with the research findings discussed by Bilinski [45] and Obolewicz [46] in their papers.

All in all, this paper proposes a framework for quantifying the risk of time variation in a project based on Monte Carlo simulation and a probabilistic Time-at-risk analysis. It is an approach that explicitly quantifies the uncertainty in the duration of the whole project as well as its individual stages. The possibilities of the proposed approach are explained using a simple example of the construction of a housing estate in Warsaw-Bemowo, which was carried out in the period 1999–2012.

7. Patents

Another important practical contribution of this paper is the proprietary investment project management model (extended by a comprehensive map of investment process management). On 28 December 2012 it was granted a patent protection in the Patent Office of the Republic of Poland under the number P-402301, entitled Method of managing a technical project, particularly a construction investment project. It describes in detail the development phases of the project life cycle and is an alternative to other models known from the literature on the subject. The model constitutes an important practical contribution to the field of construction project management (the model was also elaborated in a digital version).

Author Contributions: Conceptualization, J.S.; methodology, J.S. and D.M.; validation, J.S.; investigation, J.S. and D.M.; resources, J.S.; data curation, D.M.; writing—original draft preparation, J.S. and D.M.; writing—review and editing J.S. and D.M.; visualization, D.M.; supervision, J.S. All authors have read and agreed to the published version of the manuscript.

Funding: This research received no external funding.

Institutional Review Board Statement: Not applicable.

Informed Consent Statement: Not applicable.

Data Availability Statement: Raw data were generated from the Department of Building Engineering (Warsaw University of Technology). We confirm that the data, models and methodology used in the research are proprietary, and the derived data supporting the findings of this study are available from the first author on request.

Conflicts of Interest: The authors declare no conflict of interest.

Appendix A

In this section, we provide a more detailed rationale for the relevance of the stages of investment process.

Table A1. Proprietary investment process model. Description of all its stages.

Phases/Stages of the Investment Project	Description	Cited by Authors
Opportunity study	First, an investment opportunity study and a preliminary economic assessment must be carried out, considering various investment opportunities and project proposals, which will then be pursued in further steps. An opportunity study is carried out at the earliest stage of a project. The scope of the study includes an initial exploration of a project idea or identification of opportunities. Project ideas in a specific region, for a specific industry or based on available raw materials are explored.	Behrens and Hawranek [101]; Armaneri [102]; Tamošiūnienė and Angelov [103].
Participation in local spatial development plan approvals	As part of this stage, the possibilities of preparing, approving and adopting a local spatial development plan are examined. It is considered that the adoption of a local spatial development plan should be the starting point for further investment activities. The next step is to elaborate a number of urban and architectural concepts, preliminary conceptual studies, as well as provisional, alternative media requirements for the entire area and for individual investment tasks.	Leśniak et al. [104]; Sobieraj [9,10,48]
Conceptual stage	The conceptual stage is about conceptualising and institutionalising the project's design processes. It boils down to the following activities: - Conceptualisation of the idea behind the project; - Elaboration of the main goals/objectives and expectations of the initiators; - Formation of a team of initiators (finding a compromise between the initiators); - Identification of the project participants; - Building the tree of goals of the initiators; - Formation of the team of initiators; - Creation of the project charter; - Elaboration of a detailed project definition and its alternatives; - Definition of the understanding of project success; - Definition of the rules for communication and information transfer; - Search for partners for project implementation (for large projects); - Elaboration of the concept for a general investment plan.	Tizani [105]; Biliński [45]; Obolewicz [46], Sobieraj [9,10]
Pre-feasibility study	For each project, a pre-feasibility study must be carried out specifying how the project will be implemented, which allows the selection of the final version of the project. The study ends with the formulation of the final version of the project feasibility study, covering all technical, economic, commercial and financial aspects of the investment process. The final version of the project (feasibility study) is then submitted to banks, potential foreign contractors, developers and investors to assure them of the future success and expediency of the intended project. When acquiring the information necessary to prepare both the feasibility study and the prefeasibility study, particular attention should be paid to its reliability, timeliness, completeness and usefulness. At this stage the experience and assistance of the most reputable experts is used, relying on their specialist knowledge. Therefore, in this phase of the investment process, the following activities should be carried out: - Preliminary concept of the functional and utilisation programme; - Gathering general information for feasibility studies; - Preparation of the sketches of the investment plan; - Preparation of the pre-feasibility study; - Report on the pre-feasibility study; - Decision on the part of project initiators (after evaluation of the analysis results) with regard to the commencement of the project implementation; - Development of a general investment programme; - Preliminary identification of financing methods for the investment process; - Initiation of negotiations with financial institutions on the terms of financing; - Obtainment of a financing commitment for the investment activity under conditions specified in the pre-feasibility study or securing a financing option after meeting the conditions of the financial institutions.	Escobar-García et al. [106]; Sobieraj [48]; Kim et al. [107]

Table A1. *Cont.*

Phases/Stages of the Investment Project	Description	Cited by Authors
Implementation planning stage—phase I	Investment planning is such an important element of each investment process that mistakes made at this stage/phase determine the final project costs (those that will be known retrospectively when the project is completed) to a greater extent than those made during project implementation [45,46]. Therefore, each project should include a separate planning phase, preferably consisting of two stages, each of which should integrate a number of different activities that significantly facilitate the implementation of the preparation phase, followed by the pre-implementation, implementation and investment closure phases [9,10]. The implementation planning stage (phase I) entails: - Obtainment of necessary approval documents that will allow the investment process to proceed; - Initiation of the process of adopting the local spatial development plan (LSDP); optional solution, provided that more favourable construction parameters are obtained than in the decision on construction conditions (also known as individual construction permit); - Carrying out the procedure for obtaining a decision on building conditions if the LSDP has not been adopted and it is possible to obtain more favourable building conditions than those set out in the LSDP; - Signing memorandum of understanding (MoU) with selected partners (option); - Acquisition of property rights if a decision on construction conditions has been reached or the LSDP is in force and a decision is made on the implementation of the project. Otherwise, the decision to acquire the title can be made at a later date, or instead of acquiring the title, a commitment agreement for the sale of the title may be entered into in the form of a notarial deed; - Determination of required qualifications and skills of the project manager and project team; - Selection of the project manager responsible for the phase prior to the implementation of the investment; - Decision on the method of implementation of the investment process based on the analysis of the scope and difficulty of the project; - If necessary, when new information becomes known, possible correction of the master plan.	Biliński [45]; Ebi [108] Obolewicz [46]; Sobieraj [9,10,48]
Implementation planning stage—phase II	The implementation planning stage (phase II) entails: - Appointment of the project team managers dealing with the project in the structures of an investor or a substitutive investor; - Development of a functional-utility programme (final version); - Decision on the preparation and implementation of a tender; - Elaboration of the procedure of conducting the competition for an architectural office (open or restricted call for proposals); - Preparation and approval of the competition documentation including a draft contract; - Selection of the design office (signing a contract); - Elaboration of detailed architectural concepts as an element of the tender/competition; - Selection and approval of the final architectural concept (for the whole project or its part, if there are several investors); - Preparation of the land development design; - Elaboration of the preliminary budget; - Development of a detailed investment programme.	Biliński [45]; Obolewicz [46]; Sobieraj [9,10,48]

Table A1. *Cont.*

Phases/Stages of the Investment Project	Description	Cited by Authors
Arrangements stage and execution of construction documentation	In the realisation of every project, a strong emphasis should be placed on the design and execution of construction documentation including a description, calculations and construction drawings [9,109]. The list of individual tasks to be completed during this stage is as follows: - Analysis of the possibility to choose the methods of obtaining permits for construction works (decision with regards to a building permit or notification procedure); - Obtainment of a decision on the conditions for connecting the investment facility to the existing infrastructure; - Preparation of a multi-sectoral construction design and accompanying studies; - Review of project documentation and submission of comments; - Preliminary acceptance of the project documentation (construction design) for filing an application for its approval with the competent authority (in accordance with the procedure for obtaining a building permit); - Notification and approval of demolitions; - Commencement of the procedure for obtaining a building permit; - Obtainment of a valid building permit decision; - Preparation of the feasibility study; - Continuation of talks with financial institutions, funds and exploration of the possibilities of obtaining funds from the EU, Provincial Fund for Environmental Protection, National Fund for Environmental Protection, etc.; - Verification and updating of the feasibility study.	Elnagar and Yates [98]; Roy et al. [110]; Adriańczyk [109]; Levy [111]; Sobieraj [9,10]
Developing a detailed investment management map	The list of individual tasks to be completed during this stage is as follows: - Approval of the final composition of the project team and supplementary training for individual team members; - Approval of the list of key experts who agreed to cooperate with the project manager; - Approval of the building documentation and, if necessary (e.g., new information emerges, introduction of new requirements, etc.) preparation of an alternative building documentation—a return to stage 5; - Verification, updating and approval of a detailed plan for the implementation of the project (directive schedules, preferably in the form of a Gantt diagram, bills of quantities, in the form of an annex attached to the materials supplementing the tender documentation) - Final assessment of the project manager before starting an implementation of the project.	Yakura [55]; Kim et al. [112]; Maya [113]; Houston [114]; Sobieraj [9,10]
Obtaining and securing financing (feasibility study)	The obtaining and securing financing stage entails: - Submission of the feasibility study documentation to financial institutions (after its verification); - Negotiations with financial institutions on final funding terms; - Approval of the final feasibility study; - Conclusion of investment financing agreements or obtainment of investment financing stand-by agreements.	Strzelecka et al. [81]; Sobieraj [48]
Executing executive documentation	- Elaboration on an executive design by the design office or, possibly at a later stage, by the general contractor (after its selection); - Submission of comments and amendments with regards to the executive documentation; - Incorporation of the observations and amendments to the project documentation by the project office (design office).	Biliński [45]; Sobieraj [9,10]

Table A1. *Cont.*

Phases/Stages of the Investment Project	Description	Cited by Authors
Selecting of general contractor(s) and verification of executive documentation	This stage involves: - Selection of the method of conducting the competition for the general contractor of construction works (open or closed competition); - Preparation and approval of competition materials together with a detailed contract draft(s); - Multi-stage tender (competition) for the selection of a general contractor (or general contractors), under the conditions—enabling the elimination of poor companies with bad financial condition (every candidate entity is obliged to prepare a substantive and financial schedule of the project or part of the project consistent with the tender specifications); - Multiple comments on the documentation and proposals for replacement solutions submitted by tender participants and replies to them made by the project design office - Negotiation of the contract with potential general contractors and obtainment of bank guarantee promises by the finalists of the tender; - Selection of the general contractor or, in the case of larger projects, more than one or even several general contractors; - Signing a contract with the general contractor(s); - Approval of the executive project plans or individual stages of the project if partial acceptance of the executive documentation is stipulated in the contract terms (for all sectors according to the time schedule).	Sobieraj [9,10]
Project implementation stage	At this stage, the efficiency of project (contract) management directly influences the smooth implementation of each objective [115,116]. Project implementation stage includes: - Appointment of a team of supervisors for the whole undertaking; - Appointment of site manager(s); - Establishment of the site's organisational structure by the site manager; - Preparation of the Health and Safety Plan (HSP) for each construction site; - Submitting a declaration of taking up the site manager's position to the District Construction Supervision Inspectorate; - Notification about demolition (construction site) to the National Construction Supervision Inspectorate; - Establishment of the demolition log (development of a technology and organisation scheme for dismantling/demolition, if required); - Establishment of the construction log; - Filing a declaration by the head of the team of supervision inspectors on undertaking supervision to the District Construction Supervision Inspectorate; - Registration of the construction log in the municipality's local architecture and construction department; - Registration of the functions of site manager and investor's supervision inspectors in the construction logbook; - Commencement of declared demolition/construction works; - Geodetic demarcation of the facility/construction site including surveyor's entry in the construction log with a graphic attachment; - Indication of the place of supply of the utilities (by the investor) for the entire duration of the construction works; - Protocolar introduction of the contractor to the construction site; - Preparatory work for the demolition/construction process according to the approved site development plan; - Development by the occupational health and safety (OHS) coordinator of a detailed safety plan for construction site works (for each construction site); - Implementation of a detailed OHS plan; - Approval by the health and safety coordinator of the compliance of the site preparation/demolition work with regard to occupational health and safety (OHS); - Acceptance of preparatory works by the site manager and investor's supervision; - Implementation of organisational, technological and workshop projects by professional contractors; - Approval of organisational, technological designs and acceptance of them for implementation by the site manager and investor supervision; - Verification of the completeness and validity of the project documentation forwarded to the general contractor;	Zhao-xia [115]; Yan and Chen [116]; Sobieraj [9,10]

Table A1. Cont.

Phases/Stages of the Investment Project	Description	Cited by Authors
	- Development by the general contractor of a comprehensive organisational and technological approach to construction processes; - Launch of the project implementation and common parts of the project infrastructure as a whole; - Implementation processes of the demolition/construction/works; - Updates of the time schedules during the implementation of works; - Project Implementation Management involving: strategic project management (for projects lasting more than 5 years), operational project management (for projects lasting between 3 and 5 years), scope management, project documentation management, value management, quality management, cost management, financial management, risk management, time management, site management, change management, procurement management, resource management, communication management, human resources management, stakeholder management, graphic design management, safety management (OSH), environmental and sustainable development management, configuration management, knowledge management, and integration management.; - Monitoring, control and coordination of the entire project.	
Commissioning stage	Many projects go well until they enter the commissioning phase and then even comparatively minor problems can cause a disproportionate amount of trouble and delay [117]. This stage involves the following steps: - Notification of subcontractors/suppliers about their readiness for technical acceptance of works/services performed; - Preparation of technical acceptance protocol(s); - Signing of the final acceptance of works protocol between the general contractor and subcontractors with a clause on the conditions of final acceptance of all construction works by the investor/substitute investor; - Procedures for notification by the general contractor towards final acceptance by the investor; - Examination by the general contractor of the completeness of the work performed and the correctness/completeness of the collected acceptance documents; - Site manager's entry in the construction log indicating preparedness of commissioned works for final acceptance; - Confirmation in the form of an official entry into the construction logbook with regards to the preparedness of the commissioned works for acceptance by the investor's supervision inspectors; - Establishment of the final acceptance committee; - Activities performed by the final acceptance committee; - Establishment of a deadline for the removal of defects by the general contractor; - Preparation of the list of errors/defects/faults encountered during the investment process; - Final settlement between the general contractor and subcontractors/suppliers; - Final financial settlement of the contract; - Completion of the documentation for the operation permit application or, if such a permit is not required—notification to the District Construction Supervision Inspectorate about the completion of construction works; - Verification of the completeness of collected documents and submission of the application (with a set of annexes) for an operation permit; - Analysis of the application and attached documents by the District Construction Supervision Inspectorate; - Construction site inspection (or its lack) by the District Construction Supervision Inspectorate; - Issuance of the operation permit; - Legal validity of the operation permit decision.	Barnes [118]; Covey et al. [117]; Połoński [82]; Baryłka and Baryłka [83]; Sobieraj [9,10]

Table A1. Cont.

Phases/Stages of the Investment Project	Description	Cited by Authors
Evaluation stage of obtaining results and effects of the project (1st phase of project closing)	This stage addresses the following activities: - General contractor's analysis of the parameters/results established by the investor following project launch; - Measurements, analyses and results achieved during this stage; - Evaluation of the achieved results; - Comparison of the achieved results with the parameters established in the conceptual stage and approved in the final version of the project directed to implementation; - Conclusions and recommendations on the basis of the achieved results; - Analysis—verification of the results obtained by the investor (i.e., the type and quality of added value obtained); - Comments, suggestions and decisions made by the investor following the performance analysis; - In case of unsatisfactory results, an investigation should be initiated between the investor and the general contractor; - Where necessary, implementation of corrective procedures; - Re-examination of the results following corrective procedures; - In case of a satisfactory result, an application for the financial settlement of the completed stage (return of the contract performance bond and 50% of the quality/performance bond with regards to the construction works) or in case of a negative result, a legal action.	Trocki and Wyrozębski [119]; Sobieraj [9,10]
Phase of drawing up proposals for future implementation after Project Closure Phase I	This project closure (Phase I) involves: - Archiving of all analytical results; - Statistical analysis of the achieved results; - Preparation of the list of errors/defects/faults encountered during the investment process; - Preparation of the facility card containing all facility parameters, including financial parameters that can be used for comparisons when creating pre-feasibility studies for similar facilities; - Implementation of investment process correction procedures for similar facilities.	Trocki and Wyrozębski [119]; Sobieraj [9,10]
Initial operation stage (usually 3 years of warranty and guarantee)	The initial operation stage requires: - Inspection of the general contractor's compliance with the instructions for the use and operation of the facility - Approval of the instructions by the investor and application of any changes/amendments that have occurred during 3 years of operation; - Investor's decision as to how to administer the undertaking (sale of the facility, lease/rental, own investment) - Appointment of the facility administrator; - Establishment of the facility logbook; - Instructions for the facility administrator; - Implementation of the facility operation procedures; - Periodic reports prepared by the facility administrator; - Report verification.	Biliński [45]; Zabielski [79]; Grzywiński [78]; Sobieraj [9,10]
The stage of final evaluation of the results and effects of the project (II stage of project closing)	This stage involves the following activities: - General contractor's analysis of the parameters/results established by the investor when launching the project (assuming full production capacity); - Update of the schedule for achieving the target indicators; - Detailed plan of the nodal points, whose passage will be followed by measurements and analyses, i.e., after 3, 5, 7, 10 years; - Re-audit which should be conducted by the investor; - Projections concerning target results, performed on the basis of the stage results; - Measurements and analysis of the completed production/service process; - Evaluation of the achieved results; - Comparison of the achieved results with the parameters established in the conceptual stage and approved in the final version of the project directed to implementation; - Conclusions and recommendations on the basis of the achieved results - Analysis-verification of the achieved results by the investor; - Investor's decisions following results analysis;	Trocki and Wyrozębski [119]; Sobieraj [9,10]

Table A1. *Cont.*

Phases/Stages of the Investment Project	Description	Cited by Authors
Phase of drawing up proposals for future implementation after Phase II of the project closure	- In case of unsatisfactory results, an investigation should be initiated between the investor and the general contractor; - Where necessary, implementation of corrective procedures; - Re-examination of the results following corrective procedures; - In case of a satisfactory result, an application for the financial settlement of the completed stage or in case of a negative result, a court action (or a discount/compensation granted by the general contractor). This stage involves: - Archiving of all analytical results; - Statistical analysis of the achieved results; - Preparation of the list of errors/defects/faults encountered during the investment process; - Preparation of the facility card containing all facility parameters, including financial parameters that can be used for comparisons when creating pre-feasibility studies for similar facilities; - Implementation of the investment process correction procedures for similar facilities.	Trocki and Wyrozębski [119]; Sobieraj [9,10]

Table A2. Different schedule management techniques—practitioners' perspective.

Method	Descriptions	Authors
Critical Path Method (CPM)	The planning method most commonly used in large construction projects is the critical path method (CPM). CPM is based on the assumption that the completion of each activity depends on a few critical resources or constraints. Because of the critical importance of boundary conditions, project managers study CPM as part of their project management certification (PMP). For this reason, it is also the legal standard for measuring delays when project-related disputes arise. CPM creates a graphical view of a project and calculates how much time and resources are needed to complete each activity. It also determines the critical activities that need attention to ensure the project is completed on time.	Yamin and Harmelink [11]; East [12]; Galloway [120]
Program Evaluation and Review Technique (PERT)	PERT is one of the most accessible tools for building design and scheduling. It provides a visual representation of the key project activities and the order in which they must be completed. Each of these steps represents the commitment of time or resources. The diagram can be thought of as a roadmap for the completion of the project; only when all milestones have been reached has the building project reached its final phase. PERT Diagrams are often built from back to front, as many projects have a pre-determined deadline, but contractors have some flexibility in the early phases and stages of the project [9,10].	Kirytopoulos et al. [93]; Liu [121]; Sobieraj [9,10]; Galloway [120]
Critical Chain Project Management (CCPM)	Critical Chain Project Management (CCPM) was developed by Goldratt [122]. This method emphasises the appropriate use of resources required to complete project tasks, i.e., people, equipment and physical space. In contrast to more traditional methods such as CPM or PERT (which take into account task sequencing and rigid scheduling), the CCPM method relies on resource alignment, which of course involves some flexibility in their allocation (as well as in start times). In other words, the appropriate use of resources plays a key role in the CCPM method. With the critical chain method (CCM), one can perform an analysis of the scheduling network, addressing dependencies between tasks, resource availability and appropriate buffers. Thus, the CCM allows for better (more efficient) planning (scheduling) when the execution of project tasks is accompanied by some uncertainty related to resource management, namely to their availability or constraints.	Goldratt [122]

Table A2. Cont.

Method	Descriptions	Authors
Line of Balance (LOB)	LOB is a construction scheduling tool that relies on thoughtful planning of projects through repeated iterations. It is a management control process where the project contains blocks of repetitive work activities. LOB collects, measures and presents information in terms of time, cost and completion and compares it to a specific plan. It helps identify where projects are going off track by identifying the specific moments when deviations occur. It reflects project objectives as a single line on a chart in terms of completed activities/time that teams are expected to adhere to in order to stay on track.	Arditi and Albulak [22]; Arditi et al. [24]; Soini et al. [26]; Tokdemir et al. [23]; Damci et al. [25]
Q Scheduling	Q Scheduling Quantitative scheduling is a planning approach that uses a bar chart to indicate the quantities of materials that will be used at different locations and times during the project. This type of scheduling allows companies to clearly see the amount and type of material needed at different times and places. It also includes a hierarchical component so that staff and managers can see what materials they need at what time, order them accordingly, carry out the activities/tasks in the right order and not disrupt the work of others—all while controlling costs.	Sulbaran and Ahmed [28]; Majumder et al. [27]
Resource Oriented Scheduling	Resource Oriented Scheduling method focuses on project resources, prioritising the most efficient use of those resources. As limited resources increase the likelihood of delays due to different teams fighting over them. Without a smart approach to determine who gets them and when, a construction company may be powerless to prioritise. A resource-driven schedule takes into account everyone who needs the resources in advance and then assigns them to an orderly use throughout the project.	Trimble [29]; Venkatesh et al. [30]
Last Planner System	The Last Planner system is a short-term collaborative planning process, and in this sense, it is not a typical construction scheduling tool compared to other techniques. Even though it is not a stand-alone tool, it can be used well with planning techniques such as CPM. LPS brings together those who will do the work (the team) to plan when and how the work will be done through a series of conversational processes. This allows teams to assess and remove likely obstacles in advance, promoting the timely completion of each task.	AlSehaimi et al. [31];
Gantt Chart	A Gantt Chart is a type of bar chart that encourages stakeholders to structure the project with several levels of detail and consider dependencies between tasks [32]. Gant Charts help them estimate the duration of the project and identify the critical path to take during construction. A Gantt chart is a bar chart used to illustrate a project schedule, that includes some milestones, and it is not as detailed as a full CPM. It normally includes start/end dates of activities and a summary of activities of a project. However, it lacks the complexity of more comprehensive approaches and doesn't include the resources or materials needed to complete it. Gantt Charts are excellent for creating a hierarchy among projects, showing which ones require immediate attention and which must be completed before other, dependent projects can follow.	Maylor [33]; Geraldi and Lechter [32];

References

1. Poh, P.; Lam, Y.M. Confidence based scheduling procedure (CBSP): A pragmatic approach to manage project schedule uncertainty. *Int. J. Constr. Proj. Manag.* **2014**, *6*, 119.
2. Sobieraj, J.; Metelski, D. Identification of the key investment project management factors in the housing construction sector in Poland. *Int. J. Constr. Manag.* **2020**, 1–12. [CrossRef]
3. Sobieraj, J.; Metelski, D. Quantifying Critical Success Factors (CSFs) in Management of Investment-Construction Projects: Insights from Bayesian Model Averaging. *Buildings* **2021**, *11*, 360. [CrossRef]
4. Pinto, J.K.; Slevin, D.P. Critical factors in successful project implementation. *IEEE Trans. Eng. Manag.* **1987**, *34*, 22–27. [CrossRef]

5. Cleland, D.I.; King, W.R. *Project Management Handbook*, 3rd ed.; REI: New York, NY, USA, 1983.
6. Sobieraj, J.; Metelski, D.; Nowak, P. The View of Construction Companies' Managers on the Impact of Economic, Environmental and Legal Policies on Investment Process Management. *Arch. Civ. Eng.* 2021, 67, 111–129.
7. Herroelen, W.; Leus, R. Project scheduling under uncertainty: Survey and research potentials. *Eur. J. Oper. Res.* 2005, 165, 289–306. [CrossRef]
8. Bruni, M.E.; Beraldi, P.; Guerriero, F.; Pinto, E. A scheduling methodology for dealing with uncertainty in construction projects. *Eng. Comput.* 2011, 28, 1064–1078. [CrossRef]
9. Sobieraj, J. *Wpływ Polityki Gospodarczej, Środowiskowej i Prawnej na Zarządzanie Procesem Inwestycyjnym w Budownictwie Przemysłowym*; ITE-PIB: Radom, Poland, 2019.
10. Sobieraj, J. *Investment Project Management on the Housing Construction Market*; Aurum Universitas Grupo Hespérides: Madrid, Spain, 2020.
11. Yamin, R.A.; Harmelink, D.J. Comparison of linear scheduling model (LSM) and critical path method (CPM). *J. Constr. Eng. Manag.* 2001, 127, 374–381. [CrossRef]
12. East, E. *Critical Path Method (CPM) Tutor for Construction Planning and Scheduling*; McGraw-Hill Education: New York, NY, USA, 2015.
13. Mohamed, Y. A Framework for Systematic Improvement of Construction Systems. Ph.D. Thesis, University of Alberta, Edmonton, AB, Canada, 2002.
14. Koskela, L. *An Exploration Towards a Production Theory and Its Application to Construction*; VTT Technical Research Centre of Finland: Espoo, Finland, 2000.
15. Howell, G.; Ballard, G. Lean production theory: Moving beyond "Can-Do". *Lean Constr.* 1997, 17–23.
16. Kong, Z.; Zhang, J.; Li, C.; Zheng, X.; Guan, Q. Risk assessment of plan schedule by Monte Carlo simulation. In Proceedings of the 4th International Conference on Information Technology and Management Innovation (ICITMI), Shenzhen, China, 12–13 September 2015; pp. 509–513.
17. Frein, J. *Handbook of Construction Management and Organization*; Springer Science & Business Media: Berlin, Germany, 2012.
18. Hulett, D. *Practical Schedule Risk Analysis*; Routledge: London, UK, 2016.
19. Ruogang, L.; Guoxiang, W.; Yue, L.; Jialin, R. Study on uncertainty of activity duration in PERT. *Syst. Eng. Electron.* 1997, 19, 40–45.
20. Petroutsatou, K. A proposal of project management practices in public institutions through a comparative analyses of critical path method and critical chain. *Int. J. Constr. Manag.* 2019, 1–10. [CrossRef]
21. Gonçalves-Dosantos, J.C.; García-Jurado, I.; Costa, J. ProjectManagement: An R Package for Managing Projects. *R J.* 2020, 12, 419–436. [CrossRef]
22. Arditi, D.; Albulak, M.Z. Line-of-balance scheduling in pavement construction. *J. Constr. Eng. Manag.* 1986, 112, 411–424. [CrossRef]
23. Tokdemir, O.B.; Erol, H.; Dikmen, I. Delay Risk Assessment of Repetitive Construction Projects Using Line-of-Balance Scheduling and Monte Carlo Simulation. *J. Constr. Eng. Manag.* 2019, 145, 1–19. [CrossRef]
24. Arditi, D.; Tokdemir, O.B.; Suh, K. Challenges in line-of-balance scheduling. *J. Constr. Eng. Manag.* 2002, 128, 545–556. [CrossRef]
25. Damci, A.; Arditi, D.; Polat, G. Resource leveling in line-of-balance scheduling. *Comput. -Aided Civ. Infrastruct. Eng.* 2013, 28, 679–692. [CrossRef]
26. Soini, M.; Leskelä, I.; Seppänen, O. Implementation of line-of-balance based scheduling and project control system in a large construction company. In Proceedings of the 12th Annual Conference of the International Group for Lean Construction, Helsingor, Denmark, 3–5 August 2004.
27. Majumder, S.; Majumder, S.; Biswas, D. Impact of effective construction planning in project performance improvement. *Qual. Quant.* 2021, 1–12. [CrossRef]
28. Sulbaran, T.; Ahmed, F. Expert System for Construction Scheduling Decision Support Based on Travelling Salesman Problem. In Proceedings of the 53rd ASC Annual International Conference Proceedings, Seattle, WA, USA, 5–8 April 2017.
29. Trimble, G. Resource-oriented scheduling. *Int. J. Proj. Manag.* 1984, 2, 70–74. [CrossRef]
30. Venkatesh, M.P.; Malathi, B.; Umarani, C. Factors Affecting Implementation of Resource Scheduling in Indian Construction Projects. *Appl. Mech. Mater.* 2012, 174, 2782–2786. [CrossRef]
31. AlSehaimi, A.O.; Fazenda, P.T.; Koskela, L. Improving construction management practice with the Last Planner System: A case study. *Eng. Constr. Archit. Manag.* 2014, 21, 51–64. [CrossRef]
32. Geraldi, J.; Lechter, T. Gantt charts revisited: A critical analysis of its roots and implications to the management of projects today. *Int. J. Manag. Proj. Bus.* 2012, 5, 578–594. [CrossRef]
33. Maylor, H. Beyond the Gantt chart: Project management moving on. *Eur. Manag. J.* 2001, 19, 92–100. [CrossRef]
34. Khodakarami, V.; Fenton, N.; Neil, M. Project Scheduling: Improved approach to incorporate uncertainty using Bayesian Networks. *Proj. Manag. J.* 2007, 38, 39–49. [CrossRef]
35. Khodakarami, V.; Abdi, A. Project cost risk analysis: A Bayesian networks approach for modeling dependencies between cost items. *Int. J. Proj. Manag.* 2014, 32, 1233–1245. [CrossRef]
36. Gondia, A.; Siam, A.; El-Dakhakhni, W.; Nassar, A.H. Machine learning algorithms for construction projects delay risk prediction. *J. Constr. Eng. Manag.* 2020, 146, 04019085. [CrossRef]

37. Kim, S.Y.; Van Tuan, N.; Ogunlana, S.O. Quantifying schedule risk in construction projects using Bayesian belief networks. *Int. J. Proj. Manag.* **2009**, *27*, 39–50.
38. Anastasopoulos, P.C.; Labi, S.; Bhargava, A.; Mannering, F.L. Empirical assessment of the likelihood and duration of highway project time delays. *J. Constr. Eng. Manag.* **2012**, *138*, 390–398. [CrossRef]
39. Azaron, A.; Ghomi, S.F. Lower bound for the mean project completion time in dynamic PERT networks. *Eur. J. Oper. Res.* **2008**, *186*, 120–127. [CrossRef]
40. Sobieraj, J.; Metelski, D.; Mihi Ramírez, A. Pivotal project management factors in the context of polish residential construction projects. In Proceedings of the International Conference on Industry, Business and Social Sciences (IBSS), Osaka, Japan, 28–30 August 2019.
41. Bolgorian, M.; Raei, R. A quantile-based Time at Risk: A new approach for assessing risk in financial markets. *Phys. A Stat. Mech. Appl.* **2013**, *392*, 5673–5677. [CrossRef]
42. Kovalenko, T.; Sornette, D. Dynamical Diagnosis and Solutions for Resilient Natural and Social Systems. *arXiv* **2012**, arXiv:1211.1949.
43. Sobieraj, J.; Metelski, D.; Mihi Ramírez, A. The view of SME construction companies' managers on the impact of economic, environmental and legal policies on investment process management. Case of Polish Companies. In In Proceedings of the XXVIII AEDEM International Conference, Tokyo, Japan, 3–4 September 2019; pp. 1046–1069.
44. Cooper, K.G. The $2000 hour: How managers influence project performance through the rework cycle. *Proj. Manag. J.* **1994**, *25*, 11–24.
45. Biliński, T. Struktura i uwarunkowania współczesnego procesu inwestycyjno-budowlanego. *Przegląd Bud.* **2010**, *81*, 46–52.
46. Obolewicz, J. Koordynacja budowlanego procesu inwestycyjnego. *Bud. I Inżynieria Śr.* **2016**, *7*, 153–163.
47. Laufer, A.; Howell, G.A. *Construction Planning: Revising the Paradigm*; Project Management Institute: Newtown Square, PA, USA, 1993.
48. Sobieraj, J. Impact of spatial planning on the pre-investment phase of the development process in the residential construction field. *Arch. Civ. Eng.* **2017**, *63*, 113–130. [CrossRef]
49. Project Management Institute. *A Guide to the Project Management Body of Knowledge*; Project Management Institute: Newtown Square, PA, USA, 2004.
50. Basu, A. *Practical Risk Analysis in Scheduling*; AACE International Transactions: Morgantown, WV, USA, 1998.
51. Rubinstein, R.Y.; Kroese, D.P. *Simulation and the Monte Carlo Method*; John Wiley & Sons: New York, NY, USA, 2016.
52. Sarma, M.; Thomas, S.; Shah, A. Selection of Value-at-Risk models. *J. Forecast.* **2003**, *22*, 337–358. [CrossRef]
53. Lindkvist, L.; Soderlund, J. Managing product development projects: On the significance of fountains and deadlines. *Organ. Stud.* **1998**, *19*, 931–951. [CrossRef]
54. Dille, T.; Söderlund, J. Managing inter-institutional projects: The significance of isochronism, timing norms and temporal misfits. *Int. J. Proj. Manag.* **2011**, *29*, 480–490. [CrossRef]
55. Yakura, E.K. Charting time: Timelines as temporal boundary objects. *Acad. Manag. J.* **2002**, *45*, 956–970.
56. Vaughan, D. *The Challenger Launch Decision: Risky Technology, Culture, and Deviance at Nasa*; University of Chicago Press: Chicago, IL, USA, 1996.
57. Sobieraj, J. *Review of Knowledge on Strategy Development, Strategic Management and Strategic Analysis*; Wydawnictwo ITEE: Radom, Poland, 2017.
58. Jugdev, K.; Müller, R. A retrospective look at our evolving understanding of project success. *Proj. Manag. J.* **2005**, *36*, 19–31. [CrossRef]
59. Flyvbjerg, B.; Garbuio, M.; Lovallo, D. Delusion and deception in large infrastructure projects: Two models for explaining and preventing executive disaster. *Calif. Manag. Rev.* **2009**, *51*, 170–194. [CrossRef]
60. Flyvbjerg, B.; Garbuio, M.; Lovallo, D. Delusion and deception in large infrastructure projects: Two models for explaining and preventing executive disaster. *Def. AR J.* **2017**, *24*, 583–585. [CrossRef]
61. Beckers, F.; Chiara, N.; Flesch, A.; Maly, J.; Silva, E.; Stegemann, U. A risk-management approach to a successful infrastructure project. *Mckinsey Work. Pap. Risk* **2013**, *52*, 18.
62. Bertsimas, D.; Sim, M. The price of robustness. *Oper. Res.* **2004**, *52*, 35–53. [CrossRef]
63. Jaśkowski, P. Methodology for enhancing reliability of predictive project schedules in construction. *Eksploat. I Niezawodn.* **2015**, *17*, 470–479. [CrossRef]
64. Al-Fawzan, M.A.; Haouari, M. A bi-objective model for robust resource-constrained project scheduling. *Int. J. Prod. Econ.* **2005**, *96*, 175–187. [CrossRef]
65. Van de Vonder, S.; Demeulemeester, E.; Herroelen, W.; Leus, R. The use of buffers in project management: The trade-off between stability and makespan. *Int. J. Prod. Econ.* **2005**, *97*, 227–240. [CrossRef]
66. Namazian, A.; Yakhchali, S.H.; Yousefi, V.; Tamošaitienė, J. Combining Monte Carlo simulation and Bayesian networks methods for assessing completion time of projects under risk. *Int. J. Environ. Res. Public Health* **2019**, *16*, 5024. [CrossRef]
67. Fitzsimmons, J.; Hong, Y.; Brilakis, I. Improving Construction Project Schedules before Execution. In Proceedings of the 37th International Symposium on Automation and Robotics in Construction (ISARC), Aarhus, Denmark, 27–28 October 2020.
68. Ökmen, Ö.; Öztaş, A. Construction project network evaluation with correlated schedule risk analysis model. *J. Constr. Eng. Manag.* **2008**, *134*, 49–63. [CrossRef]

69. Levin, J. Relational incentive contracts. *Am. Econ. Rev.* **2003**, *93*, 835–857. [CrossRef]
70. Mitkus, S.; Mitkus, T. Causes of conflicts in a construction industry: A communicational approach. *Procedia-Soc. Behav. Sci.* **2014**, *110*, 777–786. [CrossRef]
71. Kaming, P.F.; Olomolaiye, P.O.; Holt, G.D.; Harris, F.C. Factors influencing construction time and cost overruns on high-rise projects in Indonesia. *Constr. Manag. Econ.* **1997**, *15*, 83–94. [CrossRef]
72. Moselhi, O.; Gong, D.; El-Rayes, K. Estimating weather impact on the duration of construction activities. *Can. J. Civ. Eng.* **1997**, *24*, 359–366. [CrossRef]
73. Honek, K.; Azar, E.; Menassa, C.C. Recession effects in United States public sector construction contracting: Focus on the American Recovery and Reinvestment Act of 2009. *J. Manag. Eng.* **2012**, *28*, 354–361. [CrossRef]
74. Bragadin, M.A.; Kähkönen, K. Safety, space and structure quality requirements in construction scheduling. *Procedia Econ. Financ.* **2015**, *21*, 407–414. [CrossRef]
75. ElZomor, M.; Burke, R.; Parrish, K.; Gibson, G.E., Jr. Front-end planning for large and small infrastructure projects: Comparison of project definition rating index tools. *J. Manag. Eng.* **2018**, *34*, 04018022. [CrossRef]
76. Ortiz-González, J.I.; Pellicer, E.; Howell, G. Contingency management in construction projects: A survey of Spanish contractors. In Proceedings of the IGLC-22, Oslo, Norway, 25–27 June 2014; pp. 195–206.
77. Mubarak, S.A. *Construction Project Scheduling and Control*; John Wiley & Sons: Hoboken, NJ, USA, 2015.
78. Grzywiński, J. *Proces Inwestycyjny Zgodnie z Polskim Prawem Budowlanym*; Wyd. Kancelarii Furtek Komosa Aleksandrowicz: Warsaw, Poland, 2015.
79. Zabielski, J. *Proces Inwestycyjno-Budowlany. Materiały Dydaktyczne Wydziału Prawa i Administracji UW*; Wyd. CRE Edukacja: Warsaw, Poland, 2014.
80. Dzierżewicz, Z.; Dylewski, J. *Proces Budowlany w Świetle Ustawy Prawo Budowlane*; Wyd. Grupa APEXnet: Lublin, Poland, 2011.
81. Strzelecka, E.; Glinkowska, B.; Maciejewska, M.; Wiażel-Sasin, B. *Zarządzanie Przedsięwzięciami Budowlanymi. Podstawy, Procedury, Przykłady*; Wyd. Politechniki Łódzkiej: Łódź, Poland, 2014.
82. Połoński, M. *Kierowanie Budowlanym Procesem Inwestycyjnym*; Wyd. SGGW: Warsaw, Poland, 2009.
83. Baryłka, A.; Baryłka, J. *Funkcje Techniczne w Budownictwie. Przewodnik Po Inwestycyjnym i Eksploatacyjnym Procesie Budowlanym*; Polcen: Warsaw, Poland, 2015.
84. Liu, Y.; Li, Y. Risk management of construction schedule by PERT with Monte Carlo simulation. *Appl. Mech. Mater.* **2014**, *548*, 1646–1650. [CrossRef]
85. Hendradewa, A.P. Schedule Risk Analysis by Different Phases of Construction Project Using CPM-PERT and Monte-Carlo Simulation. In Proceedings of the 11th International Seminar on Industrial Engineering & Management (ISIEM), Technology and Innovation Challenges towards Industry 4.0 Era, Makasar, Indonesia, 27–29 November 2018.
86. Abdelkader, Y.H. Evaluating project completion times when activity times are Weibull distributed. *Eur. J. Oper. Res.* **2004**, *157*, 704–715. [CrossRef]
87. Clark, C.E. Letter to the editor—The PERT model for the distribution of an activity time. *Oper. Res.* **1962**, *10*, 405–406. [CrossRef]
88. Johnson, N.L.; Kotz, S.; Balakrishnan, N. *Continuous Univariate Distributions*; John Wiley & Sons: New York, NY, USA, 1995.
89. Kay, M. ggdist: Visualizations of Distributions and Uncertainty. R Package Version 3.0.1. Available online: https://mjskay.github.io/ggdist/ (accessed on 18 January 2022).
90. Vose, D. *Risk Analysis: A Quantitative Guide*; John Wiley & Sons: New York, NY, USA, 2008.
91. Buchsbaum, P. *Modified PERT Simulation*; Great Solutions: Rio de Janeiro, Brazil, 2012.
92. Barraza, G.A.; Bueno, R.A. Probabilistic control of project performance using control limit curves. *J. Constr. Eng. Manag.* **2007**, *133*, 957–965. [CrossRef]
93. Kirytopoulos, K.A.; Leopoulos, V.N.; Diamantas, V.K. PERT vs. Monte Carlo Simulation along with the suitable distribution effect. *Int. J. Proj. Organ. Manag.* **2008**, *1*, 24–46. [CrossRef]
94. Vanhoucke, M. Measuring the efficiency of project control using fictitious and empirical project data. *Int. J. Proj. Manag.* **2012**, *30*, 252–263. [CrossRef]
95. Koulinas, G.K.; Xanthopoulos, A.S.; Tsilipiras, T.T.; Koulouriotis, D.E. Schedule delay risk analysis in construction projects with a simulation-based expert system. *Buildings* **2020**, *10*, 134. [CrossRef]
96. Rumelt, R.P. Good Strategy/Bad Strategy: The Difference and Why It Matters. *Strateg. Dir.* **2012**, *28*, 8. [CrossRef]
97. Metelski, D.; Mihi-Ramirez, A.; Arteaga-Ortiz, J. Research and development projects upon real options view. *Eng. Econ.* **2014**, *25*, 283–293. [CrossRef]
98. Elnagar, H.; Yates, J.K. Construction documentation used as indicators of delays. *Cost Eng.* **1997**, *39*, 31.
99. Sobieraj, J.; Metelski, D.; Nowak, P. PMBoK vs. PRINCE2 in the context of Polish construction projects: Structural Equation Modelling approach. *Arch. Civil Eng.* **2021**, *67*, 551–579.
100. Kapliński, O.; Dziadosz, A.; Zioberski, J.L. Próba standaryzacji procesu zarządzania na etapie planowania i realizacji przedsięwzięć budowlanych. *Zesz. Nauk. Politech. Rzesz. Bud. I Inżynieria Sr.* **2011**, *58*, 11.
101. Behrens, W.; Hawranek, P.M. *Manual for the Preparation of Industrial Feasibility Studies*; United Nations Industrial Development Organization: Vienna, Austria, 1991; pp. 176–181.
102. Armaneri, Ö. An Integrated Multi-Criteria Decision Making Methodology for Risky Investment Projects Evaluation. Ph.D. Thesis, DEÜ Fen Bilimleri Enstitüsü, Buca/İZMİR, Turkey, 2009.

103. Tamošiūnienė, R.; Angelov, K. *Project and Programme Management and Evaluation*; Publishing House of Technical University–Sofia: Sofia, Bulgraria, 2011.
104. Leśniak, A.; Plebankiewicz, E.; Kozik, R.; Amanowicz-Marcinkowska, K. Impact of the current local spatial development plans on the activity of investor on the Polish residential real estate market. *Earth Environ. Sci.* **2021**, *656*, 012004. [CrossRef]
105. Tizani, W. Collaborative Design in Virtual Environments at Conceptual Stage. In *Collaborative Design in Virtual Environments*, 2nd ed.; Wang, X., Tsai, J., Eds.; Springer: Dordrecht, The Netherlands, 2011; pp. 67–76.
106. Escobar-García, D.A.; Younes-Velosa, C.; Moncada-Aristizábal, C.A. Application of a prefeasibility study methodology in the selection of road infrastructure projects: The case of Manizales (Colombia). *Dyna* **2015**, *82*, 204–213. [CrossRef]
107. Kim, H.Y.; Shin, Y.H.; An, J.H.; Vinh, B.T.; Dung, T.Q. Prefeasibility Study on the Construction and the Operation of the Underground Cold Storage in Lam Dong Province, Vietnam. *Tunn. Undergr. Space* **2021**, *31*, 184–197.
108. Ebi, U. Implementation of projects in the housing sector: A view of world bank assisted low income housing project in Aba and umuahia in retrospect. *Sch. J. Sci. Res. Essay* **2015**, *6*, 90–95.
109. Adriańczyk, A.K. The Project Fire Water Storage Tank with a Capacity of 6000 m^3. Ph.D. Thesis, Insitute of Civil Engineering, Warsaw University of Technology, Warsaw, Poland, 2016.
110. Roy, R.; Low, M.; Waller, J. Documentation, standardization and improvement of the construction process in house building. *Constr. Manag. Econ.* **2005**, *23*, 57–67. [CrossRef]
111. Levy, S.M. *Project Management in Construction*; McGraw-Hill Education: New York, NY, USA, 2018.
112. Kim, N.; Park, M.; Lee, H.S.; Roh, S. Performance management method for construction companies. In Proceedings of the 24th International Symposium on Automation and Robotics in Construction Companies, Kochi, India, 19–21 September 2007; pp. 523–529.
113. Maya, R.A. Performance management for Syrian construction projects. *Int. J. Constr. Eng. Manag.* **2016**, *5*, 65–78.
114. Houston, D.A. Knowledge Management and Positive Deviance: A Study of Construction Project Outcomes. Ph.D. Thesis, Capella University, Minneapolis, MN, USA, 2019.
115. Zhao-xia, M.A. Research on Contract Management in Project Implementation Stage. *Nonferrous Met. Des.* **2010**, *2*, 70–77.
116. Yan, W.Z.; Chen, P. Based on the System Dynamics construction phase of the project cost control Study. *Appl. Mech. Mater.* **2014**, *501*, 2691–2694. [CrossRef]
117. Covey, G.; Shore, D.; Harvey, R.; Faber, G. Preparing for commissioning. *Appita Technol. Innov. Manuf. Environ.* **2011**, *64*, 314–322.
118. Barnes, M. Construction project management. *Int. J. Proj. Manag.* **1988**, *6*, 69–79. [CrossRef]
119. Trocki, M.; Wyrozębski, P. *Planowanie Przebiegu Projektów*; Oficyna Wydawnicza SGH: Warsaw, Poland, 2015.
120. Galloway, P.D. Survey of the construction industry relative to the use of CPM scheduling for construction projects. *J. Constr. Eng. Manag.* **2006**, *132*, 697–711. [CrossRef]
121. Liu, M. Program evaluation and review technique (PERT) in construction risk analysis. *Appl. Mech. Mater.* **2013**, *357*, 2334–2337. [CrossRef]
122. Goldratt, E.M. *Critical Chain*; The North River Press: Great Barrington, MA, USA, 1997.

Article

Resilient Scheduling as a Response to Uncertainty in Construction Projects

Martina Milat *, Snježana Knezić and Jelena Sedlar

Faculty of Civil Engineering, Architecture and Geodesy, University of Split, 21000 Split, Croatia; snjezana.knezic@gradst.hr (S.K.); jelena.sedlar@gradst.hr (J.S.)
* Correspondence: mmilat@gradst.hr

Abstract: Complex construction projects are developed in a dynamic environment, where uncertainty conditions have a great potential to affect project deliverables. In an attempt to efficiently deal with the negative impacts of uncertainty, resilient baseline schedules are produced to improve the probability of reaching project goals, such as respecting the due date and reaching the expected profit. Prior to introducing the resilient scheduling procedure, a taxonomy model was built to account for uncertainty sources in construction projects. Thence, a multi-objective optimization model is presented to manage the impact of uncertainty. This approach can be described as a complex trade-off analysis between three important features of a construction project: duration, stability, and profit. The result of the suggested procedure is presented in a form of a resilient baseline schedule, so the ability of a schedule to absorb uncertain perturbations is improved. The proposed optimization problem is illustrated on the example project network, along which the probabilistic simulation method was used to validate the results of the scheduling process in uncertain conditions. The proposed resilient scheduling approach leads to more accurate forecasting, so the project planning calculations are accepted with increased confidence levels.

Keywords: resilience; baseline schedule; uncertainty; taxonomy; construction project

Citation: Milat, M.; Knezić, S.; Sedlar, J. Resilient Scheduling as a Response to Uncertainty in Construction Projects. *Appl. Sci.* **2021**, *11*, 6493. https://doi.org/10.3390/app11146493

Academic Editors: Mariusz Szóstak, Jarosław Konior and Marek Sawicki

Received: 2 June 2021
Accepted: 12 July 2021
Published: 14 July 2021

Publisher's Note: MDPI stays neutral with regard to jurisdictional claims in published maps and institutional affiliations.

Copyright: © 2021 by the authors. Licensee MDPI, Basel, Switzerland. This article is an open access article distributed under the terms and conditions of the Creative Commons Attribution (CC BY) license (https://creativecommons.org/licenses/by/4.0/).

1. Introduction

Large construction projects are characterized by their complexity in terms of organization, as they consist of hundreds of activities and require numerous resources. To successfully manage important project objectives, scheduling efforts must be applied to ensure that a project is completed within the contract requirements [1]. Schedule management concentrates on the processes that are essential to appropriately deliver critical project aspects, such as time, cost, resources, etc. [2] Scheduling methods which are used to establish reliable construction plans can be broadly classified into exact [3–6], heuristic [7–9], or metaheuristic approaches [10–14]. Current practices in the domain of construction planning and scheduling are oriented towards automatic schedule development and the application of specialized optimization techniques [15]. Although modern technologies such as BIM have already been applied to the optimization problems in the realm of construction scheduling [16–21], additional development and possible extensions are still needed to effectively automatize scheduling practices and improve both customizability and user-friendliness of emerging technologies for practical use [22,23].

As construction projects take place in a dynamic environment, they are consequently prone to the negative impacts of internal and external sources of uncertainty. Due to the fact that uncertainty has been recognized as a cause of risks that can influence the final outcome of a project [24], there is an indisputable need to manage the uncertainty which is present in almost all of the construction activities [25]. However, there is still a lack of mechanisms that would leverage state-of-the-art technologies such as BIM with the uncertainty management frameworks in complex construction projects [26]. BIM-based uncertainty management frameworks are currently emerging at the theoretical level, where the automatization and practical integration of existing frameworks with new technologies remain the main challenges thus far [27].

To diminish unfavorable impacts stemming from the construction project environment, a relevant strategy is to prepare for uncertainty from the early stages of the project life cycle. To account for uncertainty as early as in the project planning phase, a resilient scheduling approach has been developed recently [28–30]. Resilient scheduling is a procedure to develop the optimal baseline makespan for a project, considering the trade-off between schedule stability or robustness, and other important objectives, such as makespan minimization [29]. Moreover, additional trade-offs can be considered when defining the equilibrium state of the project, such as the limited budget of the project, expected net present value, as well as adequate risk management, among other equilibrium facets [28].

Resilient schedules are defined by their multidimensionality: they tend to be robust, flexible, and adaptable. Previous studies in construction scheduling have predominantly focused on the robustness aspect, which is already a complex concept. On the one hand, robustness depicts the insensitivity of objective function in the optimization model, may it be project minimization or NPV maximization, for example; on the other hand, it tends to minimize the deviations between baseline schedule and realized state. For instance, Zhao et al. [31] propose the framework for integrated robustness evaluation, considering composite robustness measure. Authors have been using improved subjective and objective weights in order to evaluate the schedule robustness. In another research, Zhao et al. [32] investigated the importance of schedule robustness by use of metaheuristic optimization techniques, considering both activities' starting-time deviation, as well as a structural deviation in a schedule.

On the other hand, information about the uncertainty in a construction project is usually scattered, as it arises from various sources, and due to the dynamics of complex construction projects, it is extremely demanding to organize, collect, and reuse that knowledge. Lack of information or ambiguous data can have undesirable consequences on project success and cause a negative impact to project objectives. According to Reference [33], significant efforts have been undertaken to consider more general sources of uncertainty in the project management domain. Because of the inherent complexity in large construction projects, there is a need to manage a considerable amount of information [34].

In the realm of construction management, different researchers have attempted to integrate knowledge about uncertainty into formal conceptualization. This way, domain knowledge can be accessed and reused by users in a form of computer-readable data [35]. For example, Tah and Carr [36] proposed a knowledge-based approach to facilitate effective risk management procedures in a construction project. Ping Tserng et al. [37] developed an ontology-based risk management framework to improve the overall effectiveness of risk management practices for a construction contractor. The study of Ding et al. [38] coupled ontology and semantic web technology in a BIM environment to manage construction risk knowledge. Apart from enhancing general risk management procedures, other practical applications of ontology in the construction domain include knowledge sharing [39–41], information extraction [42–44], and performance analysis [45–48].

So far, however, there has been little discussion about modeling a comprehensive knowledge base by considering general sources of uncertainty in construction projects. Therefore, the first aim of this research was to structure uncertainty sources related to complex construction projects in a faceted taxonomy, as a basis for the analysis and uncertainty management in construction projects from the early planning stages. Comprehensive identification and characterization of uncertainties in the construction domain is a first step towards increasing the probability of reaching project goals during the execution phase.

Considering the nature of complex construction projects, which are financially extremely demanding undertakings, appropriate cash procurement is of vital importance. The major source of financing for construction projects is the establishment of the bank overdraft [49,50]. If the cash deficit occurs during the project realization period, contractors will encounter difficulties related to the implementation of the project activities in accordance with a baseline plan [51]. Therefore, the development of a schedule where the cash flows will be suitable for the established bank overdraft is an important subject, since large construction projects require extensive investments and rarely depend solely on the savings of the contractor [49,50].

The main objective of this research is to propose a comprehensive mathematical model for resilient scheduling as a trade-off between project robustness, project duration, and contractor's profit. Previous studies on resilient project scheduling [28–30] have not dealt with the finance flow modeling, so the mathematical problem presented in this study aims to contribute to the growing area of research on resilient baseline scheduling by introducing the additional objective function for profit maximization at the end of the project. The present research contributes to a more practical setting in the context of resilient construction scheduling by including the financial aspect in the existing problem. In this way, we assess the issue of financial feasibility in resilient scheduling for construction projects.

2. Resilience Approach to Manage Uncertainty in Construction Projects

The methodological approach taken in this research is presented as an algorithm for resilient scheduling to manage uncertainty in construction projects, as depicted in Figure 1. At the start, uncertainty sources in a construction project are analyzed to clarify how uncertainty impacts project objectives. From here, the multi-objective optimization (MOO) model is developed to serve as a response to uncertainty in construction projects. The objective functions in the model stem from the uncertainties related to variability in the durations of activities. In such a way, the optimization method overcomes problems of the deterministic approach which cannot account for activities' completion times with absolute certitude. The optimization problem consists of three objective functions and complementary constraints. Decision variables in the MOO model define the start times for activities in a project once the optimization problem is solved by the use of a hierarchical approach. The solution is obtained in a form of a resilient baseline schedule. Finally, the baseline schedule is validated by the simulation analysis, where recognized uncertainty sources are modeled as stochastic variables.

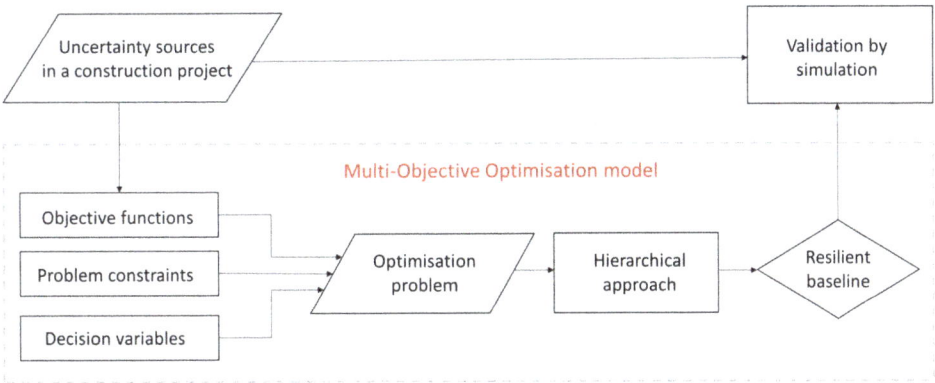

Figure 1. Algorithm for the resilience approach to manage uncertainty.

The structure of this study is organized as follows. The following subsection begins with a discussion on the nature of uncertainty in construction projects. In Section 2.2, the taxonomy development process is described. Resilience framework is discussed in more detail in Section 2.3. In Section 2.4, the MOO model for resilient scheduling is presented. The Section 3 consists of two parts, where the solving process is discussed in more detail in Section 3.1, while the resilient scheduling approach is validated on the construction project example by simulating two baseline cases in Section 3.2. The overall conclusions and suggestions for future research are given in Section 4.

2.1. Understanding Uncertainty

During the execution phase, many undesirable events might affect project constraints, such as budget overrun or timely execution of significant project activities. Disruptions in a project schedule due to mechanical malfunctions or unfavorable meteorological conditions on site could cause delays in project makespan unless managed properly. Occurrence, as well as the impact of disruptive events on project objectives, is uncertain; therefore, it is important to understand sources of uncertainty in a construction project.

Some authors consider uncertainty as various undesirable conditions or events, e.g., unpredictable resource unavailability, changing environmental conditions, various user needs, and system intrusions or faults [52]. Further examples of uncertain events include an increase in material prices, an accident during construction, a decision-maker's change of mind, etc. [53]. The corresponding perspective was adopted in this study, defining uncertainty as a high variety of possible events and conditions that could occur and affect project objectives.

Previous research has revealed certain characteristics considering the uncertainty in construction projects. For example, Gündüz et al. [54] reported that most of the uncertainty sources can be traced to conditions and events on site. Authors discovered that from 83 identified uncertainty factors that could delay construction performance, over 90% were traceable to the activities within the construction site. On the other hand, considering the time distribution of uncertainty through the project's life cycle, Ustinovičius [55] concluded that in the initiation phase the majority of uncertain events and conditions could be expected. When it comes to the classification of uncertainty sources, various perspectives are emerging from different studies in the project management domain. Most of the research work categorizes uncertainties in two major categories: internal or external, while both of the categories could be further extended into subcategories, such as technical, commercial, environmental, socioeconomic, political, etc. [56]. Since most studies in the construction domain have only focused on certain aspects of uncertainty analysis, such as uncertainty related to the specific phase of a project [55], or uncertainty related to the particular objective of a construction project [56], this paper aims to discuss a conceptual method based on the comprehensive literature review to accumulate, process, and reuse the dispersed knowledge on uncertainty, considering different aspects of a construction project. The following section aims to explain the process of development for the structured and systematic knowledge base of uncertainties, considering construction projects.

2.2. Taxonomy

Herein, a voluminous body of information is deliberately omitted, aiming to present only the fundamental principles in the process of creating the taxonomy model. Taxonomy is modeled by the use of the open-source platform Protégé. The taxonomy model serves as a basis for modeling the impact of uncertainty on baseline schedules in the final step of resilient schedule validation. In this study, the taxonomy model is a representation of domain knowledge by the use of concepts organized into facets and/or hierarchical structures. In the multifaceted taxonomy, every facet represents different features of the concept which is classified [57]. For example, one facet embodies uncertainties considering project life-cycle phases, and the other depicts uncertainties regarding stakeholders in a construction project, etc. Each of the facets could be further described by an independent taxonomy [57], either in hierarchical, faceted, or combined form.

The taxonomy model is developed to account for a high variety of possible events and conditions that could occur and affect project objectives. As the main objective is the development of construction schedules that are resilient to disruptions stemming from uncertainties, the starting point, i.e., universe of discourse of the taxonomy, is determined as "Uncertainty sources in a construction project". The main concept is further depicted by sub-concepts, i.e., facets, which ought to exhaustively describe the universe of discourse.

Faceted systems are in their nature general knowledge management models based on a multidimensional classification of heterogeneous data [57]. Since different authors argued sources of uncertainty in construction projects in a constrained way, e.g., solely with impact on transaction cost [56], multifaceted classification was imposed on the representation of the gathered knowledge.

The main concept is depicted by sub-concepts, i.e., facets, which ought to exhaustively describe the universe of discourse. To maintain integrity while modeling sub-dimensions (facets) of the main concept, fundamental categories were developed by following the core principles of knowledge organization, referring to five fundamental categories of knowledge classification [58], as presented in Table 1.

Table 1. Fundamental facets for the uncertainty sources in a construction project.

Abstraction Category	Facet Question	Facet	Description
Personality	Who	Member	Uncertainty related to the member in a construction project.
Matter	What	Project	Uncertainty related to a construction project itself.
Energy	How	Impact	Uncertainty impact on the project objective.
Space	Where	Cause	Cause of uncertainty source in a construction project.
Time	When	Life cycle	The time dimension of a construction project presenting a stage when uncertainty may arise.

The facets are revealed through the literature review, and they tend to describe the main concept exhaustively. To prevent ambiguity or vagueness of any kind, we offer five main facets:

- Member facet tends to describe all aspects of human attributes together with their mutual relationships which could lead to project perturbations. Members in a construction project could be individuals as well as groups of people, i.e., organizations.
- The project is described as features of a construction project which represent substantial means of how a project affects uncertainty existence.
- Impact means all the direct effects of uncertainty sources on traditionally determined project objectives (schedule-budget-quality goals), with additional sustainability aspects [59].
- Causes are states, conditions, and events in which sources of uncertainty are triggered.
- The life cycle presents stages of a construction project in which uncertainties may arise. This facet is the time dimension of the universe of discourse.

The structure of five main facets further depicted with the second-level entities (sub-facets) is shown in Table 2. Currently, the taxonomy model counts more than 300 entities in the form of mutually categorized terms, accompanied by the definition for each term to prevent any ambiguity. While the main idea of classification is to systematically represent the integrity of the source term, compounding terms is the process of composing different concepts to form a distinct term. For example, the term "Crane failure due to mechanical malfunctions or breakdowns" is compounded from concepts "Resources" and "Breakdowns", so there is no need to explicitly denote the existence of the compound term. Due to the complexity of the task, it is on the final users of taxonomy to validate possible combinations of uncertainty sources in construction projects.

The presented taxonomy model in a comprehensive, yet restricted manner describes uncertainty sources in a construction project. The uncertainty assessment plays an important role in the overall approach to model resilient schedules since the detailed examination of uncertainty sources is needed to recognize the impact on project objectives. Therefore, this taxonomy reveals the fundamental need to cope with unfavorable disruptions in the process of creating a baseline schedule. As a next step in the overall methodology, resilient scheduling is deployed to produce stable makespans for construction projects under the impact of uncertainty.

Table 2. The second level of faceted hierarchy for uncertainty sources in a construction project, where the primary node is "Uncertainty sources in a construction project".

Member	Project	Impact	Cause	Life Cycle
Behavior	Activities	Cost	Location	Conceptualization
Capacity	Complexity	Quality	Threats	Planning
Decision making	Novelty	Schedule	Type	Execution
Interactions	Type	Sustainability	-	Termination
Nature	-	-	-	-

Full taxonomy model is available upon request from the corresponding author.

2.3. Resilience as a Response to Uncertainty in Construction Projects

Herein, we propose the procedure based on the resilient scheduling concept and optimization modeling to manage uncertainty in a construction project and develop a construction schedule resilient to disruptions. The resilience strategy to cope with uncertainties is to recognize them as early as possible, namely in the project planning phase. By doing this we reduce the number and intensity of disruption shocks during the project implementation. Resilient scheduling, together with early identification of the uncertainties, plays a crucial role in reducing the risks of the construction projects. The main goal of resilient project scheduling is to maximize the preservation of the project baseline schedule at the equilibrium state during project execution. The concept of equilibrium state generally refers to timely project completion and minimized project tardiness amount, considering both project due date and activities' start times when executed activities are compared to a baseline schedule [28]. In this study, we extend the concept of equilibrium state by examining the degree of credit limit preservation and achievement of the baseline profit goal.

The goal of the proposed approach is to obtain the proactive baseline schedule as a response to the impact of uncertainty, from the contractors' perspective. As illustrated in Figure 2, the research procedure consists of three main parts. The starting point is identification, evaluation, and interpretation of uncertainty sources by the systematic literature review and also by using the developed taxonomy to avoid ambiguity and vagueness. The systematic literature review in this study was done by following the process for uncertainty identification and classification from Warmink et al. [60]. This way, the taxonomy serves as an essential tool for contractors to manage uncertainty in the planning phase, in view of the knowledge extraction and reuse during the optimization process. However, by finding new concepts during the identification process, the taxonomy is being constantly updated. The second phase in the proposed method is oriented toward solving the multi-objective optimization problem, where objective functions capture uncertainties. In this paper, we show an optimization model for resilient scheduling problem with three objective functions that strive to simultaneously (i) minimize project duration, (ii) maximize the level of resilience built in the baseline schedule, and (iii) maximize profit from the contractors' perspective.

The detailed finance framework is incorporated into the problem of resilient baseline scheduling so the cash flows can be calculated throughout the duration of a project. The model is built from the contractors' perspective to ensure the financial feasibility of a baseline schedule. In the case of a cash deficit, the contractor is threatened with the possibility of impeded work on the construction site. The interrupted workflow could lead to significant delays and eventually to additional overheads and liquidated damages and, hence, lower profit [51]. Since the appropriate financing of the construction project is of the utmost importance when it comes to the timely execution of project activities, the research hypothesis is that the contractor will be able to reliably commit to the baseline schedule if the final profit is maximized. Project profit maximization indicates reduced overhead expenses, so the overall project cost is minimized; therefore, project performances are improved to benefit all stakeholders. This way, the additional layers of adaptability and flexibility are implicitly included in the mathematical model by realistically calculating cumulative cash flows throughout the project makespan.

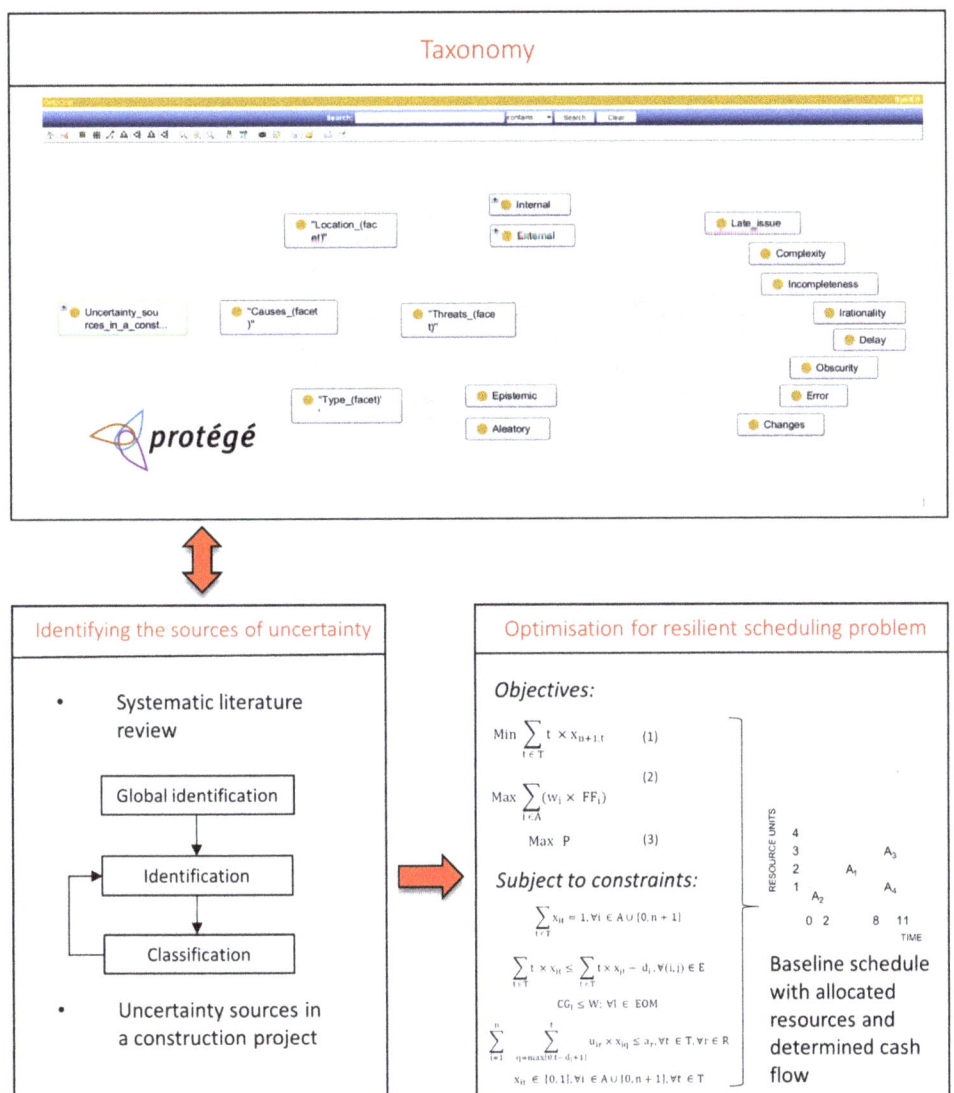

Figure 2. Layout of the proposed resilience procedure for uncertainty management in construction projects.

In the following step of the proposed procedure, the optimization process is conducted for the resilient scheduling problem, and the result is obtained in the form of a baseline makespan, by determining the optimal start time for each activity.

2.4. Optimization Model for Resilient Scheduling Problem

This section describes the optimization model for creating a feasible, resilient baseline schedule. Details related to the resilience improvement through surrogate measure maximization are discussed in the following subsection, followed by the statement of the mathematical problem in Section 2.4.2.

2.4.1. Surrogate Measure and Activity Weights

The concept of resilient scheduling is built on the idea of providing enough time intervals in the initial baseline structure so the schedule is able to absorb disturbances caused by the impact of uncertainty. A common approach is to provide enough time floats in the initial solution by the use of different surrogate measures [28–30,61,62]. This way, the stability of a schedule is approximated under the low cost of computational complexity, providing a simple solution in the overall computationally demanding optimization process.

The surrogate measure in this study is modeled as a weighted sum of resource-technology free floats [30] of all activities in a project. The resource-technology free float (FF_i) calculates the amount of time for which an activity can be prolonged without postponing start times of succeeding activities, in a way that both precedence and resource relations are considered. On the other hand, activity weights (w_i) are significant for appropriate float distribution, since they express the relative importance of the FF_i. For activities with the higher weight (w_i), it is more important to provide bigger FF_i, so the baseline schedule is as resilient as possible. Activity weight is calculated based on four indices:

- AS_i—Equation (1) states the relative number of successors per activity. The number of direct and indirect successors (N_{succ}) for the activity i is divided by the number of all activities in a project, n (both dummy start and dummy end included);
- DP_i— Equation (2) calculates relative duration of the activity, i;
- AC_i— Equation (3) determines relative cost of the activity, i;
- RU_i— Equation (4) counts relative resource usage as required by the activity, i.

The assumptions are as follows: (i) with an increased number of successors, there is a higher probability to propagate time disruptions through the baseline schedule; (ii) activities that are expected to be executed during a longer time have a higher probability of disruption; (iii) the higher the cost of the activity, the bigger is the impact to the cash flow of a project; and (iv) with higher resource consumption, there is a higher probability of resource breakdown. Therefore, the activity weight (w_i) is calculated as follows:

$$AS_i = \frac{N_{succ}}{n} \quad (1)$$

$$DP_i = \frac{d_i}{\sum_{i=1}^{n} d_i} \quad (2)$$

$$AC_i = \frac{c_i}{\sum_{i=1}^{n} c_i} \quad (3)$$

$$RU_i = \frac{\sum_{r=1}^{k} u_{ir}}{\sum_{i=1}^{n} \sum_{r=1}^{k} u_{ir}} \quad (4)$$

Finally,

$$w_i = \frac{AS_i + DP_i + AC_i + RU_i}{4} \quad (5)$$

Multiplication of FF_i by the activity weight (w_i) accounts for proper float distribution in the baseline schedule, aiming to preserve the project at the equilibrium state. Therefore, by maximizing the weighted sum of the resource-technology free floats, the resilience of the initial makespan is improved.

2.4.2. Optimization Problem

The optimization model presented in this study is based on the existing bi-objective resilient scheduling problem [30], which is improved by introducing the additional objective function to maximize the profit from a contractors' perspective. This way, an additional layer of financial stability is provided in the project baseline structure. With the net profit increase, the hypothesis is that contractor is fit to make financial compensation for perturbed activities. This means that adaptability to project disruptions is enhanced by maximizing the third objective function in the model. Moreover, the credit limit is included as one of the constraints in the optimization model, thus ensuring the financial feasibility of the optimal baseline schedule. A credit limit is imposed to assure that a financial flow is bounded to an allocated cash amount through the entire execution of the project.

In addition, the methodology of activity weight calculation is introduced. This way, the surrogate measure is appropriately quantified so the resilience of a schedule can be approximated in the optimization process. Activity weight calculation is inspired by the work of Torabi Yeganeh and Zegordi [29], where the weight of each activity is determined based on resource usage, network complexity, and time overlap. For this study, the additional layer concerning the direct cost of each activity was combined with the existing factors to quantify the activity weights. Table 3 summarizes the notation used in the optimization model. In the optimization model, binary decision variables were used to calculate activity start times, where $x_{it} = 1$ if activity, i, starts at the time, t; otherwise, $x_{it} = 0$.

Table 3. Parameters and sets used in the optimization problem.

Symbol	Description
T	Length of the planning horizon (t = 1, 2, ... , m)
A	Set of project activities (i = 1, 2, ... , n), including dummy start 0 and dummy end n + 1
E	Set of precedence relations
R	Set of project resources (r = 1, 2, ... , k)
w_i	Weight of activity i
d_i	Expected duration for activity i
c_i	Deterministic cost of activity i
u_{ir}	Consumption of resource r as required by activity i
a_r	Availability of resource r during project time T
FF_i	Resource-technology free float for activity i
P	Final profit at the end of a project
EOM	End of the month considering project timeline (time step used when calculating Cash Flow), (EOM = 1, 2, ... , l)
CG_l	Cumulative cash flow value at the end of the month l
W	Credit limit for the project

The multi-objective optimization model (MOO) for resilient baseline scheduling in construction projects is expressed as follows:

$$\text{Min} \sum_{t \in T} t \times x_{n+1,t} \tag{6}$$

$$\text{Max} \sum_{i \in A} (w_i \times FF_i) \tag{7}$$

$$\text{Max } P \tag{8}$$

Subject to:

$$\sum_{t \in T} x_{it} = 1, \forall i \in A \cup \{0, n+1\} \tag{9}$$

$$\sum_{t \in T} t \times x_{it} \leq \sum_{t \in T} t \times x_{jt} - d_i, \forall (i,j) \in E \tag{10}$$

$$CG_l \leq W; \forall l \in EOM \tag{11}$$

$$\sum_{i=1}^{n} \sum_{q=\max\{0, t-d_i+1\}}^{t} u_{ir} \times x_{iq} \leq a_r, \forall t \in T, \forall r \in R \tag{12}$$

$$x_{it} \in \{0, 1\}, \forall i \in A \cup \{0, n+1\}, \forall t \in T \tag{13}$$

The first objective function, shown in Equation (6), minimizes the start time for dummy end activity, minimising the project duration in total. Multiplication of binary variable $x_{it} = 1$ with specific time step (*t*) defines the start of activity (*i*) at the start time, *t*, supposing that time, *t*, corresponds to the start time of activity (*i*) from the baseline solution. For all other time steps, the binary variable x_{it} equals 0, as stated in the constraint part of the model. The second objective function, as stated in Equation (7) maximizes resilience through the surrogate measure (SM), which is represented by the sum of weighted resource-technology free float [30] for each activity, as explained in the previous subsection. The third objective function, shown in Equation (8), maximizes profit for the contractor at the end of the project by maximising the cumulative cash flow of dummy end activity. Constraint expressed in Equation (9) ensures that each activity, including the dummy start and end, can be started only once. Precedence constraint is expressed in Equation (10), defining that no activity can start before the precedent one is finished, for all activities that belong to the set of precedence related pairs. Financial constraint is shown in Equation (11) and it states that at the end of each month, the cumulative cash gap (before receiving the payment from the investor) must not exceed the permitted credit limit. Resource constraints are specified in Equation (12) by considering all possible start times, x_{iq}, for all activities (*i*) such that activity is in progress in period *q*, for every time step, *t*, and every resource, *r*. Multiplication of binary variable x_{iq} with resource consumption u_{ir} must not exceed resource availability, a_r. Constraint which is stated in Equation (13) defines the decision variable as a binary one.

Financial flow is modeled from the contractors' perspective by equations used in the work of Elazouni and Metwally [49] that, in turn, rely on the financial terminology clarified by Au and Hendrickson [63]. Since the bank overdraft has been recognized as a prevailing method for financing construction projects [49,64,65], this study also considers the bank overdraft as a single source of financing the construction project. For this reason, the cash-flow model in this study adopted established calculation [49] and incorporated two adjustments to improve the validity of financial flows. The first modification refers to the inclusion of interest rate, *h*, which rewards the contractor with an interest paid for positive cash flow. The interest rate (*h*) applicable on positive balances is less than the interest rate, *ir*, charged for borrowing in the case of a negative cash balance for the contractor [66]. The second adjustment is applied to explicitly display the return of the retainage amount at the end of the project. By modeling the retainage amount as an absolute value, it is assured that a contractor receives complete payment from an investor. Notation for required input variables to model financial flow is presented in Table 4.

Table 4. Input variables required for financial flow modeling.

Symbol	Description
d_i, $i \in A$	Duration of activities
s_i, $i \in A$	Start times for activities from baseline schedule
c_i, $i \in A$	Total direct cost per each activity in thousand of financial units
OP	Overhead percentage in decimal form
MP	Mobilization percentage in decimal form;
TP	Tax percentage in decimal form
MP	Markup percentage in decimal form
BP	Bond percentage in decimal form
N	Negotiated duration of the project in months
S	Realized duration of the project in months
ADV	Advance payment (percentage of TBP in decimal form)
D	Late completion penalty in thousand of cost units
RET	Percentage of retainage for investors' payments (decimal)
ir	Interest percentage in decimal form
h	Surplus percentage in decimal form (h < ir)
k	Percentage for interest on the unused portion of a credit
W	Specified credit limit of the overdraft

It is assumed that receipts from the investor lag one month after realized expenses. Effects of financing costs are considered and calculated as seen in Reference [49], with two adjustments as previously described. The goal of finance flow modeling is to determine accurate values for the maximal cash gap (CG) and final profit for the contractor (P) for a particular instance of the baseline schedule. In the work of Elazouni and Metwally [49], the mentioned output variables of interest are denoted as $\hat{F}_t$ for the cash gap amount at the end of the period t, and G for the net profit achieved at the end of the project. The direct cost of an activity is linearly distributed over the length of its duration. For implementation purposes, finance calculation is coded by using Python programming language. Any readers interested in the detailed calculation process considering cash-flow analysis can contact the corresponding author for additional information.

3. Application of Resilient Scheduling on a Test Problem Instance

3.1. Solving the Multi-Objective Optimization Problem

The multi-objective optimization model was applied to the test case adjusted from Reference [30] to provide a detailed description of the solving process for the multi-objective optimization problem on a single problem instance. The activity list for the problem instance is given in Table 5. The project network is shown in Figure 3, along with activities' durations in months, their resource requirement per activity, and activities' direct costs in thousand of financing units. Only one resource type is required during project execution, of which 7 units are available at any time. Since the model proposed herein requires cash-flow analysis, financial data, and contract terms are provided in Table 6. All integer values considering financial input are given in thousand of financial units.

Table 5. Activity list for the problem instance.

Activity Index	Activity Description
A_0	Dummy start
A_1	Procurement
A_2	Field mobilization and site work
A_3	Landscape equipment mobilization
A_4	Systems work
A_5	Structure work
A_6	Construction finishing operations
A_7	Landscape earthwork
A_8	Landscape surfaces
A_9	Dummy end

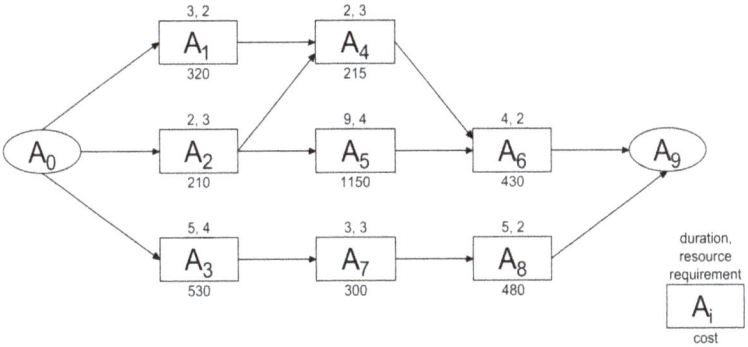

Figure 3. The network of a test problem instance.

Table 6. Financial data and contract terms for the test instance.

Symbol	Data	Value
OP	Overhead percentage	0.15
MP	Mobilization percentage	0.05
TP	Tax percentage	0.02
MP	Markup percentage	0.20
BP	Bond percentage	0.01
ADV	Advance	0.10
D	Penalty	2
RET	Retainage	0.05
ir	Interest	0.008
h	Surplus	0.005
k	Interest on unused credit	0.002

In the test case, we assume recognized uncertainty sources as follows:
- Unfavorable weather conditions might interfere with the execution of Activity 2, which is performed on the construction site;
- Architectural innovation involves an element of uncertainty while implementing the Activity 4;
- Change orders in Activity 6 might cause technical difficulty for accomplishing the execution in the estimated time frame.

Having in mind optimization problems with multiple objectives, a single optimal solution cannot always be found in an unambiguous manner. For example, one baseline schedule may provide a shorter makespan, with lower values considering resilience surrogate measure and profit, while another baseline schedule may have a longer duration and significantly higher values of surrogate measure and profit. Therefore, from the mathematical point of view, one optimal solution generally is not guaranteed when solving optimization problems with more than one objective function.

To date, different methods have been introduced to solve the MOO problems. One of them is to combine or weigh different objective functions [67], so the range of solutions can be found as Pareto optimal points. Following this approach, it is hard to obtain a single optimal schedule, since neither of the Pareto solutions can be improved without compromising other objective values [68]. As can be seen in Reference [69], although Pareto solutions can be considered broadly comparable, it is not possible to select only one dominant solution. Another problem then arises, as it is not always clear how to combine objective functions with different measurement units. For these reasons, a hierarchical approach was used to proceed with the solving process in the current study.

Since the initial assumption was that the decision-maker's preferences in obtaining the final solution were known in advance, a hierarchical approach is employed to find a single baseline solution that can be implemented for a construction project. Priorities for all objectives are assigned in decreasing order of importance as follows: timeline minimization, surrogate measure maximization, and profit maximization. This way, project duration is optimized on the uppermost level of the solving process, and the following preference is to find the highest surrogate measure value amongst all solutions that satisfy the first hierarchical objective. At the final step of solving process, a profit value is maximized for the baseline schedule, without causing degradation in earlier objective functions.

The test problem was solved to optimality with an exact algorithm written in the Mathematica programming language. The overall process led to obtaining a single baseline schedule, with a makespan duration of 18 months, SM value of 0.487, and deterministic profit value evaluating to 904.12 thousand of financial units. This way, the stable baseline schedule, which is considered to be resilient under the impact of uncertainty, was produced as a result of the optimization process.

3.2. Validating Resilient Scheduling Process

To validate the resilience of the proposed scheduling process, a probabilistic simulation analysis was conducted for different baseline solutions of the previously presented test problem. Due to the computational complexity of the underlying scheduling problem, the simulation analysis was bounded on two distinctive instances. The first baseline schedule refers to the single makespan instance obtained by solving the MOO test problem as previously described, while the other solution is found when the second and the third objective function were shifted in their hierarchy. For the second case, timeline minimization is still on the highest level of hierarchy in the optimization process; however, the profit maximization is in the second place of importance, and the surrogate measure maximization is at the bottom considering the hierarchical importance of objective functions. This way, the evaluation is made to compare the simulated behavior of two different baseline schedules: both of them have a duration of globally optimal 18 months, but in the first case, the SM value is higher and the baseline profit value is lower compared to SM value and baseline profit value from the second case. Objective function values for both cases are given in Table 7, along with the baseline start times for all activities, including dummy start and dummy end. The credit limit was set to 280 thousand of financial units in both cases.

Table 7. Objective function values for two test instances.

Instance	Baseline	Duration	SM	Profit
Case 1	0, 5, 0, 0, 8, 5, 14, 10, 13, 18	18	0.487	904.12
Case 2	0, 0, 3, 0, 5, 5, 14, 7, 10, 18	18	0.353	911.81

Since the goal of the simulation process is to analyse the project behavior under the impact of uncertainty, some of the activities' durations in the process of validation are allowed to take stochastic values. For this reason, the duration of activities 2, 4, and 6 in the project network are allowed to take stochastic values. Uncertain activity times are modeled with beta distribution, where shape parameters are set to $\alpha = 2$ and $\beta = 5$ in order to calculate realised activity durations in the Monte Carlo analysis. Stochastic activity durations in the simulation process are bounded between 0.7 and 2.2 of their estimated length, while the mode of the simulated activities is equal to the planned duration, d_i. Considering the simulated direct cost for each activity, it is assumed that the new cost is proportionally related to the realized duration of activity at a ratio, R, which is used to delineate if the cost of activity (C_a) at a simulated duration (T_a) consists mainly of material or equipment and labour. For example, if the cost of activity consists mainly of material, it will have a lower ratio value, R, while higher dependence on equipment and labour will be represented with a higher R-value [70]. In the simulation analysis, R is set to 0.6 for all activities. To calculate activities' cost C_a at a simulated duration T_a, Equation (14) is employed from the Reference [70]:

$$C_a = C_m + \frac{(T_a - T_m)}{T_m} \times R \times C_m, \tag{14}$$

where C_m represents the original cost of activity at the deterministic duration, T_m.

In the simulation analysis, additional resource relations are appended to accompany existing precedence connections between activities, so the baseline resource flow can be preserved throughout the entire simulation process. This way, resource conflicts are avoided through the simulation process where activities' durations take stochastic values. Additional resource relations are shown for each baseline schedule in Figure 4. Finally, the enhanced Monte Carlo simulation approach, which takes into consideration resource constraints, is taken on 10,000 iterations for each baseline case.

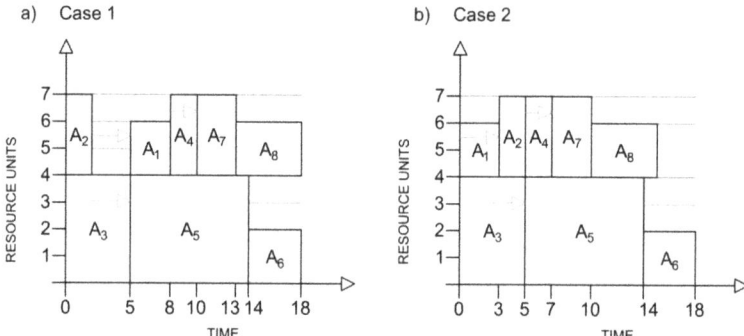

Figure 4. Baseline schedules with additional resource constraints for (**a**) Case 1 with the highest SM value and lower profit than in Case 2, and (**b**) Case 2 with the highest profit and lower SM value than in Case 1.

Equilibrium state was examined throughout 6 resilience dimensions: (Eq_i) probability of reaching baseline due date, (Eq_{ii}) probability of reaching baseline due date multiplied with coefficient 1.1, (Eq_{iii}) average tardiness amount for the proposed baseline, (Eq_{iv}) average tardiness amount for activities' start times, (Eq_v) the percentage of simulations where credit limit was broken, and (Eq_{vi}) probability of reaching or exceeding the baseline profit. The last equilibrium dimension (Eq_{vi}) is calculated by considering only those simulations for which credit limit, W, was not broken. From the simulation results, as shown in Table 8, it is evident that the first case (single optimal solution from the previous section) generates better results in all equilibrium calculations than the second case, for which profit maximization was considered as a more important objective than maximizing SM value. With an increase in *SM* value, there is a constant rise in the probability of reaching both strict (Eq_i) and relaxed (Eq_{ii}) baseline duration, as well as a constant decrease of average project tardiness (Eq_{iii}), and a decrease of average tardiness for activities' start times (Eq_{iv}). Moreover, the percentage of simulations for which the credit limit is broken (Eq_v) is slightly lower in the first case, where the SM value was higher. Finally, the probability of reaching baseline profit or attaining even higher profit values (Eq_{vi}) is significantly better for the test case with a higher SM value.

Table 8. Simulation results.

Instance	SM Value	Profit	W	Eq_i	Eq_{ii}	Eq_{iii}	Eq_{iv}	Eq_v	Eq_{vi}
Case 1	0.487	904.12	280	0.3	0.9135	0.66	0.12	1.65%	0.3769
Case 2	0.353	911.81	280	0.2046	0.8249	0.93	0.42	1.73%	0.0978

According to the simulation results, the initially suggested approach, where SM value maximization was taken as a more important objective function than maximizing the baseline profit, leads to better results considering the overall resilience of the baseline solution. Validation results indicate a positive correlation between increased SM value and resilience of baseline schedule with allocated resources. As a result of enhanced Monte Carlo analysis, probability distribution charts for makespan duration, considering both cases, are shown in Figure 5a. Probability distribution charts for final profit are shown in Figure 5b.

Since the average deviation of the maximal cash gap is negligible in both baseline cases (average deviation from the negotiated CL was around 0.01% in the simulations where the CL was broken), the probability distribution for profit is shown by considering all simulations, including those where violation of credit limit has occurred. After simulating two different baseline schedules with the Monte Carlo approach, where additional resource relations were appended, it was shown that a schedule with a higher SM value produces a better response to uncertainty.

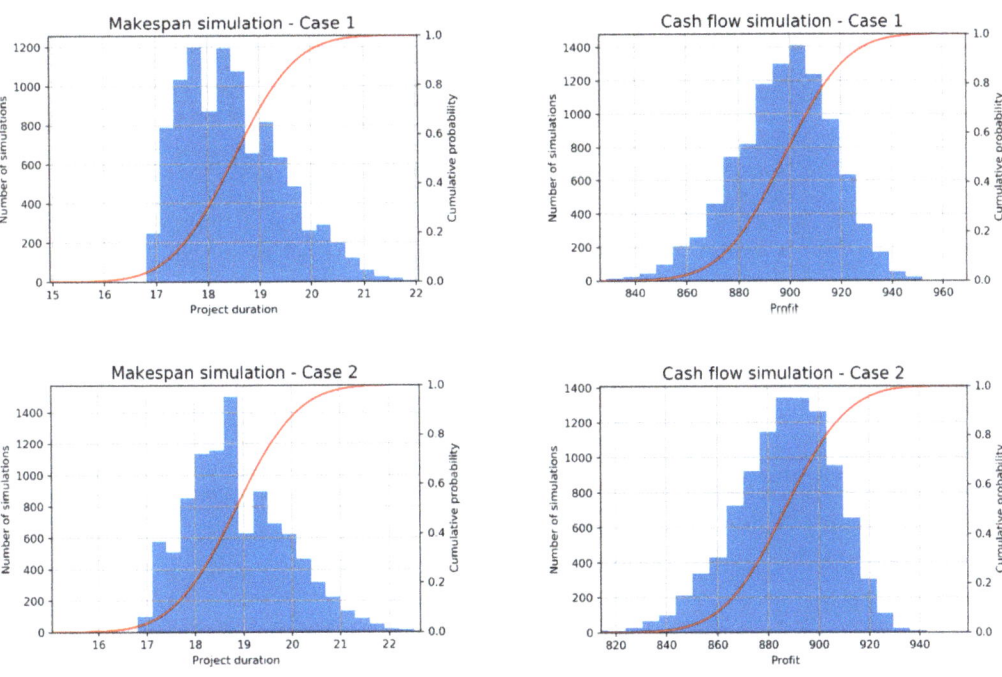

Figure 5. Simulation results: (**a**) total project duration (months); (**b**) total project profit for all iterations in simulation, CL may be broken (thousand of financing units).

The results of the equilibrium analysis showed improved performances in all equilibrium dimensions ($Eq_i - Eq_{vi}$) when comparing the baseline schedule with higher SM value to another solution in which profit maximization dominated SM improvement. The optimal trade-off between project duration, stability, and profit is obtained for a baseline schedule calculated as a result of the resilient scheduling approach presented in this study. According to validation results, it can be stated that the contractor will benefit from the enhanced resilience of the baseline schedule, since its schedule is improved for the case with the higher SM value. Therefore, this research contributes to the project-management body of knowledge by exploring an approach to develop resilient baseline schedules which will maximize the probability of reaching project goals. With the introduction of the financing aspects in the underlying resilience scheduling problem, the essential issues in construction-management reality are considered. The final output of the MOO problem is given in a form of a baseline schedule which simultaneously minimizes duration of a project, maximizes its resilience, and maximizes final profit from for a contractor. This way, project performances are improved, since the baseline calculations can be accepted with improved confidence levels.

4. Conclusions

In this study, a novel resilience procedure was proposed as a response to prevailing uncertainty in construction projects. In the initial stage of the research, a taxonomy was built to identify uncertainty sources during the life cycle of a project. The formal conceptualization of the domain knowledge, which was modeled in the open-source software Protégé, enables the reuse of information considering the uncertainty in construction projects. The main contribution of the research is the development of the optimization model with three objective functions. This way, a proactive baseline schedule is produced as a result of the optimization process.

Validation results suggest that resilience of a baseline solution improves with SM value maximization. However, the scope of the research was limited to a small project network, so caution must be applied when examining larger problem instances, since the findings might not be unconditionally transferable to a construction projects based on the more complex precedence networks. Although the research has laid the theoretical foundations for resilient scheduling procedure, the study has certain limitations. For example, the resilience framework should be tested on a larger set of project data collected from real construction projects to analyze the systems' behavior. Moreover, further research might explore different surrogate measures to interpret the resilience capacity of the baseline schedules for various types of construction projects. Finally, the development of new metaheuristic algorithms will enable detailed analysis and validation of the proposed resilience framework on a larger set of problems.

The present research introduces the financing aspect into the process of resilient scheduling, so the comprehensiveness and feasibility of the initial schedule are significantly improved. The advantage of the proposed resilient project planning is enhanced stability of the baseline schedule in comparison with the simulated state of project execution. This leads to more accurate forecasting, so the project-planning calculations are accepted with higher confidence levels.

Author Contributions: Conceptualization, M.M., S.K., and J.S.; methodology, M.M., S.K., and J.S.; software, M.M.; validation, M.M.; formal analysis, M.M., S.K., and J.S.; investigation, M.M.; resources, M.M., S.K. and J.S.; data curation, M.M.; writing—original draft preparation, M.M.; writing—review and editing, M.M., S.K. and J.S.; visualization, M.M.; supervision, S.K. and J.S.; project administration, S.K.; funding acquisition, S.K. All authors have read and agreed to the published version of the manuscript.

Funding: This research received no external funding.

Institutional Review Board Statement: Not applicable.

Informed Consent Statement: Not applicable.

Data Availability Statement: Data available on request, due to restrictions, e.g., privacy or ethical. The data presented in this study are available on request from the corresponding author. The data are not publicly available due to further research to be published.

Acknowledgments: This research is partially supported through project KK.01.1.1.02.0027, a project co-financed by the Croatian Government and the European Union through the European Regional Development Fund—the Competitiveness and Cohesion Operational Programme.

Conflicts of Interest: The authors declare no conflict of interest.

References

1. Derbe, G.; Li, Y.; Wu, D.; Zhao, Q. Scientometric review of construction project schedule studies: Trends, gaps and potential research areas. *J. Civ. Eng. Manag.* **2020**, *26*, 343–363. [CrossRef]
2. Faghihi, V.; Nejat, A.; Reinschmidt, K.F.; Kang, J.H. Automation in construction scheduling: A review of the literature. *Int. J. Adv. Manuf. Technol.* **2015**, *81*, 1845–1856. [CrossRef]
3. Cajzek, R.; Klanšek, U. Cost optimization of project schedules under constrained resources and alternative production processes by mixed-integer nonlinear programming. *Eng. Constr. Archit. Manag.* **2019**, *26*, 2474–2508. [CrossRef]
4. García-Nieves, J.; Ponz-Tienda, J.; Ospina-Alvarado, A.; Bonilla-Palacios, M. Multipurpose linear programming optimization model for repetitive activities scheduling in construction projects. *Autom. Constr.* **2019**, *105*, 102799. [CrossRef]
5. Zou, X.; Fang, S.; Huang, Y.; Zhang, L. Mixed-Integer Linear Programming Approach for Scheduling Repetitive Projects with Time-Cost Trade-Off Consideration. *J. Comput. Civ. Eng.* **2017**, *31*, 06016003. [CrossRef]
6. Klanšek, U. Mixed-Integer Nonlinear Programming Model for Nonlinear Discrete Optimization of Project Schedules under Restricted Costs. *J. Constr. Eng. Manag.* **2016**, *142*, 04015088. [CrossRef]

7. Liu, Z.; Zhang, Y.; Yu, M.; Zhou, X. Heuristic algorithm for ready-mixed concrete plant scheduling with multiple mixers. *Autom. Constr.* **2017**, *84*, 1–13. [CrossRef]
8. Sonmez, R.; Iranagh, M.; Uysal, F. Critical Sequence Crashing Heuristic for Resource-Constrained Discrete Time–Cost Trade-Off Problem. *J. Constr. Eng. Manag.* **2016**, *142*, 04015090. [CrossRef]
9. Li, H.; Xu, Z.; Demeulemeester, E. Scheduling Policies for the Stochastic Resource Leveling Problem. *J. Constr. Eng. Manag.* **2015**, *141*, 04014072. [CrossRef]
10. Tran, D.; Chou, J.; Luong, D. Multi-objective symbiotic organisms optimization for making time-cost tradeoffs in repetitive project scheduling problem. *J. Civ. Eng. Manag.* **2019**, *25*, 322–339. [CrossRef]
11. Agdas, D.; Warne, D.; Osio-Norgaard, J.; Masters, F. Utility of Genetic Algorithms for Solving Large-Scale Construction Time-Cost Trade-Off Problems. *J. Comput. Civ. Eng.* **2018**, *32*, 04017072. [CrossRef]
12. Aminbakhsh, S.; Sonmez, R. Pareto Front Particle Swarm Optimizer for Discrete Time-Cost Trade-Off Problem. *J. Comput. Civ. Eng.* **2017**, *31*, 04016040. [CrossRef]
13. Sroka, B.; Rosłon, J.; Podolski, M.; Bożejko, W.; Burduk, A.; Wodecki, M. Profit optimization for multi-mode repetitive construction project with cash flows using metaheuristics. *Arch. Civ. Mech. Eng.* **2021**, *21*, 1–17. [CrossRef]
14. Tao, S.; Wu, C.; Hu, S.; Xu, F. Construction project scheduling under workspace interference. *Comput.-Aided Civ. Infrastruct. Eng.* **2020**, *35*, 923–946. [CrossRef]
15. Amer, F.; Koh, H.; Golparvar-Fard, M. Automated Methods and Systems for Construction Planning and Scheduling: Critical Review of Three Decades of Research. *J. Constr. Eng. Manag.* **2021**, *147*, 03121002. [CrossRef]
16. ElMenshawy, M.; Marzouk, M. Automated BIM schedule generation approach for solving time–cost trade-off problems. *Eng. Constr. Archit. Manag.* **2021**. Epub ahead of printing.
17. Wang, Z.; Azar, E.R. BIM-based draft schedule generation in reinforced concrete-framed buildings. *Constr. Innov.* **2019**, *19*, 280–294. [CrossRef]
18. Abbasi, S.; Taghizade, K.; Noorzai, E. BIM-Based Combination of Takt Time and Discrete Event Simulation for Implementing Just in Time in Construction Scheduling under Constraints. *J. Constr. Eng. Manag.* **2020**, *146*, 04020143. [CrossRef]
19. Dasović, B.; Galić, M.; Klanšek, U. A Survey on Integration of Optimization and Project Management Tools for Sustainable Construction Scheduling. *Sustainability* **2020**, *12*, 3405. [CrossRef]
20. Nusen, P.; Boonyung, W.; Nusen, S.; Panuwatwanich, K.; Champrasert, P.; Kaewmoracharoen, M. Construction Planning and Scheduling of a Renovation Project Using BIM-Based Multi-Objective Genetic Algorithm. *Appl. Sci.* **2021**, *11*, 4716. [CrossRef]
21. Xie, L.; Chen, Y.; Chang, R. Scheduling Optimization of Prefabricated Construction Projects by Genetic Algorithm. *Appl. Sci.* **2021**, *11*, 5531. [CrossRef]
22. Wang, H.; Lin, J.; Zhang, J. Work package-based information modeling for resource-constrained scheduling of construction projects. *Autom. Constr.* **2020**, *109*, 102958. [CrossRef]
23. Sbiti, M.; Beddiar, K.; Beladjine, D.; Perrault, R.; Mazari, B. Toward BIM and LPS Data Integration for Lean Site Project Management: A State-of-the-Art Review and Recommendations. *Buildings* **2021**, *11*, 196. [CrossRef]
24. Perminova, O.; Gustafsson, M.; Wikström, K. Defining uncertainty in projects—A new perspective. *Int. J. Proj. Manag.* **2008**, *26*, 73–79. [CrossRef]
25. Project Management Institute. *Construction Extension to the PMBOK Guide*; Project Management Institute, Inc.: Newtown Square, PA, USA, 2016; pp. 29–30.
26. Ahmad, Z.; Thaheem, M.; Maqsoom, A. Building information modeling as a risk transformer: An evolutionary insight into the project uncertainty. *Autom. Constr.* **2018**, *92*, 103–119. [CrossRef]
27. Badran, D.; AlZubaidi, R.; Venkatachalam, S. BIM based risk management for design bid build (DBB) design process in the United Arab Emirates: A conceptual framework. *Int. J. Syst. Assur. Eng. Manag.* **2020**, *11*, 1339–1361. [CrossRef]
28. Xiong, J.; Chen, Y.; Zhou, Z. Resilience analysis for project scheduling with renewable resource constraint and uncertain activity durations. *J. Ind. Manag. Optim.* **2016**, *12*, 719–737.
29. Yeganeh, F.T.; Zegordi, S.H. A multi-objective optimization approach to project scheduling with resiliency criteria under uncertain activity duration. *Ann. Oper. Res.* **2020**, *285*, 161–196. [CrossRef]
30. Milat, M.; Knezic, S.; Sedlar, J. A new surrogate measure for resilient approach to construction scheduling. *Proc. Comp. Sci.* **2021**, *181*, 468–476. [CrossRef]
31. Zhao, M.; Wang, X.; Yu, J.; Xue, L.; Yang, S. A construction schedule robustness measure based on improved prospect theory and the Copula-CRITIC method. *Appl. Sci.* **2020**, *10*, 2013. [CrossRef]
32. Zhao, M.; Wang, X.; Yu, J.; Bi, L.; Xiao, Y.; Zhang, J. Optimization of Construction Duration and Schedule Robustness Based on Hybrid Grey Wolf Optimizer with Sine Cosine Algorithm. *Energies* **2020**, *13*, 2015. [CrossRef]
33. Chapman, C.; Ward, S. *Project Risk Management: Processes, Techniques and Insights*, 2nd ed.; John Wiley & Sons Ltd.: Chichester, UK, 2003; pp. 1–15.
34. Zhang, J.; El-Diraby, T.E. Social semantic approach to support communication in AEC. *Int. J. Proj. Manag.* **2012**, *26*, 90–104. [CrossRef]
35. Elghamrawy, T.; Boukamp, F.; Kim, H.S. Ontology-based, semi-automatic framework for storing and retrieving on-site construction problem information—An RFID-based case study. In Proceedings of the Construction Research Congress 2009: Building a Sustainable Future, Seattle, WA, USA, 5–7 April 2009.

36. Tah, J.H.M.; Carr, V. Knowledge-based approach to construction project risk management. *J. Comput. Civ. Eng.* **2001**, *15*, 170–177. [CrossRef]
37. Tserng, H.P.; Yin, Y.L.S.; Dzeng, R.J.; Wou, B.; Tsai, M.D.; Chen, W.Y. A study of ontology-based risk management framework of construction projects through project life cycle. *Autom. Constr.* **2009**, *18*, 994–1008. [CrossRef]
38. Ding, L.Y.; Zhong, B.T.; Wu, S.; Luo, H.B. Construction risk knowledge management in BIM using ontology and semantic web technology. *Saf. Sci.* **2016**, *87*, 202–213. [CrossRef]
39. El-Diraby, T.A.; Lima, C.; Feis, B. Domain taxonomy for construction concepts: Toward a formal ontology for construction knowledge. *J. Comput. Civ. Eng.* **2005**, *19*, 394–406. [CrossRef]
40. Costa, R.; Lima, C.; Sarraipa, J. Facilitating knowledge sharing and reuse in building and construction domain: An ontology-based approach. *J. Intell. Manuf.* **2016**, *27*, 263–282. [CrossRef]
41. Niu, J.; Issa, R.R.A. Developing taxonomy for the domain ontology of construction contractual semantics: A case study on the AIA A201 document. *Adv. Eng. Inform.* **2015**, *29*, 472–482. [CrossRef]
42. Fidan, G.; Dikmen, I.; Tanyer, M.A.; Birgonul, T.M. Ontology for relating risk and vulnerability to cost overrun in international projects. *J. Comput. Civ. Eng.* **2011**, *25*, 302–315. [CrossRef]
43. Zhang, L.; Issa, R.R.A. Ontology-based partial building information model extraction. *J. Comput. Civ. Eng.* **2013**, *27*, 576–584. [CrossRef]
44. Baudrit, C.; Taillandier, F.; Tran, T.T.P.; Breysse, D. Uncertainty processing and risk monitoring in construction projects using hierarchical probabilistic relational models. *Comp. Aid. Civ. Inf. Eng.* **2019**, *34*, 97–115. [CrossRef]
45. Jiang, S.; Wang, N.; Wu, J. Combining BIM and ontology to facilitate intelligent green building evaluation. *J. Comput. Civ. Eng.* **2018**, *32*. [CrossRef]
46. Xing, X.; Zhong, B.; Luo, H.; Lic, H.; Wu, H. Ontology for safety risk identification in metro construction. *Comp. Ind.* **2019**, *109*, 14–30. [CrossRef]
47. Zhong, B.; Li, H.; Luo, H.; Zhou, J.; Fang, W.; Xing, X. Ontology-based semantic modeling of knowledge in construction: Classification and identification of hazards implied in images. *J. Constr. Eng. Manag.* **2020**, *146*, 04020013. [CrossRef]
48. Zhong, B.; Gan, C.; Luo, H.; Xing, X. Ontology-based framework for building environmental monitoring and compliance checking under BIM environment. *Build. Environ.* **2018**, *141*, 127–142. [CrossRef]
49. Elazouni, A.M.; Metwally, F.G. Finance-Based Scheduling: Tool to Maximize Project Profit Using Improved Genetic Algorithms. *J. Constr. Eng. Manag.* **2005**, *131*, 400–412. [CrossRef]
50. Fathi, H.; Afshar, A. GA-based multi-objective optimization of finance-based construction project scheduling. *KSCE J. Civ. Eng.* **2010**, *14*, 627–638. [CrossRef]
51. El-Abbasy, M.; Elazouni, A.; Zayed, T. Finance-based scheduling multi-objective optimization: Benchmarking of evolutionary algorithms. *Autom. Constr.* **2020**, *120*, 103392. [CrossRef]
52. Damian, D.; Knauss, A.; Zavala, E.; Marco, J.; Franch, X. SACRE: Supporting contextual requirements' adaptation in modern self-adaptive systems in the presence of uncertainty at runtime. *Exp. Syst. Appl.* **2018**, *98*, 166–188.
53. Taillandier, F.; Taillandier, P.; Tepeli, E.; Breysse, D.; Mehdizadeh, R.; Khartabil, F. A multi-agent model to manage risks in construction project (SMACC). *Autom. Constr.* **2015**, *58*, 1–18. [CrossRef]
54. Gündüz, M.; Nielsen, Y.; Özdemir, M. Quantification of Delay Factors Using the Relative Importance Index Method for Construction Projects in Turkey. *J. Manag. Eng.* **2012**, *29*, 133–139. [CrossRef]
55. Ustinovičius, L. Uncertainty analysis in construction project's appraisal phase. In Proceedings of the 9th International Conference Modern Building Materials, Structures and Techniques, Vilnius, Lithuania, 16–18 May 2007.
56. Ali, Z.; Zhu, F.; Hussain, S. Identification and assessment of uncertainty factors that influence the transaction cost in public sector construction projects in Pakistan. *Buildings* **2018**, *8*, 157. [CrossRef]
57. Sacco, G.; Tzitzikas, Y. *Dynamic Taxonomies and Faceted Search*; Springer: Berlin, Germany, 2013.
58. Ranganathan, S.R. *The Colon Classification*; Rutgers University Press: New Brunswick, ME, Canada, 1965.
59. Rafindadi, A.D.; Mikić, M.; Kovačić, I.; Cekić, Z. Global Perception of Sustainable Construction Project Risks. *Procedia–Soc. Behav. Sci.* **2014**, *119*, 456–465. [CrossRef]
60. Warmink, J.J.; Janssen, J.A.E.B.; Booij, M.J.; Krol, M.S. Identification and classification of uncertainties in the application of environmental models. *Environ. Model. Soft.* **2010**, *25*, 1518–1527. [CrossRef]
61. Hazir, O.; Haouari, M.; Erel, E. Robust scheduling and robustness measures for the discrete time/cost trade-off problem. *Eur. J. Oper. Res.* **2010**, *207*, 633–643. [CrossRef]
62. Zahid, A.; Agha, M.H.; Schmidt, T. Investigation of surrogate measures of robustness for project scheduling problems. *Comput. Ind. Eng.* **2019**, *129*, 220–227. [CrossRef]
63. Au, T.; Hendrickson, C. Profit Measures for Construction Projects. *J. Constr. Eng. Manag.* **1986**, *112*, 273–286. [CrossRef]
64. Ahuja, H. *Construction Performance Control by Networks*; Wiley: New York, NY, USA, 1976.
65. Al-Shihabi, S.; AlDurgam, M. A max-min ant system for the finance-based scheduling problem. *Comput. Ind. Eng.* **2017**, *110*, 264–276. [CrossRef]
66. Hendrickson, C. Project Management for Construction: Fundamental Concepts for Owners, Engineers, Architects and Builders. Available online: https://www.cmu.edu/cee/projects/PMbook/ (accessed on 18 May 2021).
67. Demeulemeester, E.; Herroelen, W. *Project Scheduling*; Kluwer Academic Publishers: New York, NY, USA, 2002; p. 67.

68. Hapke, M.; Jaszkiewicz, A.; Słowiński, R. Interactive analysis of multiple-criteria project scheduling problems. *Eur. J. Oper. Res.* **1998**, *107*, 315–324. [CrossRef]
69. Chaturvedi, S.; Rajasekar, E.; Natarajan, S. Multi-objective Building Design Optimization under Operational Uncertainties Using the NSGA II Algorithm. *Buildings* **2020**, *10*, 88. [CrossRef]
70. Al-Sadek, O.; Carmichael, D. On simulation in planning networks. *Civ. Eng. Syst.* **1992**, *9*, 59–68. [CrossRef]

Article

Automated Extraction and Time-Cost Prediction of Contractual Reporting Requirements in Construction Using Natural Language Processing and Simulation

Parinaz Jafari, Malak Al Hattab, Emad Mohamed and Simaan AbouRizk *

5-080 NREF, Department of Civil and Environmental Engineering, University of Alberta, Edmonton, AB T6G 2W2, Canada; parinaz@ualberta.ca (P.J.); elhattab@ualberta.ca (M.A.H.); ehmohame@ualberta.ca (E.M.)
* Correspondence: abourizk@ualberta.ca

Featured Application: The approach rapidly extracts reporting requirements from construction contracts and predicts overhead costs and durations associated with report preparation. Application of the approach is anticipated to provide the insights necessary to enhance contract negotiations, reporting workflow processes, and submittal procedures between clients and contractors.

Abstract: Due to a lack of suitable methods, extraction of reporting requirements from lengthy construction contracts is often completed manually. Because of this, the time and costs associated with completing reporting requirements are often informally approximated, resulting in underestimations. Without a clear understanding of requirements, contractors are prevented from implementing improvements to reporting workflows prior to project execution. This study developed an automated reporting requirement identification and time–cost prediction framework to overcome this challenge. Reporting requirements are extracted using Natural Language Processing (NLP) and Machine Learning (ML), and stochastic simulations are used to predict overhead costs and durations associated with report preparation. Functionality and validity of the framework were demonstrated using real contracts, and an accuracy of over 95% was observed. This framework provides a tool to rapidly and efficiently retrieve requirements and quantify the time and costs associated with reporting, in turn providing necessary insights to streamline reporting workflows.

Keywords: construction reports; construction contracts; natural language processing; machine learning; simulation modeling

Citation: Jafari, P.; Al Hattab, M.; Mohamed, E.; AbouRizk, S. Automated Extraction and Time-Cost Prediction of Contractual Reporting Requirements in Construction Using Natural Language Processing and Simulation. *Appl. Sci.* **2021**, *11*, 6188. https://doi.org/10.3390/app11136188

Academic Editor: Mariusz Szóstak

Received: 17 June 2021
Accepted: 1 July 2021
Published: 3 July 2021

Publisher's Note: MDPI stays neutral with regard to jurisdictional claims in published maps and institutional affiliations.

Copyright: © 2021 by the authors. Licensee MDPI, Basel, Switzerland. This article is an open access article distributed under the terms and conditions of the Creative Commons Attribution (CC BY) license (https://creativecommons.org/licenses/by/4.0/).

1. Introduction

Work within the construction industry is allocated through construction contracts [1], which include information such as instructions, definitions, supporting statements, and contractual requirements that detail the standards and project specifications of the client [2]. A core component of construction contracts is reporting and information requirements, which require contractors to periodically submit various reports detailing different aspects of project progress to the client [3]. As construction projects and contracts are becoming increasingly complex, clients are demanding that contractors provide more information and reports on different project aspects [4,5]. Reporting has quickly become a laborious procedure, with construction personnel spending as much as 40% of their time gathering field data, organizing and analyzing data, preparing reports, and verifying report accuracy [6].

Although a large administrative burden for contractors [7], the time and resources needed to complete reporting requirements—as well as the precise reporting requirements themselves—are often unknown and unaccounted for during the preliminary planning stages of a project. An integral component of project success, preliminary planning involves, among other activities, the selection of the project management team and the creation of the project documentation system. Consideration of specific reporting requirements during the preliminary planning stage ensures that (1) an adequate number of personnel is available to complete reporting requirements on time and within budget, (2) efficient project reporting systems are implemented, and (3) redundant and/or overlapping reporting requirements are addressed prior to the execution phase of a project. Contract documentation, however, remains an immature area of practice, with the identification of reporting requirements involving the manual reading, interpretation, and analysis of hundreds of unstructured textual contract pages to differentiate between statements related to requirements and other unimportant texts, such as instructions and definitions. Due to the time and effort involved, the manual extraction of reporting requirements is often not completed during the preliminary planning stages of a project, with project managers informally approximating reporting costs and resource requirements. Indeed, it has been reported that office-related processes, such as project reporting, continue to suffer from low reliability, where planned durations are often underestimated [8].

The poor estimation of project reporting costs and resource requirements during the preliminary planning stages of construction can result in a number of challenges for contractors [4,9–11]. For example, an inadequate number of available project management personnel may result in project reports that are submitted late or with errors. In the case of certain types of contracts (e.g., cost-plus), reporting costs that exceed the preliminary estimate can result in disputes between the contractor and client. Furthermore, a lack of understanding of the reporting requirements in the preliminary stages of a project may prevent contractors from increasing reporting efficiency during the construction phase through the consolidation of redundant reporting requirements or by optimizing the composition of the project management team. As such, the ability to quickly, accurately, and automatically extract reporting requirements and predict associated costs is expected to have a notable impact on project performance [12]. Although text-mining techniques, such as information extraction, text classification, and other predictive analytics, have been used by researchers to develop requirement extraction models [4,5,13], existing models are not fully automated, do not provide high requirement extraction accuracy, and lack a cost-and-time prediction component. Methods capable of automatically extracting contractual reporting requirements and predicting the time and costs associated with report preparation, therefore, remain relatively unexplored.

To address this challenge, this study has developed a framework capable of (1) automating the identification and extraction of reporting requirements and (2) predicting and analyzing the overhead costs and durations associated with report preparation. The framework employs Natural Language Processing (NLP) and Machine Learning (ML) techniques to automate the extraction of reporting requirements, and uses stochastic simulation to predict the durations and costs associated with report preparation using historical project data. Real contractual documents from an actual case study were used to (1) develop and refine the reporting requirement extraction module and (2) demonstrate the functionality and validity of the complete framework. This framework provides practitioners and researchers with an automated tool to more efficiently identify reporting requirements and quantify the time and costs associated with report preparation. Practical application of this approach is anticipated to provide decision makers with the insights necessary to enhance contract negotiations, reporting workflow processes, and submittal procedures between clients and contractors, in turn increasing value for all project stakeholders.

2. Research Background

2.1. Construction Reporting

Many problems in the construction industry involve communication and reporting procedures, with ineffective reporting systems leading to poor project management [11,14,15]. Construction reports, therefore, are often required by clients as a means of monitoring project progress, estimating production rates, and resolving disputes and claims [6]. Project reporting involves the collection and structuring of large volumes of site data from numerous field management activities by many site personnel on a frequent—even daily—basis [16,17]. Given the amount of preparation work required together with the frequency of submittals, reporting has become a time- and effort-intensive procedure that can result in notable increases in overhead costs of the project.

Various construction field management tools have been developed to establish project reporting systems tailored to the needs of contractors, while ensuring the reporting requirements of projects are met [2,6,16]. For example, Russell [18] developed a daily construction project management system that rapidly reports and shares site information and project progress status between project participants. Similarly, Shiau and Wang [19] developed a construction management information system consisting of daily reports as well as cost management and design-change management modules to compile daily site management information. El-Omari and Moselhi [3] proposed a model to facilitate automated data acquisition from construction sites by deploying an information technology platform. Their goal was to integrate automated data acquisition technologies to collect required data for progress measurement purposes to support efficient time–cost tracking and control of construction projects [3]. Following the same line of work, Lee et al. [16] proposed an approach to automatically generate daily reports from text messages exchanged through a commonly used text messaging system.

It is important to note that the aforementioned models were primarily focused on effective data acquisition, information flow, and communication to facilitate the monitoring of site work, incurred costs, and potential challenges [3,20]. Although these studies have addressed the downstream aspect of reporting, they have been developed with the assumption that reporting requirements are already defined and known in advance. In practice, however, reporting requirements for complex types of construction, such as oil and gas or infrastructure projects, often differ between projects and from contract-to-contract, making the time, resources, and costs associated with report preparation difficult to approximate. Methods for automating the extraction of contractual reporting requirements or the estimation of time and cost implications associated with reporting, however, remain relatively unexplored.

Importantly, the lack of research literature in the area of contract documentation is not indicative of the practical importance of this issue. Discussions with experienced professionals at a construction company in Alberta, Canada, revealed that contractors are very interested in techniques that can support the extraction, management, and time–cost prediction of reporting requirements. Once considered an obligatory and static activity, contractors are beginning to explore methods capable of enhancing the planning, and therefore efficiency and cost, of project reporting—particularly for complex types of construction where contracts are often specific to each individual project.

2.2. NLP Applications in Construction

To avoid unnecessary changes, rework, and potential claims, contractors must thoroughly analyze construction contracts and specifications to ensure that client requirements are identified, managed, and fulfilled [13]. The traditional approach of identifying reporting requirements involves the manual reading, interpretation, and analysis of hundreds of unstructured textual contract pages to differentiate between statements related to requirements and other extraneous text (e.g., instructions and definitions). Techniques capable of accelerating the reporting requirement extraction process, therefore, represent a key prerequisite for the development of an automated time–cost prediction model.

Natural Language Processing (NLP) is an area of Artificial Intelligence (AI) that focuses on the development of techniques to analyze, process, and extract information from natural human language. Applications include machine translation, speech recognition, and automated content analysis [21]. In construction, a large number of project documents are generated in text format [22]. The use of NLP techniques to organize and improve access to information contained in these types of documents is becoming ever more essential for effective construction management [7], with NLP techniques being increasingly applied in construction research [22–24]. Caldas et al. [7], for instance, employed NLP techniques to automate the classification of construction documents to improve organization of and access to information within interorganizational systems. Al Qady and Kandil [25] also developed an automated classification system of construction documents according to their semantic relationships. Fan and Li [23] used NLP for the automatic retrieval of similar cases from an electronic case repository of construction accidents.

Text classification, a subfield of NLP, is an automated process for classifying text into categories [26,27]. Text classification is divided into rule-based and Machine Learning (ML)-based methods [26]. Rule-based text classification categorizes text using a manually defined pattern to create rules; in contrast, under ML-based text classification, a machine learns how to classify text on its own using data. A variety of text classification models have been developed for the construction domain. For example, Salama and El-Gohary [28] developed a hybrid semantic, multilabel ML-based text classification algorithm for classifying clauses and subclauses of general conditions to support automated compliance checking. Lee et al. [5] proposed a rule-based model to automatically detect risk-related sentences of contracts to support contract risk management for construction contractors. Zhong et al. [29] combined NLP and convolutional neural networks to develop a classification model capable of automatically classifying accident narratives to support safety management on site. Zhou and El-Gohary [30] proposed an ontology-based, multilabel text classification approach for classifying environmental regulatory clauses to support automated compliance checking in construction, and Hassan and Le [4] proposed a domain-specific classification model to identify client requirements from construction contracts. It is important to note that the implementation of existing text classifiers to different applications remains limited, as text classification models, text features, and performance requirements vary greatly across domains and applications [28]. Designed for a specific domain, the aforementioned text classifiers and are not well-suited for applications that require alternate classification structures.

2.3. Research Gap

Despite these advancements, research focused on enhancing the management of contractual reporting requirements remains relatively unexplored and fragmented. Most of the studies in the area of construction reporting have focused on the development of systems that improve data acquisition, information flow, and communication. While other studies, such as those mentioned previously, have deployed NLP and AI to automate information retrieval and extraction from construction contracts, the text approaches used to develop these requirement extraction models are limited by a lack of full automation, low extraction accuracy, and the absence of a cost–time prediction component [4,5,13]. Indeed, a review of construction literature could not identify any established study capable of automatically extracting reporting requirement statements from hundreds of pages of contractual and project specification documents.

3. Framework Overview

To address the gap existing in literature, this study developed a novel, NLP-based framework for the automated extraction and time–cost prediction of contractual reporting requirements in construction. The framework consists of two modules, namely the (1) automatic extraction of reporting requirements module and (2) prediction of reporting time and cost module that are linked as illustrated in Figure 1.

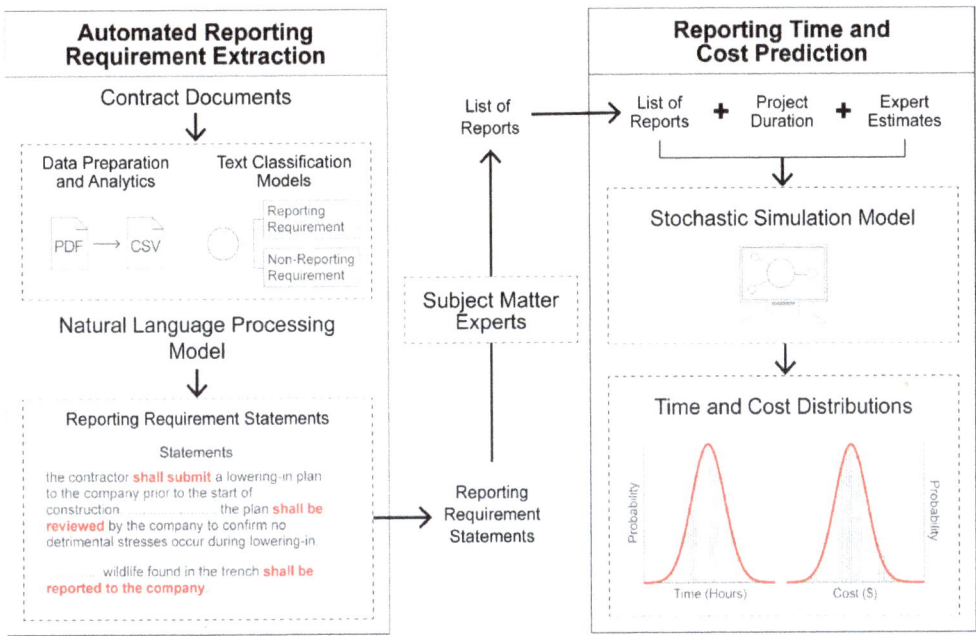

Figure 1. Proposed Framework.

The first module, hereafter referred to as the extraction module, is responsible for identifying statements in contract or project specification documents that prescribe reporting requirements. Contract documents and project specifications are input into the NLP model. Relevant text is extracted from the documents and is transformed into a format that is compatible with the text classification models.

Text classification algorithms are then used to classify contractual and project specification statements as either a (1) reporting requirement or (2) non-reporting statement. Both rule-based and ML-based text classification methods can be used to classify statements; application of either method will depend on the specific requirements of the user. While rule-based text classification is more time-consuming than ML-based classification due to the involvement of manual rule development, rule-based classification commonly results in higher precision and recall [26]. ML, on the other hand, makes it possible to automatically classify text, provided that sufficient learning opportunities are available [27].

The second module, hereafter referred to as the prediction module, is responsible for generating relevant time–cost predictions. Reporting requirements output from the extraction module are used by practitioners to prepare a list of required reports and their associated submittal frequencies that are then input into the prediction module. Estimates of the time required to prepare a specific report are used as inputs. Then, a Monte Carlo simulation model, which uses random sampling to obtain numerical results or a probability distribution [31], is used to predict the cost associated with report preparation based on project duration and historical data.

Although distinct, the practical functionality of these two modules increases considerably when used together. Manual extraction is very tedious and time-consuming, and contractors do not have enough time during the bidding stage to identify the contract requirements and plan accordingly. Because of the ability of the extraction module to quickly extract reporting requirements, the prediction module can now be applied in a more impactful stage of the project life-cycle (i.e., pre-construction bidding and planning stages). Specifically, the outputs of the prediction module can be used to (1) ensure that sufficient cost and time contingencies for report preparation are included in bid estimates, (2) engage in negotiations to reduce redundant reporting requirements before finalizing contracts, and (3) improve resource allocation.

4. Extraction Module

Development of the extraction module was completed in three main steps, namely (1) data preparation and pre-processing, (2) development and training of text classification models, and (3) evaluation of model performance. Ten contractual and specification documents of an oil-and-gas project were supplied by a private Canadian construction contracting firm and were used to develop the extraction module. Python [32], an open-source programming language, was used to automate module development steps.

4.1. Data Preparation and Preprocessing

Data preparation and pre-processing transformed raw data into a labeled dataset that was used to develop and train the text classification models. This involved the (1) extraction of textual data from documents, (2) manual assignment of labels to extracted statements, and (3) cleaning of labeled data. An overview of the data preparation process is illustrated in Figure 2.

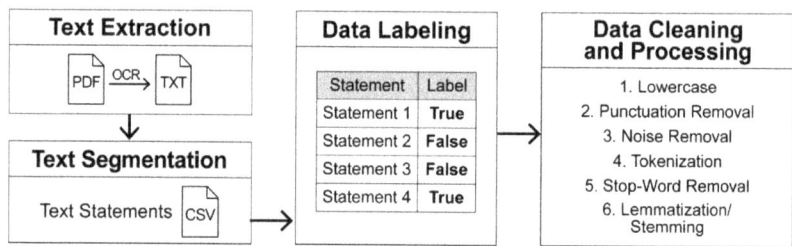

Figure 2. Data preparation and pre-processing during extraction module development.

Documents, which were provided in an imaged portable document format (i.e., .pdf), were first converted into a standard, processable text format (i.e., .txt) using Optical Character Recognition (OCR). Next, text documents were automatically segmented into individual text statements using document formatting. A total of 8943 text segments were extracted from 10 contractual and project specification documents. Since pre-labeled textual construction datasets are not as readily available as other domain applications, such as movie reviews or twitter messages [33], extracted statements were labeled manually. A label of "true" was manually applied to statements prescribing a reporting requirement, while a label of "false" was applied to non-reporting statements. A sample of the labeled dataset is presented in Figure 3.

Data were then structured into a single, comma-separated values (.csv) file, with text statements stored in the first column and the document name and associated page stored in subsequent columns. The last column contained the pair label (i.e., type: true/false) of each statement. The labeled dataset was validated by domain experts to ensure accuracy of the manual labels. The final dataset used in the study included 340 reporting requirements and 8603 non-reporting statements.

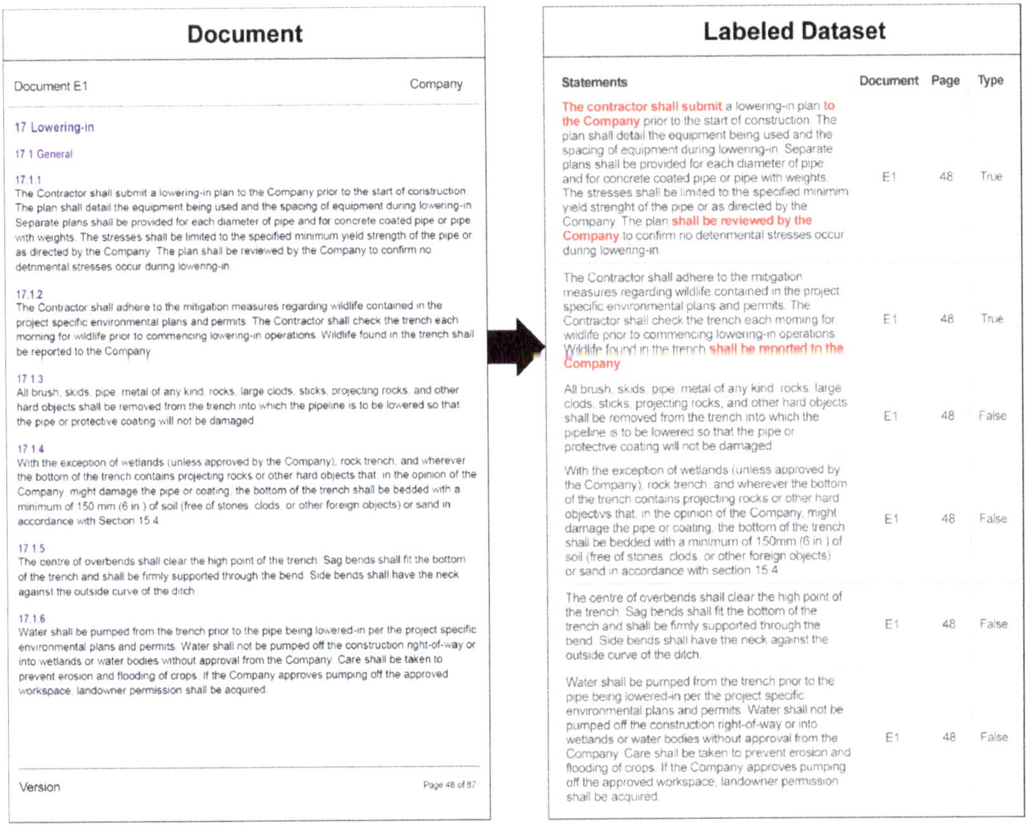

Figure 3. Sample contract document (left) and labeled dataset (right).

The final dataset was then cleaned to reduce data noise and enhance the quality of data used to train the text classification models. First, text was converted to a lowercase form to ensure identical words were treated as like terms (e.g., "Submit" and "SUBMIT"). Then, punctuation was removed from the text. In addition to text in the main body of the document, OCR extracted text from footers, page numbers, headers, annotations, and footnotes. This text acted as data noise for the text classification algorithms and was, therefore, removed. Then, tokenization was used to divide text statements into words (i.e., tokens) and to convert text into a feature vector form to prepare text for feature engineering and further analysis [34]. Stop-words, which are frequent words such as conjunctions, prepositions, and pronouns (e.g., the, for, so, is, of, and a) that do not carry relevant information for text classification, were removed using a standard English stop-word list [22]. Lemmatization and stemming were applied to reduce the number of features through word grouping. While word stemming groups words by removing prefixes and/or suffixes to conflate words to their original root [35], lemmatization groups words subsequent to a full morphological analysis. Once data cleaning was completed, the dataset was randomly split into training (80%) and testing (20%) sets, which were used to train the text classifiers and to evaluate classifier performance, respectively.

4.2. Rule-Based Classification

In the rule-based classification approach, a set of hand-coded "IF-THEN" rules that define the label assignment criteria for a certain category were prepared [5]. These rules were iteratively constructed and refined to improve accuracy of the classifier. The process used to develop the rule-based classification model is depicted in Figure 4.

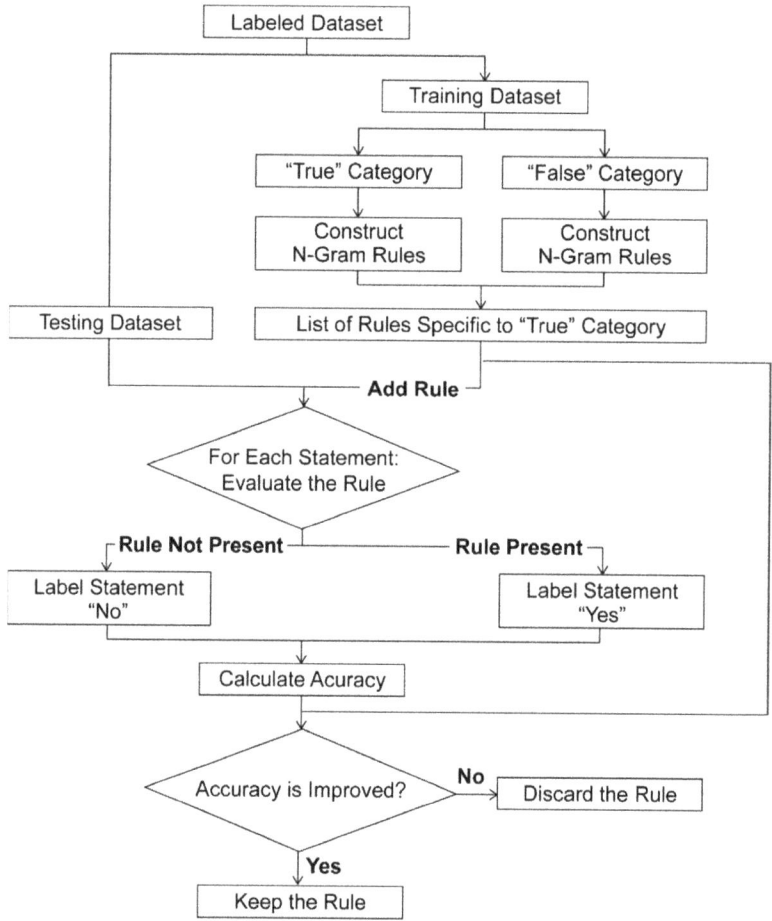

Figure 4. Rule-based classification.

Accurate filtering of reporting requirements from contractual documents requires the development of robust and comprehensive rules. Although keywords, such as "report" and "submit," may be helpful in identifying certain reporting requirements, construction contracts also contain key phrases, such as "shall be reported/submitted," which indicate that a report or deliverable must be provided contractually. It is important to note that keywords alone cannot distinguish a reporting requirement from any other contractual requirement. As such, critical phrases were extracted using text analytics. Using the training dataset, the rule-based model was used to extract n-grams (i.e., a sequence of co-occurring words as a single token) from the textual statements. The most common n-grams (i.e., phrases) appearing in the reporting requirement statements are summarized in Figure 5.

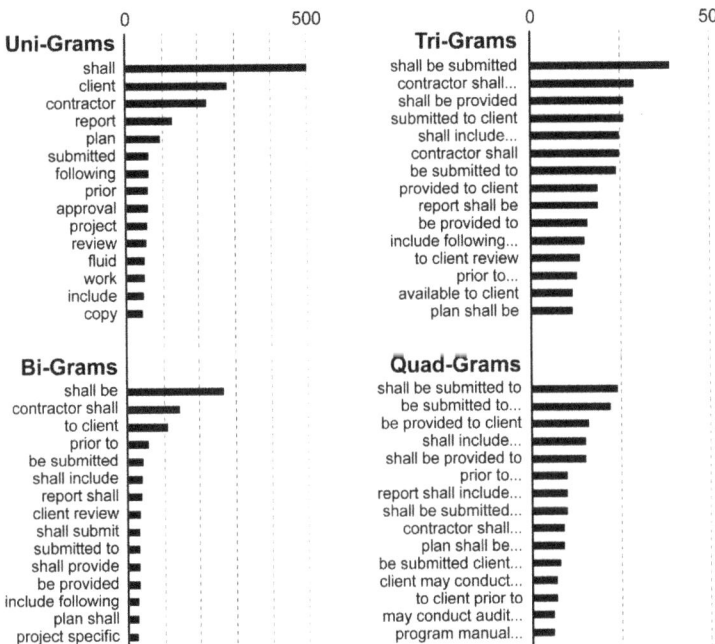

Figure 5. Most common n-grams in reporting requirement statements.

To avoid errors, a list of n-grams specific only to the "true" category was prepared. N-grams common to both the "true" and "false" categories were removed. The final rules consisted of four different sets of n-grams capable of discerning between "true" and "false" statements, namely uni-grams, bi-grams, tri-grams, and quad-grams representing single-, two-, three-, or four-word phrases, respectively. These four sets of n-grams were developed to evaluate the effect of each n-gram set on the performance of the rule-based text classifier. N-grams were flagged as rules, with each rule consisting of a pattern and a predicted category. N-grams in each N-gram set were closely monitored, and rules for each statement were evaluated. Each rule was added, one-by-one, to the text classification model. Predicted labels were then compared to actual labels, and classifier performance was calculated. If the performance (i.e., accuracy) increased with the addition of the rule, the rule was retained. If not, the rule was removed. This was repeated for each rule of each n-gram until a final list was created.

4.2.1. Performance of Rule-Based Classification Models

Performance of the text classification models were evaluated using accuracy, precision, recall, and F1-score, which were calculated using Equation (1) through Equation (4), respectively.

$$\text{Accuracy} = \frac{TP + TN}{TP + FP + TN + FN} \quad (1)$$

$$\text{Recall} = \frac{TP}{TP + FN} \quad (2)$$

$$\text{Precision} = \frac{TP}{TP + FP} \quad (3)$$

$$\text{F1} - \text{Score} = \frac{2 \times Precision \times Recall}{Precision + Recall} \quad (4)$$

where *TP* are true positives (i.e., statements correctly labeled "true"), *FP* are false positives (i.e., incorrectly labeled "true"), *TN* are true negatives (i.e., correctly labeled "false"), and *FN* are false negatives (i.e., incorrectly labeled "false").

Accuracy (Equation (1)) is defined as the percentage of correctly classified statements over the total number of statements in the testing set. Recall (Equation (2)), is defined as the percentage of true positives identified by the model. Precision (Equation (3)) is defined as the percentage of positives that are correctly labeled [36]. Finally, the F1-score (Equation (4)), combines precision and recall to provide an overall assessment of model effectiveness.

Recall is considered to be the most critical performance metric in the context of requirement extraction, where the extraction of all reporting requirements is the primary objective. For instance, a model may have low performance accuracy because it results in a larger number of false positives (i.e., non-reporting statements labeled as requirements). However, the model may have high recall results (i.e., 100%) if it is able to correctly label all reporting requirements as "true".

Specific rules for text processing were developed and applied to improve results of the rule-based classification model. Initial tests were conducted on different n-gram sets. The testing approach was conducted in an iterative manner, and results from 24 different combinations of n-grams and text pre-processing techniques (e.g., stop-word removal, lemmatization, etc.) were compared. The four sets of n-grams extracted from the training set are summarized in Table 1. The total number of rules generated increased with the number of n-grams (Table 1). Using the process summarized in Figure 4, the number of rules maintained for each n-gram was considerably reduced for all n-gram sets (Table 1). For example, of the 2363 rules generated for the bi-grams set, only 38 rules were retained. Accuracy was increased from 97%, using uni-grams, to 99%, using bi-grams, yet was decreased to 98% and 97% using tri- and quad-grams, respectively. Although differences between n-gram sets were minimal, optimal accuracy was achieved using the retained bi-gram rules. The impact of adding the first 10 and the last bi-gram rule on model accuracy is visualized in Figure 6.

Table 1. Number of generated and retained bi-gram rules and associated accuracy.

N-Gram Set	Number of Generated Rules	Number of Retained Rules	Maximum Accuracy (%)
Uni-Grams	118	8	97
Bi-Grams	2363	38	99
Tri-Grams	3762	38	98
Quad-Grams	4268	34	97

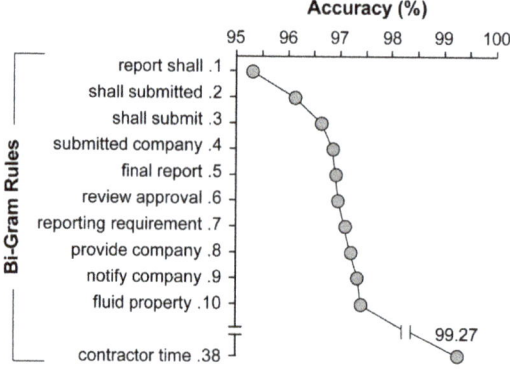

Figure 6. Impact of adding bi-gram rules on the accuracy of the rule-based text classifier.

The first bi-gram rule, "report shall", resulted in an accuracy of 95.3%. The third bi-gram, "shall submit", further increased the accuracy of the classifier to 96.7%. The 37 rules added after "shall submit" collectively increased performance by 3.97% to 99.3%.

The impact of stop-word removal, lemmatization, and stemming on the performance of text classification models is known to differ based on the textual context and application. As such, the impact of stop-word removal and lemmatization/stemming were evaluated. Various experimental scenarios examining the impact of n-gram sets, stop-word removal, and lemmatization on model performance are summarized in Table 2.

Table 2. Effect of n-gram sets and data pre-processing on performance of rule-based classification for two experimental scenarios.

N-Gram Set	Class Label	Scenario 1: without Lemmatization Stop-Words Retained			Scenario 2: with Lemmatization Stop-Words Removed		
		Precision (%)	Recall (%)	F1-Score (%)	Precision (%)	Recall (%)	F1-Score (%)
Uni	True	91	56	69	93	55	69
	False	98	100	99	98	100	99
Bi	True	100	86	92	96	88	92
	False	99	100	100	99	100	100
Tri	True	99	86	92	98	71	83
	False	99	100	100	98	100	99
Quad	True	100	74	85	100	49	49
	False	99	100	99	97	100	99

Uni-grams had the lowest performance for both experimental scenarios (Table 2), and bi-grams demonstrated the highest performance in all three metrics in the base condition (Scenario 1).

When stop-words were removed and lemmatization was applied, bi-grams had the highest recall and F1-score, with precision only differing marginally from other n-grams. Interestingly, lemmatization and stop-word removal resulted in a 2% increase in the recall of the bi-gram classifier, while the recall of the other n-gram sets decreased (Table 2). Notably, the F1-score of tri-grams (without lemmatization and with stop-words retained) was equal to the F1-score of bi-grams (with lemmatization and with stop-words removed). This result is expected as, in some cases, removing stop-words from tri-grams transforms them into bi-grams. For example, when the stop-word "be" is removed from the tri-gram "shall be submitted", the bi-gram "shall submitted" results. Given the importance of the recall measurement when extracting reporting requirements, and based on the findings that bi-grams resulted in the highest model accuracy (Table 1) and recall (Table 2), bi-grams are selected as the optimal classifier for rule-based text classification.

4.3. Machine Learning-Based Classification

In contrast to rule-based classification, the alternate classification approach used in the present study was supervised ML models, with the learning process driven by previous knowledge of the data [28]. In ML algorithms, a general inductive process automatically builds a classification model for each class by observing the characteristics of a set of manually classified statements. The ML-based text classification approach is summarized in Figure 7.

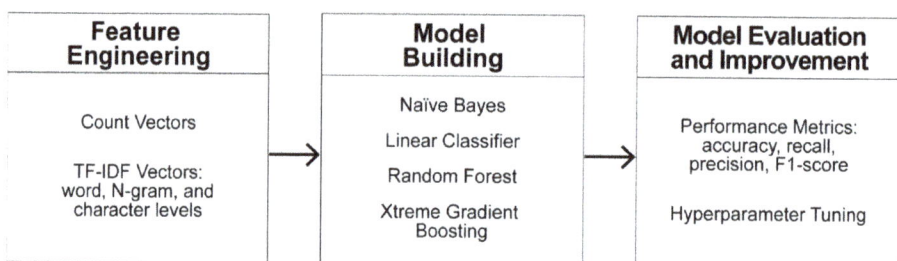

Figure 7. Machine learning-based text classification method for extraction module development.

To ensure compatibility with computer processors, words were first converted into a numeric format using feature engineering. Here, raw text data were transformed into feature vectors, and new features were created using the dataset. Different methods were used to create relevant dataset features prior to input into the text classification algorithm [37].

Two approaches for constructing representation vectors, namely count vectors and term frequency-inverse document frequency (TF-IDF) vectors, were implemented. Count vectors are a matrix representation of the dataset, where every row represents a statement, every column represents a word, and every cell represents the frequency count of a particular word (i.e., either zero or a real number) in a particular statement [38]. Words that appear in many textual statements are considered less meaningful and, therefore, each vector component (i.e., a word) can be weighed based on the number of statements in which the word appears. Another approach for constructing representation vectors is TF-IDF, which is a technique designed to identify important terms in a dataset by weighing a term's frequency (TF) together with its inverse document frequency (IDF), which weighs down high-frequency domain-specific terms while scaling up rare terms [38]. In TF-IDF vectors, terms can be extended to include characters and n-gram-level models, such as uni-gram (i.e., words), bi-grams (i.e., pairs of words), as well as tri and quad-grams (i.e., phrases). The TF-IDF of terms are calculated using Equation (5) [38],

$$\text{TF} - \text{IDF} = \frac{tf_i}{T} \times \left(1 + \log\left(\frac{N}{N_i}\right)\right) \quad (5)$$

where tf_i is the frequency of term i in the statement, T is the total number of words in the statement, N is the total number of statements, and N_i is the number of statements containing term i.

For the proposed method to be feasible in practice, retraining and prediction [39] must be completed within a relatively short period of time. As such, models that required longer than an hour to be fine-tuned and retrained (e.g., deep learning algorithms) were excluded from this study to ensure applicability of this research. Based on this criterion, four popular supervised ML algorithms, which have been shown to perform differently depending on the application and domain [4,28], were implemented to build the ML-based text classification model. Characteristics of the ML algorithms are summarized as follows:

Naïve Bayes (NB) is a simple algorithm, based on the Naïve Bayes Theorem, that is used for solving practical domain problems including text classification [40]. Because it assumes that every feature is conditionally independent of other features for a given class label, the computational cost of applying the NB algorithm is comparatively low.

Logistic Regression (LR) is a linear statistical ML algorithm that correlates discrete categorical dependent features with a set of target variables [40]. It is a complex form of linear regression that can predict data probability for predefined categories.

Random Forest (RF) is a supervised ML method based on ensemble learning that involves the construction of multiple decision trees during training. Outputs are classes that are averaged or voted the most by individual trees [41]. Decision Tree algorithms, such as the RF classifier, are often used to combat imbalanced classes, such as the scenario described here, where the number of non-reporting statements considerably exceeds the number of reporting requirements.

Extreme Gradient Boosting (XGBOOST) is a scalable ML system based on gradient boosting [42]. It generates a strong classifier by iteratively updating parameters of the former classifier to decrease the gradient of loss function. XGBOOST has superior performance in supervised ML, with high accuracy and low risk of overfitting.

4.3.1. Performance of ML-Based Classification Models

The final step in the development of the extraction module was the evaluation of the various ML-based text classification models. The performance of ML-based text classification algorithms is highly dependent on feature selection (i.e., domain dependent), type of ML techniques, and training datasets [28]. Therefore, all possible combinations of text pre-processing, feature engineering, and ML algorithms—resulting in 160 exhaustive combinations—were evaluated. Various conditions of text pre-processing approaches, such as stop-word retention or removal with or without the implementation of stemming and/or lemmatization, were tested to evaluate the effect on model performance. While the methodology was conducted in an iterative manner to allow for the detailed comparison of results, only a subset of the results is presented to maintain brevity.

The effect of using lemmatization or stemming is illustrated in Figure 8. Stemming improved classification accuracy of LR and XGBOOST algorithms, while lemmatization marginally improved the accuracy of NB and RF algorithms. Notably, the difference in classification performance accuracy between the two text pre-processing techniques was negligible, ranging from 0.03% to 0.4% (Figure 8). The XGBOOST algorithm with stemming applied resulted in the highest accuracy (98.4%).

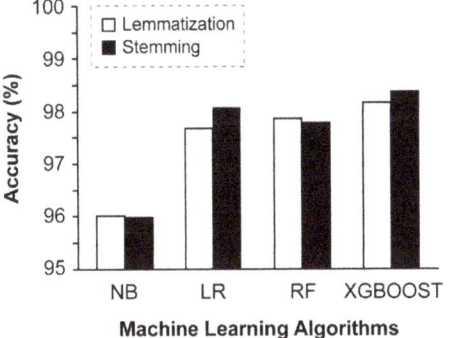

Figure 8. Impact of lemmatization and stemming on performance of Naïve Bayes (NB), Logistic Regression (LR), Random Forest (RF), and Extreme Gradient Boosting (XGBOOST)-based machine learning algorithms.

The recall, precision, and F1-score of the different ML algorithms were evaluated (Figure 9). Given that XGBOOST was found to have the highest accuracy with stemming, stemming was applied to all ML techniques prior to performance metric evaluation. The ML algorithms exhibited relatively similar recall values of over 98% for non-reporting statements (i.e., "false").

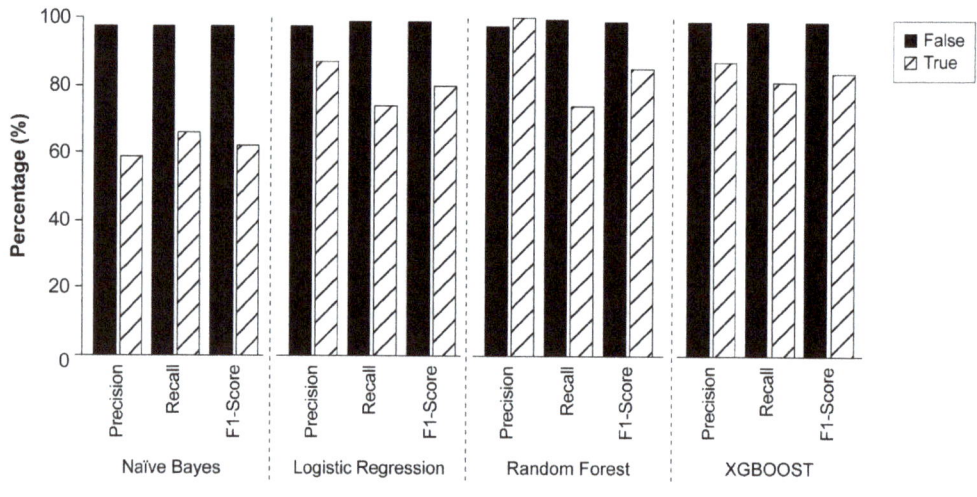

Figure 9. Performance measures of machine learning algorithms using uni-gram text classifications and word stemming.

In contrast, recall values for reporting requirements (i.e., "true") varied considerably amongst the various ML algorithms. The NB algorithm resulted in the lowest "true" recall value (66%), whereas LR, RF, and XGBOOST algorithms resulted in "true" recall values of 74%, 74%, and 81%, respectively. The lower recall values and increased variability observed for the "true" class is likely due to the imbalanced distribution of statements in the contractual documents used (340 "true" requirements versus 8603 "false" statements). In terms of precision, RF, LR, and XGBOOST resulted in "true" precision results of 100%, 87%, and 87%, respectively. The XGBOOST algorithm exhibited the highest recall (Figure 9) and accuracy (Figure 8) results for both the "true" and "false" classes and the second-highest F1-score and precision measurements.

Variations in recall when using different n-gram sets for both the "true" and "false" class were evaluated and illustrated in Figure 10. As discussed previously, higher recall values were observed for non-reporting statements (i.e., "false" class). Uni-grams resulted in higher "true" recall values compared to bi-grams for all classification algorithms except the NB algorithm. The combined use of uni-grams and bi-grams with the LR and XGBOOST classification models yielded the highest "true" recall performance, with values of 77% and 87%, respectively.

It is important to note, however, that a number of factors, such as dataset size, can influence the performance of ML models. The hyperparameters of the ML models, therefore, must be tuned to specific data [43]. Here, hyperparameters were objectively changed, one-by-one, to mitigate overfitting and improve classifier performance. After identifying optimal hyperparameters (i.e., a single set of well-performing hyperparameters), the model was retrained with the full training dataset, and the testing dataset was re-evaluated.

The two models that were examined were the RF and XGBOOST algorithms, as they have many parameters, and the impact of their hyperparameters is significant. Table 3 summarizes the values of the four performance metrics of these two classifiers before and after fine-tuning of their hyperparameters. Fine-tuning parameters improved recall, precision, and F1-score measurements for both classifiers under both classes. The largest improvement for both the RF and XGBOOST models was observed for the "true" class. XGBOOST achieved the highest recall and F1-scores after fine-tuning for both the "true" and "false" classes at 89% and 100% for recall and 92% and 99% for F1-score, respectively. The results demonstrated that fine-tuning hyperparameters to optimize parameter value by analyzing their impact, in terms of over- and underfitting, results in increased model robustness.

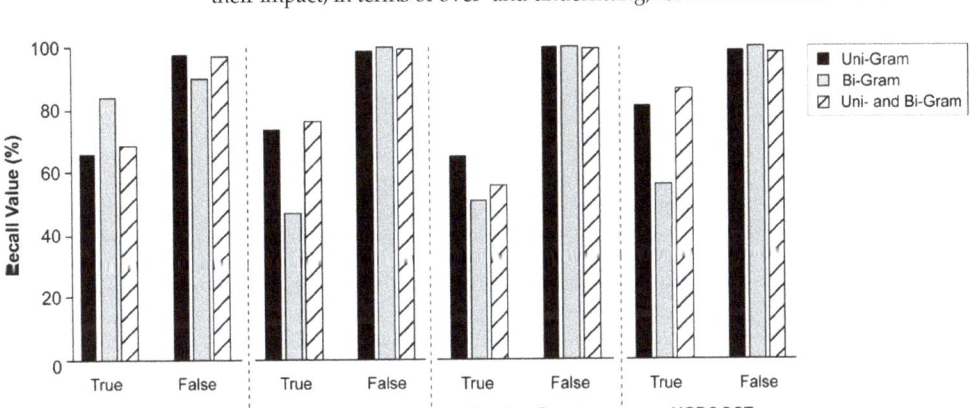

Figure 10. Impact of n-gram sets on recall performance of machine learning models.

Table 3. Effect of fine-tuning hyperparameters on performance metrics of Random Forest (RF) and Extreme Gradient Boosting (XGBOOST)-based algorithms.

Metrics	Class Label	RF		XGBOOST	
		Before	After	Before	After
Accuracy	-	97.8	98	98.4	98
Precision	True	99	100	87	96
	False	98	98	99	99
Recall	True	59	74	81	89
	False	100	100	99	100
F1-Score	True	73	85	84	92
	False	99	99	99	99

Altogether, ML-based performance measurements revealed that the XGBOOST model outperformed the other ML algorithms in terms of accuracy (Figure 8) and recall (Figure 10) performance. Accordingly, the XGBOOST model is selected as the optimal classifier for ML-based text classification.

4.4. Comparison of Classification Models

Performance results achieved by the best-performing rule-based and ML-based classifiers are summarized in Table 4. Under the rule-based classifier, application of the bi-gram rule set with lemmatization and stop-word removal resulted in accuracy and "true" recall values of 99% and 88%, respectively (Table 4). Comparatively, application of the XGBOOST-based machine learning algorithm resulted in accuracy and "true" recall values of 98% and 89%, respectively.

Table 4. Performance of rule-based versus machine learning-based text classification models.

Metrics	Class Label	Rule-Based	ML-Based
Accuracy	-	99	98
Precision	True	96	96
	False	99	99
Recall	True	88	89
	False	100	100
F1-Score	True	92	92
	False	100	99

The patterns used to construct the rules in the rule-based model were manually defined, requiring more effort in terms of rule preparation. It is also important to note that the results of the rule-based model are quite sensitive to input rules: adding or removing a specific rule may have a considerable impact on classifier performance. Alternatively, the ML-based text classification model learns the classification process by using training data. In this regard, the results of the ML model are sensitive to the number of training sets, performing best in the presence of more data. Given the results provided in Table 4, the choice of classification model depends on the availability of training data for the ML-based model or the level of effort able to be invested for rule construction in the rule-based model.

5. Prediction Module

The prediction module is used to estimate the time, resources, and cost needed to fulfill the reporting requirements. Module inputs include (1) the list of reports prepared by subject-matter experts using outputs of the rule-based or ML-based extraction module that describe the types and submittal frequencies of the reporting requirements, (2) the resources, time, and hourly rate associated with each reporting requirement, and (3) estimated project duration. To account for underlying uncertainties in model inputs and outputs, a Monte Carlo simulation model is employed, with uncertain parameters (e.g., report preparation time and project duration) input as probabilistic distributions derived from historical data. If sufficient historical data are unavailable, probabilistic distributions, such as a triangular distribution with minimum, most likely, and maximum values reported by experts, can be input into the model instead [44]. The Monte Carlo simulation is then run for multiple iterations, with each iteration randomly selecting a value from each parameter's distribution. Outputs of the model include the predicted (1) time, (2) cost, and (3) distribution among the various personnel types to complete the reporting requirements.

6. Case Study

An oil-and-gas project led by a private Canadian construction contracting firm was used to demonstrate the proposed framework. The project was considered a small-size project by the contractor and was awarded by the client to the contractor through a cost-plus contract type. This project was completed before the initiation of this research study. Actual durations of report preparation were not recorded by the contracting firm.

6.1. Data Collection

While manual extraction of reporting requirements is not required for future construction contracts, manual extraction was required, here, for initial development of the extraction module. As such, and for this case study only, the list of manually extracted reporting requirements from the set of contract documents detailed in Section 4.1 were input into the model. Notably, outputs of the rule-based or ML-based extraction module can be used to prepare a list of required reports and their submittal frequencies for input into the simulation model for resource prediction of future contracts.

Since report preparation times were not recorded by the project team, the minimum, most likely, and maximum values for the preparation time of each report were provided by company experts based on prior experience. Individual labor rates for each personnel type were not provided by the industrial partner; therefore, an average labor rate of 60 CAD per hour was input into the model. A subset of the data is summarized in Table 5.

Table 5. Sample of report preparation-associated input data.

Report Name	Frequency	Resources	Time (Minutes) Min, Most Likely, Max
Daily Update: work plan and estimated progress	Daily	1 Safety Coordinator 1 Project Manager 4 Superintendents 1 Quality Controller	45, 60, 75
Equipment Log	Bi-Weekly	1 Project Coordinator	90, 120, 150
Installation Work Package Report	Bi-Weekly	1 Project Controller 1 Scheduler 1 Project Manager	210, 240, 270
3-Week Look-Ahead Schedule	Bi-Weekly	1 Project Control 1 Scheduler 1 Project Manager	360, 420, 480

6.2. Results and Discussion

6.2.1. Extraction Module

A sample of the extracted reporting requirements is illustrated in Figure 3. As detailed in Section 6.1, 340 individual reporting requirements and their submittal frequencies were identified from over 500 contract pages. Although quite high (88%, Table 4), the recall of the current extraction module is not 100%. Sufficient for planning purposes, the extraction module should not be used as the only means of requirement extraction during the execution phase of a project. A manual review during the execution phase should continue to be performed until the ability of the framework to consistently extract 100% of reporting requirements is achieved. Failing to determine the exhaustive list of submittals and information deliverables required by the client can result in claims and litigations, subjecting both parties to disputes and conflicts that could have been prevented. Nevertheless, manual extraction is also prone to error, and the use of the extraction module as an adjunct tool during the execution phase of a project is strongly recommended. The list of reporting requirements extracted manually and by the automated extraction module should be compared to identify requirements that may be missing from the manually extracted list.

The 340 reporting requirements output by the extraction module were reviewed by the project team. It was determined that some of the reporting requirements were repetitive, requiring submission of the same report. A total of 70 distinct reports and information submittals was identified. A list of these reports was prepared and input into the prediction module.

6.2.2. Prediction Module

The Monte Carlo simulation model was run for 100,000 iterations, as increasing the number iterations beyond 100,000 slowed the execution speed without resulting in a notable impact on output results. The total duration and cost associated with the requirement reporting process was calculated using the probability distributions defined for each report. In each iteration, random numbers were sampled from the preparation time distributions of each report type, and a total reporting duration (or total cost) was achieved as the cumulative time (or cost) of all reports for that iteration. Total reporting duration (or cost) values of each iteration were then used to form a distribution of total reporting time, as shown in Figure 11.

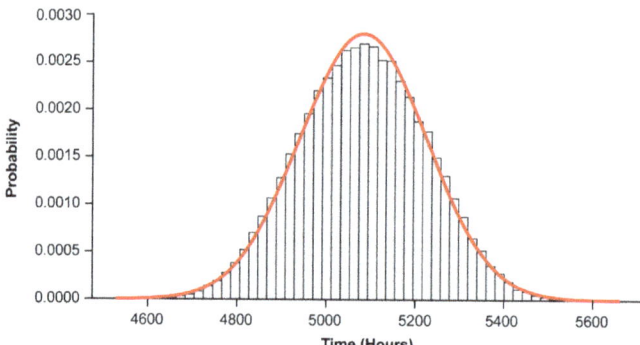

Figure 11. Predicted cumulative report preparation time as a distribution.

The mean value for the cumulative report preparation time was 5083 labor-hours ($\sigma = 142$) for the project life cycle. The simulation was then run again using an average rate of 60 CAD per hour; here, the mean value of the total cost associated with the reporting process was calculated to be $304,939 ($\sigma$ = $8538), as shown in the predicted cost distribution in Figure 12. Notably, including individual labor rates for each personnel type will increase the accuracy of the framework's results.

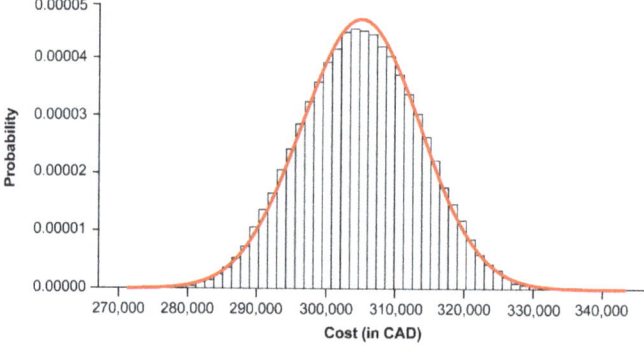

Figure 12. Predicted cumulative cost of reporting as a distribution.

The distribution of report preparation time between various personnel types is summarized in Figure 13. The plurality of the cumulative report preparation time (31%) was associated with the project manager, who must review and approve most reports. Based on the mean cumulative duration of 5083 h (Figure 11), the project manager is expected to spend an estimated 1576 h (or, assuming a 9-h work day, 175 days) completing reporting requirements. With a provided project duration of 4400 h, the project manager is estimated to be performing reporting activities 36% of the time. Similarly, two other highly utilized resources were the project control team (28%) and scheduler (28%), who are responsible for collecting, merging, and overseeing the preparation of various report types. Together, these two resources will spend an estimated 2846 h (or 316 days) completing reporting requirements—equal to 32% of their time.

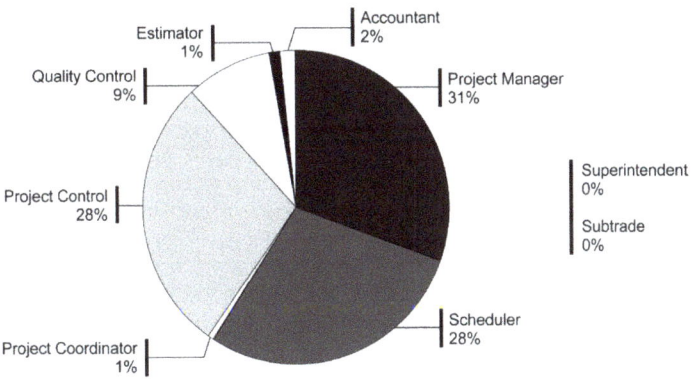

Figure 13. Distribution of report preparation time between personnel types.

6.3. Framework Validation

6.3.1. Validation of Extraction Module

The extraction module underwent extensive validation testing. Here, reporting requirements were manually extracted and compared to the list of reporting requirements identified using the extraction module. Then, a subset of real project data (different from those used for model development and training) was used to evaluate the performance of the rule-based and machine-learning-based classification models. A discussion of the validation process is detailed in Sections 4.2.1 and 4.3.1, respectively. A comparison of the models is summarized in Section 4.4.

6.3.2. Validation of Prediction Module

In contrast, the prediction module was evaluated using face validation. Since actual report preparation durations were not recorded by the contractor, face validation was performed by subject-matter experts to evaluate whether or not the simulation model results (i.e., prediction module outcomes) were accurately representing the current reporting process. Simulation results (Figures 11–13) were presented to the project management team responsible for executing the case study project. The experts confirmed that the simulated results were acceptable and were consistent with the outcomes of the actual project. Overall, face validation by the subject-matter experts confirmed that the prediction module was representative, comprehensive, and easy to use.

7. Discussion

Having a list of reporting requirements during the planning phase of a project will provide the project management team with the opportunity to enhance the reporting process, resulting in a reduction in reporting-associated costs. For example, similar or redundant reports can be consolidated, specialized data collection systems and report templates can be developed and implemented prior to project execution, and the allocation of reporting requirements to specific personnel types can be optimized.

The probability distributions output by the proposed framework allow decision makers to more accurately estimate the probability of achieving project targets, while gaining insight on potential best- and worst-case scenarios. More accurate time preparation estimates will allow contractors to ensure that a sufficient number of personnel are available to complete reporting requirements on time. Moreover, by more accurately estimating the overhead costs associated with reporting requirements for each particular project, contractors are able to enhance bid preparation to improve competitiveness, or provide more realistic direct–indirect cost ratios to avoid potential disputes for cost-plus contracts. Furthermore, these outputs can be used to optimize the composition of project management teams based on the specific requirements of each contract. Together with the list of requirements output from the extraction module, the personnel distribution results can be used to examine and potentially reallocate reporting duties to lower-wage personnel, where appropriate, thereby reducing report preparation costs.

The level of benefit achieved by considering reporting-associated costs in the planning phase of construction depends on the construction type. Repetitive types of construction, such as residential building construction, are typically associated with contracts that remain similar between projects. Due to a lack of variability in reporting requirements, project teams are able to accurately approximate the time, cost, and resources required without the need to extract reporting requirements for each contract. However, due to the increased level of risk, complexity of the work, and large project scale, contract documents for complex projects, such as those in the oil and gas industry, are typically longer, more intricate, and more variable from project to project. With these types of construction, clients tend to request additional information and detailed reporting submittals from contractors, which substantially increases overhead costs. The benefits of applying the proposed framework, therefore, are expected to expand as project complexity is increased.

8. Limitations and Future Work

An automated approach for rapidly extracting reporting requirements from contractual documents and predicting the time and cost required to complete reporting activities was developed. Although the functionality of the proposed framework was demonstrated using real contractual documents from an actual case study, the following points should be considered.

First, the extraction module was developed using a labeled dataset obtained from one set of contract and specification documents for an oil and gas project. While the extraction module is expected to be applicable—in its current form—to all construction contracts with similar characteristics (e.g., terminology, document structure, and/or report structure), the development methodology described may need to be reapplied for other contract types. Moreover, the classification models were trained using a limited amount of training data. The comparatively low performance of the classification models for the "true" class may be due to the size of the "true" dataset (i.e., an imbalanced data problem). Future work should examine the impact of increasing the training dataset through the incorporation of additional contract documents to enhance the performance of the classification models. With sufficient training data, the extraction module is anticipated to achieve the desired performance of 100% recall for the "true" class (i.e., identification of all reporting requirements).

Second, the success of the prediction module is highly dependent on accurately modeling the inputs. One of the difficulties in analyzing probabilistic processes inherent to construction is defining the probability distributions that best reflect the uncertainties associated with each variable. The more accurate the model of the inputs, the more closely the simulation model mimics real-life behavior. A primary constraint for any simulation model, therefore, is the time and effort required to collect pertinent and correct information, as well as processing it for input into the model. While the resources and time required from construction sites and administration offices to complete reporting requirements are not commonly recorded, efforts to improve data collection related to project reporting process are expected to improve model results.

Third, contract documentation remains an immature area of practice, and more reliable and efficient approaches to better and more rapidly understand contract requirements are needed. Future work should focus on providing a holistic solution to this problem, such as writing contracts using a structured-database approach. While this would provide seamless integration between clients and contractors (thereby alleviating the need for rule-based/ML-based model (re)training), achieving this ideal will require a tremendous amount of input, effort, and collaboration among all of the stakeholders involved in a project. Additionally, methods for dealing with modifications or alternate arrangements will need to be researched and developed. Consequently, the framework proposed here provides a much-needed interim solution as these more holistic solutions are pursued.

9. Conclusions

Automating the reporting requirement extraction process and estimating its associated time–cost implications are expected to reduce the effort, time, and overhead costs expended by the multiple personnel involved. To overcome the shortcomings of the traditional manual approach, this study developed a framework for reporting requirement extraction based on NLP—a domain-specific and application-oriented text classification process—that is capable of automatically identifying reporting requirements from contractual documents to considerably reduce the time and effort required to extract reporting requirements. To account for project uncertainties due to variation or unforeseeable events that may occur during execution, a Monte Carlo simulation was used to predict the time and cost needed to complete reporting requirements.

The model begins by collecting textual data, in this case the sentences and terms in contractual documents, which describe the reporting requirements mandated by the client. Rule-based and ML-based classification methods were developed, and their performances were evaluated. The performance of rule-based classification using different sets of n-grams was assessed, with an accuracy of 99.27% achieved using bi-grams as rules. Application of lemmatization to and removal of stop-words from the bi-gram rules resulted in a recall and F1-score of 88% and 92% for the "true" category, respectively. Four ML algorithms were also implemented, and their performance was assessed under different pre-processing settings and feature engineering techniques. All of the ML classification models achieved promising accuracies of over 95%; notably, XGBOOST achieved the highest recall value of 89% after parameter tuning. Then, numerical data regarding report preparation times and associated resources (based on prior experience of experts) were provided by an industrial partner and were used to predict the time and cost required to complete the reporting requirements detailed in the contractual documents. Input of these data into the Monte Carlo simulation model resulted in a mean cumulative reporting duration and cost of 5083 h and 304,939 CAD, respectively.

During the bidding and contract negotiation phase of a project, decision makers can now use the proposed framework to automatically review reporting requirements prior to accepting the contractual agreement. Not feasible using time-consuming, traditional extraction methods, the extraction speed of the framework allows decision makers to identify and subsequently negotiate difficult and/or inefficient reporting requirements prior to signing. If contract conditions are unfavorable to the contractor in terms of project reporting cost, a revision of contract conditions may be requested or a contract may be abandoned by contractors to prevent further loss. With a thorough and realistic understanding of contract reporting requirements, contractors can focus on establishing the best means, methods, pricing, and schedules for completing the proposed project.

Author Contributions: Conceptualization, P.J. and S.A.; data curation, P.J., M.A.H. and E.M.; formal analysis, P.J.; funding acquisition, S.A.; investigation, P.J.; methodology, P.J.; project administration, S.A.; software, P.J.; supervision, S.A.; validation, P.J.; visualization, P.J. and M.A.H.; writing—original draft, P.J.; writing—review and editing, M.A.H., E.M. and S.A. All authors have read and agreed to the published version of the manuscript.

Funding: This research was funded by the Natural Sciences and Engineering Research Council of Canada through a Collaborative Research and Development Grant (CRDPJ 492657).

Institutional Review Board Statement: Not Applicable.

Informed Consent Statement: Not Applicable.

Data Availability Statement: All data used in the study were provided by a third party. Direct requests for these materials may be made to the provider indicated in the Acknowledgments. Models and code that support the findings of this study are available from the corresponding author upon request and with permission from the partner indicated in the Acknowledgments.

Acknowledgments: The authors would like to thank Graham Industrial Services LP for their support and for providing contractual documents and report preparation-associated data. The authors also would like to acknowledge Catherine Pretzlaw for her assistance with manuscript editing and composition.

Conflicts of Interest: The authors declare no conflict of interest.

References

1. Shash, A.A.; Habash, S.I. Construction Contract Conversion: An Approach to Resolve Disputes. *J. Eng. Proj. Prod. Manag.* **2020**, *10*, 162–169. [CrossRef]
2. Barlow, G.; Dew, C.; Woolley, P.; Dempsey, H. *Effective Reporting for Construction Projects: Increasing the Likelihood of Project Success*; Project Advisory Leadership Series; KPMG New Zealand: Auckland, New Zealand, 2014.
3. El-Omari, S.; Moselhi, O. Integrating automated data acquisition technologies for progress reporting of construction projects. *Autom. Constr.* **2011**, *20*, 699–705. [CrossRef]
4. Hassan, F.U.; Le, T. Automated Requirements Identification from Construction Contract Documents Using Natural Language Processing. *J. Leg. Aff. Disput. Resolut. Eng. Constr.* **2020**, *12*, 04520009. [CrossRef]
5. Lee, J.; Yi, J.-S.; Son, J. Development of Automatic-Extraction Model of Poisonous Clauses in International Construction Contracts Using Rule-Based NLP. *J. Comput. Civ. Eng.* **2019**, *33*, 04019003. [CrossRef]
6. Jeong, H.D.; Gransberg, D.; Shrestha, K.J. *Framework for Advanced Daily Work Report System*; Institute for Transportation, Iowa State University: Ames, IA, USA, 2015.
7. Caldas, C.H.; Soibelman, L.; Han, J. Automated Classification of Construction Project Documents. *J. Comput. Civ. Eng.* **2002**, *16*, 234–243. [CrossRef]
8. Pestana, C.; Alves, T.; Barbosa, A. Application of Lean Construction Concepts to Manage the Submittal Process in AEC Projects. *J. Manag. Eng.* **2014**, *30*, 05014006. [CrossRef]
9. Levin, P. *Construction Contract Claims, Changes & Dispute Resolution*; American Society of Civil Engineers (ASCE): Reston, VA, USA, 1998.
10. Jeon, K.; Lee, G.; Jeong, H.D. *Classification of the Requirement Sentences of the US DOT Standard Specification Using Deep Learning Algorithms*; Springer: Berlin/Heidelberg, Germany, 2020; pp. 89–97.
11. El Gindi, M. *User Friendly Progress Reporting System for Construction Projects*; The American University in Cairo: Cairo, Egypt, 2017.
12. El Sawy, I.; Hosny, H.; Razek, M.A. A Neural Network Model for Construction Projects Site Overhead Cost Estimating in Egypt. *Int. J. Comput. Sci.* **2011**, *8*, 273–283.
13. Jallow, A.K.; Demian, P.; Anumba, C.J.; Baldwin, A.N. An enterprise architecture framework for electronic requirements information management. *Int. J. Inf. Manag.* **2017**, *37*, 455–472. [CrossRef]
14. Morgan, A. *Does Poor Project Governance Cause Delays?* Pricewaterhouse Coopers LLP: London, UK, 2010.
15. Jafari, P.; Mohamed, E.; Lee, S.; Abourizk, S. Social network analysis of change management processes for communication assessment. *Autom. Constr.* **2020**, *118*, 103292. [CrossRef]
16. Lee, G.; Cho, J.; Song, T.; Roh, H.; Jung, J.; Chung, J.; Yong, G.; Jeong, D. *Construction Field Management Using a Popular Text Messenger*; Springer: Berlin/Heidelberg, Germany, 2020; pp. 971–979.
17. Shrestha, K.J.; Jeong, H.D. Computational algorithm to automate as-built schedule development using digital daily work reports. *Autom. Constr.* **2017**, *84*, 315–322. [CrossRef]
18. Russell, A.D. Computerized Daily Site Reporting. *J. Constr. Eng. Manag.* **1993**, *119*, 385–402. [CrossRef]
19. Shiau, Y.-C.; Wang, W.-C. Daily Report Module for Construction Management Information System. In Proceedings of the 20th International Symposium on Automation and Robotics in Construction ISARC 2003—The Future Site, Budapest, The Netherlands, 21–24 September 2003; Maas, G., Van Gassel, F., Eds.; International Association for Automation and Robotics in Construction (IAARC): Eindhoven, UK, 2003; pp. 603–609.
20. Omar, T.; Nehdi, M.L. Data acquisition technologies for construction progress tracking. *Autom. Constr.* **2016**, *70*, 143–155. [CrossRef]
21. Manning, C.; Schutze, H. *Foundations of Statistical Natural Language Processing*; MIT Press: Cambridge, MA, USA, 1999; ISBN 0-262-30379-5.
22. Tixier, A.J.-P.; Hallowell, M.R.; Rajagopalan, B.; Bowman, D. Automated content analysis for construction safety: A natural language processing system to extract precursors and outcomes from unstructured injury reports. *Autom. Constr.* **2016**, *62*, 45–56. [CrossRef]
23. Fan, H.; Li, H. Retrieving similar cases for alternative dispute resolution in construction accidents using text mining techniques. *Autom. Constr.* **2013**, *34*, 85–91. [CrossRef]
24. Zhang, J.; Zi, L.; Hou, Y.; Deng, D.; Jiang, W.; Wang, M. A C-BiLSTM Approach to Classify Construction Accident Reports. *Appl. Sci.* **2020**, *10*, 5754. [CrossRef]

25. Al Qady, M.; Kandil, A. Automatic Classification of Project Documents on the Basis of Text Content. *J. Comput. Civ. Eng.* **2015**, *29*, 04014043. [CrossRef]
26. Manning, C.D.; Raghavan, P.; Schutze, H. *Introduction to Information Retrieval*; Cambridge University Press: Cambridge, UK, 2008; Volume 19, pp. 1041–4347.
27. Russell, S.; Norvig, P. *Artificial Intelligence: A Modern Approach*; Pearson: London, UK, 2020; ISBN 0-13-461099-7.
28. Salama, D.M.; El-Gohary, N.M. Semantic Text Classification for Supporting Automated Compliance Checking in Construction. *J. Comput. Civ. Eng.* **2016**, *30*, 04014106. [CrossRef]
29. Zhong, B.; Pan, X.; Love, P.E.; Ding, L.; Fang, W. Deep learning and network analysis: Classifying and visualizing accident narratives in construction. *Autom. Constr.* **2020**, *113*, 103089. [CrossRef]
30. Zhou, P.; El-Gohary, N. Ontology-Based Multilabel Text Classification of Construction Regulatory Documents. *J. Comput. Civ. Eng.* **2016**, *30*, 04015058. [CrossRef]
31. Hastings, W.K. Monte Carlo sampling methods using Markov chains and their applications. *Biomolecules* **1970**, *57*, 97–109. [CrossRef]
32. Python. *Python 3.7.0*; Python Software Foundation: Beaverton, OR, USA, 2018.
33. Priyanka, H.; Ramya, B.; Ashok, K. Classification Model to Determine the Polarity of Movie Review Using Logistic Regression. *Int. Res. J. Comput. Sci.* **2019**, *6*, 87–91. [CrossRef]
34. Grefenstette, G.; Tapanainen, P. What Is a Word, What Is a Sentence? Problems of Tokenisation. In Proceedings of the 3rd International Conference on Computational Lexicography, Budapest, Hungary, 7–10 July 1994; Research Institute for Linguistics, Hungarian Academy of Sciences: Budapest, Hungary, 1994; pp. 79–87.
35. Porter, M.F. An algorithm for suffix stripping. *Program* **1980**, *14*, 130–137. [CrossRef]
36. Buckland, M.; Gey, F. The Relationship between Recall and Precision. *J. Am. Soc. Inf. Sci.* **1994**, *45*, 12–19. [CrossRef]
37. Forman, G. An Extensive Empirical Study of Feature Selection Metrics for Text Classification. *J. Mach. Learn. Res.* **2003**, *3*, 1289–1305.
38. Sebastiani, F. Machine learning in automated text categorization. *ACM Comput. Surv.* **2002**, *34*, 1–47. [CrossRef]
39. Valieva, I.; Voitenko, I.; Björkman, M.; Åkerberg, J.; Ekström, M. Multiple Machine Learning Algorithms Comparison for Modulation Type Classification Based on Instantaneous Values of the Time Domain Signal and Time Series Statistics Derived from Wavelet Transform. *Adv. Sci. Technol. Eng. Syst.* **2021**, *6*, 658–671. [CrossRef]
40. Witten, I.; Frank, E.; Mark, A. *Hall Data Mining: Practical Machine Learning*; Elsevier: Amsterdam, The Netherlands, 2011; ISBN 9780123748560.
41. Breiman, L. Random Forests. *Mach. Learn.* **2001**, *45*, 5–32. [CrossRef]
42. Chen, T.; Guestrin, C. Xgboost: A Scalable Tree Boosting System. In Proceedings of the 22nd ACM SIGKDD International Conference on Knowledge Discovery and Data Mining, New York, NY, USA, 13–17 August 2016; pp. 785–794.
43. Bergstra, J.; Bengio, Y. Random Search for Hyper-Parameter Optimization. *J. Mach. Learn. Res.* **2012**, *13*, 281–305.
44. Abourizk, S.M.; Halpin, D.W. Statistical Properties of Construction Duration Data. *J. Constr. Eng. Manag.* **1992**, *118*, 525–544. [CrossRef]

Article

Interoperability of Digital Tools for the Monitoring and Control of Construction Projects

Luz Duarte-Vidal [1], Rodrigo F. Herrera [1,*], Edison Atencio [1] and Felipe Muñoz-La Rivera [1,2,3]

[1] School of Civil Engineering, Pontificia Universidad Católica de Valparaíso, Av. Brasil 2147, Valparaíso 2340000, Chile; luz.duarte.v@mail.pucv.cl (L.D.-V.); edison.atencio@pucv.cl (E.A.); felipe.munoz@pucv.cl (F.M.-L.R.)

[2] International Centre for Numerical Methods in Engineering (CIMNE), C/Gram Capitán S/N UPC Cambus Nord, Edifici C1, 080034 Barcelona, Spain

[3] School of Civil Engineering, Universitat Politècnica de Catalunya, Carrer de Jordi Girona 1, 080034 Barcelona, Spain

* Correspondence: rodrigo.herrera@pucv.cl

Abstract: Monitoring the progress on a construction site during the construction phase is crucial. An inadequate understanding of the project status can lead to mistakes and inappropriate actions, causing delays and increased costs. Monitoring and controlling projects via digital tools would reduce the risk of error and enable timely corrective actions. Although there is currently a wide range of technologies for these purposes, these technologies and interoperability between them are still limited. Because of this, it is important to know the possibilities of integration and interoperability regarding their implementation. This article presents a bibliographic synthesis and interpretation of 30 nonconventional digital tools for monitoring progress in terms of field data capture technologies (FDCT) and communication and collaborative technologies (CT) that are responsible for information processing and management. This research aims to perform an integration and interoperability analysis of technologies to demonstrate their potential for monitoring and controlling construction projects during the execution phase. A network analysis was conducted, and the results suggest that the triad formed by building information modeling (BIM), unmanned aerial vehicles (UAVs) and photogrammetry is an effective tool; the use of this set extends not only to monitoring and control, but also to all phases of a project.

Keywords: monitoring progress; construction phase; automated monitoring; digital tools; as-built; as-planned

1. Introduction

The construction industry is an important and dynamic economic activity that is characterized by one of the economy's reactivating mechanisms and by its contribution to the generation of employment; however, it has also been characterized by its low productivity [1,2]. Currently, the world economy is experiencing a major crisis caused by the COVID-19 pandemic. This public health emergency has presented a change in paradigms and challenges due to the lack of workers in the whole supply chain, extended closing of businesses in other areas of the economy and interruptions due to social distancing measures in existing projects [3]. These changes imply challenges for the reactivation of the economy, where strategies focus on digitalization and sustainability [4].

With the arrival of the Fourth Industrial Revolution, Industry 4.0 is a new paradigm that proposes to encourage the use of cyber physical systems; that is, technologies that enable the merging of virtual and physical worlds to create a real networked environment in which intelligent objects communicate and interact with each other [5]. In this context, the concept of Construction 4.0 has been proposed as a response to Industry 4.0 in the Architecture, Engineering, Construction and Operation (AECO) industry, which seeks to upgrade digitalization to improve the efficiency of production processes, business models and value chains. This transformation is possible due to the convergence of existing and emerging technologies, which promise to reformulate the way of designing and building the assets of the built environment [6,7].

Integration and interoperability are two key factors in Construction 4.0 [7]. Interoperability is the ability of two systems to understand each other and use the functionality of the other, which represents the ability of two systems to exchange data and share information and knowledge. Integration of information systems achieves seamless cooperation between organizations and industries [8]. The traditional method, which involves manual tracking of construction progress, still dominates the AECO industry [9]. Current monitoring practices during the execution of a construction project require multiple inspections, which are time intensive and may, by their nature, include mistakes. Therefore, a reliable monitoring system that can provide early detection and notification of project problems, whether actual or potential, is necessary [10]. In this context, the application of several technologies has great potential to improve management practices in the construction industry [11]. The advances in digital technologies that have been established in many productive sectors are permeating the AECO industry [8] at less accelerated rates [12].

The goal of this research is to perform an interoperability analysis of digital tools for monitoring and controlling construction projects during the execution phase. For this purpose, innovative technologies that are classified according to their use, in this case, field data capture technologies (FDCT) and communication and collaboration technologies (CT), which are responsible for the processing and management of information, are revised. Although these technologies have unique advantages, they often offer integration with other technologies, allowing for the formation of more powerful hybrid systems with more than a single intelligence dimension.

2. Background

Management methods are the set of actions framed in a prototype that allow for directing the activities of a public or private organization; their purpose may have one or more objectives that are aimed at achieving greater productivity in all human activities [13]. Methodologies are generally approached in different ways, many of them known and common among a high number of engineering and construction companies, with the same guidelines for decades, which have become examples to follow and/or apply [14]. Multiple problems affect the performance of projects in the construction industry, including lower labor productivity, inadequate identification of design requirements and lack of standardized construction management. Therefore, the integration of projects is essential for success, and the collaboration of different specialties in the supply chain allows for effective interaction, keeping the budget and schedule in the expected parameters [15].

The planning and control of a construction project is the process of defining, coordinating and determining the order in which activities should be carried out. The objective is to execute the most efficient and economical use of available equipment, elements and resources. For this purpose, a work plan, which must be controlled throughout the project to evaluate its compliance, is established and defined. Two actions that must be carried out in this recursive process are to carry out pertinent revisions or modifications and to eliminate unnecessary diversification of efforts. Both actions are framed in the final objective set since the beginning of a project [16]. Currently, as a result of technological advancements, new techniques that allow for remote and automated management have been incorporated to obtain information from the digitalization of the construction [8].

2.1. Traditional Construction Monitoring and Control System

The planning and control of a project is one of the key processes for the adequate development of the project and the success of each of its phases during its life cycle. Appropriate planning allows for the explicit definition of the work to be done, risk identification, consideration of different scenarios and the attainment of solutions. Adequate control offers the possibility of detecting deviations, informing on time the anomalies, allowing for their correction and ensuring the quality of a project [14]. Establishing a system to ensure the correct execution of the work is fundamental and must be adapted to the nature of each project and its environment. For this reason, there are different methodologies to carry out project management, that is, the application of knowledge, skills, tools and techniques to the activities of a project to comply with its requirements [17].

According to some studies, control tasks using traditional methods generate difficulties for information management caused by imprecise manual documentation and deferred data collection, which leads to delays in decision making [18]. It is common that a lack of opportune information about the actual state causes problems to remain unresolved and prevents field engineering staff from solving them in an opportune amount of time. An incorrect understanding of the current situation can cause errors and inappropriate adjustments by the management team, which could result in further delays and increased costs [19].

The set of monitoring and control activities consists of the processes required to monitor, analyze and regulate the progress and performance of a project, to identify areas where the plan requires changes and to make the corresponding modifications [20]. In this sense, because of the dynamism in a project, maintaining daily control can have a considerable impact, providing greater precision around the schedule and costs to comply with the scope defined in the project [21]. To carry out progress control, it is fundamental to have a work program to know when, with what and how the works will be executed. During the construction phase, it is relevant to know if the work is being performed according to the established schedule. An accurate evaluation of progress allows project managers to make adjustments to minimize costs that lead to program deviations [10] (Figure 1).

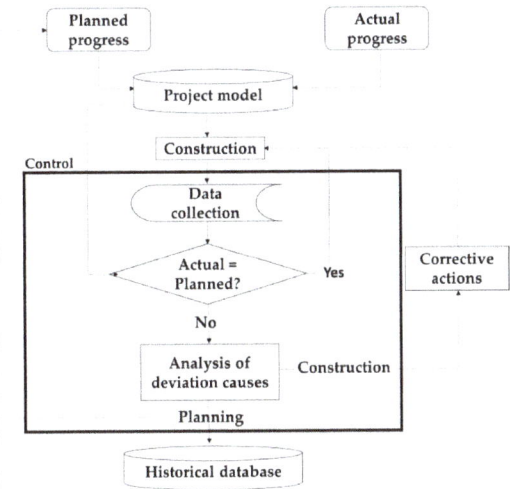

Figure 1. Schematic of model for real-time closed loop control. Adapted from [22].

Figure 1 shows a schematic of a model for real-time closed loop control. In this model, it is necessary to establish a control system that can regularly keep project managers informed of the progress made in each activity. In the case of delay or deficiency in any activity, this system enables time-sensitive corrections, either by increasing the number of workers, changing the equipment or correcting the work program. If it is verified in an opportune way that there was a planning mistake, and it is possible, although with modifications, to maintain the original program, then it will be possible to fulfill the delivery terms. If these corrections are not made on time, it is difficult to continue with the initial project schedule, and a new plan will have to be constructed, whose application will certainly mean disorders and greater project costs [23]. Therefore, the objective of monitoring is to ensure compliance with the objectives and goals proposed by the plan during its implementation, alerting any difficulties and relieving pending or delayed tasks, and allowing visualization of complementary actions by making the necessary adjustments. It is worth noting that for the monitoring of a plan, it is necessary to establish methodologically how information about projects and management measures will be obtained to understand the plan's progress and facilitate its control. This approach allows for the comparison between the baseline of the plan and its actual execution, consequently identifying the gap between what was proposed and what was achieved and defining corrective and preventive actions that minimize these gaps [17].

The practice of monitoring and control in the tracking of work, based on traditional techniques, has become indispensable for any construction project. It is common that the activities that include the management of information are based on collecting the information in the field, which is documented manually for its subsequent digitalization. The employees in charge of creating reports using this process devote between 28% and 41% of their daily time to doing so. Because information is scattered among multiple documents, which may exclude data, the traditional approach is slow and inefficient. Similarly, the construction industry makes little use of technical resources, making it difficult to track building projects due to the lack of automated processes [18].

2.2. Digitized Construction Monitoring

The current requirements presented by the AECO industry demand the application of modern monitoring and control methods and tools by companies in order to meet regulatory requirements and improve the competitiveness of companies in the construction market [24]. In this respect, increasing productivity implies the improvement in various processes by the use of new technologies and construction procedures [25]. Successful construction management requires the integration of processes, technologies and people to achieve its objectives [26], where information on the progress of a project offers a continuous diagnosis of it, allowing the different members of a team to make appropriate decisions about any measure to save the project and ensure its completion [27]. Technology is slowly paving the way for technologically supported project management practices in construction, for transforming and allowing the establishment of new tools for remote monitoring and for allowing for automation of the supervision of construction progress, improvement in data acquisition and, consequently, improvements in decision-making in project management to meet objectives and ensure productivity [28].

The continuous development of technology has made it possible to effectively solve practical problems associated with the AECO industry. [29]. Therefore, the development of digitized construction monitoring has the main purpose of developing the connection between traditional or existing methods and new technological systems. The basic theory for developing such a model is to extend the traditional approach so that construction operations become dynamic and simultaneous [30]. The World Economic Forum (2016) developed a transformational framework for the construction industry that lists 30 best practice measures. The three most important components of the transformation of its traditional approach are the following: (i) being open to innovation to take advantage of the opportunities offered by new technologies, materials and tools to reduce production costs; (ii) considering the adoption of mechanized and automated production systems with offsite construction techniques to accelerate the construction process and improve the timely completion of projects in a collaborative environment; and (iii) the role of project management and cost control in the design and planning stages [31].

It has been mentioned that the construction industry has been criticized for its slow adoption of emerging technologies. However, in recent years this trend has changed. The rapid growth of the availability and power of technologies, with their continuously decreasing cost, has allowed them to be adopted and considered an effective tool for the analysis of massive data for the purpose of monitoring and controlling the progress of projects [32]. During recent decades, research efforts have been made towards advanced 4D planning models by integrating three-dimensional (3D) models with the time parameter [30].

There is an evident need to develop an integrated model to automate the current practice, since monitoring and manual control have not produced the expected results. Efficient management can be obstructed by lost time in information recovery, poor structuring and delayed communications [33]. Manual monitoring is labor intensive and often requires a choice between monitoring based on rough estimates or spending a lot of time collecting and processing data. [34].

Introducing digitized construction monitoring in the construction phase would allow project managers and site engineers to more effectively and accurately monitor project progress. Therefore, this system will improve the decision-making process and productivity and reduce claims for delays [30]. Each team member needs to know, in a timely and accurate manner, the progress of the project and the current status to contrast such information with the originally established plans [35].

Some research has shown that monitoring of work and comparison with the project baseline can be used to assess work in progress, providing an accuracy error of less than 20% [22]. Studies in the literature indicate that greater standardization of work will make the application of automated procedures less complex [36]. For this reason, the means to represent possible discrepancies between planned progress and constructed progress is an important factor in facilitating decisions on corrective measures [10].

The results of the adoption of digital approaches in construction are increasingly showing positive results; for example, projects that would otherwise present a high risk of cost overruns are being delivered on time and within budget. The implementation of digital tools allows for the integration of teams, processes and organizations, reducing the problem of fragmentation present in traditional methods [31] (Figure 2).

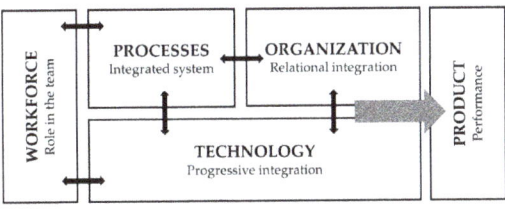

Figure 2. Integrated construction environment. Adapted from [31].

In the last few years, with the increasing level of competition in the AECO industry, research efforts have focused on the application of information technology as a way to improve the process of integration of construction supply chain management [37]. This collaboration has emerged as computer-assisted collaborative learning, facilitating interactions between two or more individuals who may be geographically and/or temporally separated [31]. As a result, there is a need to adapt technology to improve design and planning processes in a common and secure data environment (CDE). This type of environment allows for the integrated collaboration of project participants and their interaction via accessible information that supports decision making [38]. This concept could be achieved by using different technologies that provide a richness of varied and complementary information, facilitating the joint work of the teams via collaborative work and problem solving without considering the geographical distance. These technologies could also work both synchronously and asynchronously, allowing for the sharing of documents not only anywhere, but also at all times. In this way, the provided tools help communication and collaboration and provide a means of solving problems in the early stages of the project [31].

3. Materials and Methods

To achieve the objective of this study, the research was divided into three stages: (1) the design of remote monitoring and control practices—the literature review; (2) interoperability and integration; and (3) the validation of digital tools. Figure 3 shows the research tools used to perform each activity and their respective deliverables.

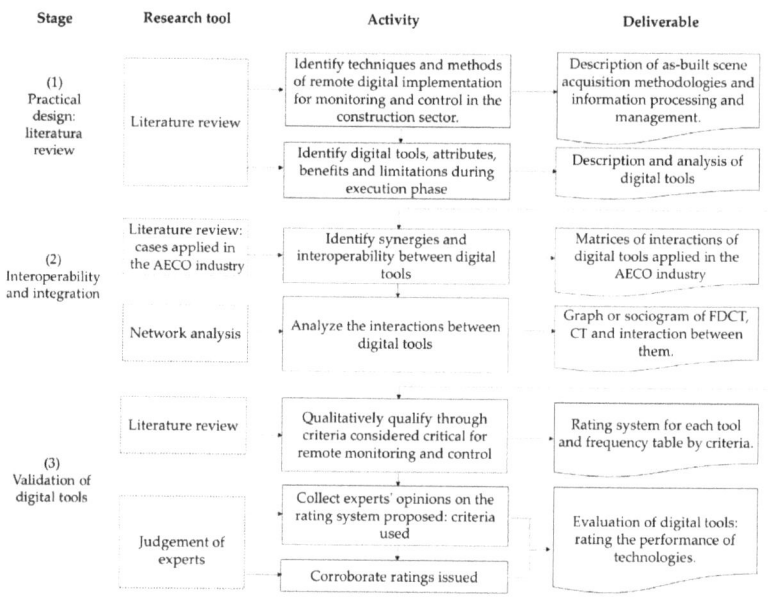

Figure 3. Three stages of the research method.

In the first stage, we identified and selected the techniques and methods that have a potential use for monitoring and control during the construction or execution phase of projects. These are presented with a brief description of their benefits, limitations and application in the AECO industry. In the second stage, the synergies and interoperability shared by different digital tools are identified, performing an exhaustive analysis of the interactions presented. In the third stage, the main researcher qualified the tools based on criteria considered critical to achieve satisfactory performance based on the deliverables of stages 1 and 2. Five experts were invited to validate these tools by means of corrections, suggestions and comments about the proposed evaluation.

3.1. Stage 1: Practice Design—Literature Review

A literature review of journals specialized in engineering and construction project management and of the proceedings of the main conferences held between 2012 and 2020 was conducted. The search was carried out in the following libraries: Google Scholar, ResearchGate, Engineering Village, Web of Science and Scopus. The search topics were automated progress monitoring, monitoring system, progress monitoring, as-built as-planned comparison, construction method into automated progress monitoring techniques, management model for construction monitoring and control, digitalizing construction monitoring, constructions phases, construction management, construction progress, interoperability, real-time monitoring and Industry 4.0.

The articles were selected by applying three inclusion/exclusion criteria: (1) innovation, (2) technology and (3) adoption of monitoring and control systems. For monitoring and information gathering, the distinction is made between (a) field data capture technologies (FDCT), which refers to sensing techniques used to capture as-built scenes, and (b) communication and collaboration technologies (CT), which are responsible for processing and managing as-built data information.

3.2. Stage 2: Interoperability and Integration

Seamless data exchange between FDCT and CT should be performed based on the search for the interoperability between them. In the search for relevant studies, technologies that have application support in the AECO industry were examined. A literature review was conducted in specialized journals in engineering and construction project management and of the proceedings of the main conferences, and a search was carried out in the following libraries: Google Scholar, ResearchGate and Scopus. The equations in the search process are presented in Table 1, where the keywords and respective Boolean operators are presented.

Table 1. Keywords and Boolean operators used to identify relevant studies.

Search	Keyword	B.O.[1]	Keyword	B.O.[1]	Keyword
FDCT + FDCT	Radio frequency identification Laser scanning Ultrawide band Wireless sensor networks Unmanned aerial vehicle Robotics Smartphones	AND	Radio frequency identification Laser scanning Ultrawide band Wireless sensor networks Unmanned aerial vehicle Robotics Smartphones	AND OR	Monitoring Progress Construction phase Execution phase
CT + CT	Mobile computing Simulations models and tools photogrammetry Mobile applications Automated regulation checking and audits Big data analytics Data mining Deep learning Embedded system Geographic information system Industrial Internet Internet of Things Machine learning Building information modeling Cloud computing Common data environment Data sharing Edge computing Social media	AND	Mobile computing Simulations models and tools photogrammetry Mobile applications Automated regulation checking and audits Big data analytics Data mining Deep learning Embedded system Geographic information system Industrial Internet Internet of Things Machine learning Building information modeling Cloud computing Common data environment Data sharing Edge computing Social media	AND OR	Monitoring Progress Construction phase Execution phase
FDCT + CT	Radio frequency identification Laser scanning Ultrawide band Wireless sensor networks Unmanned aerial vehicle Robotics Smartphones	AND	Mobile computing Simulations models and tools Photogrammetry Mobile applications Automated regulation checking and audits Big data analytics Data mining Deep learning Embedded system Geographic information system Industrial Internet Internet of Things Machine learning Building information modeling Cloud computing Common data environment Data sharing Edge computing Social media	AND OR	Monitoring Progress Construction phase Execution phase

[1] Boolean operator.

The articles were selected by applying three inclusion/exclusion criteria: (1) the case of monitoring and control during the execution stage; (2) being linked to one of the following areas—progress, machinery operation, intelligent construction or productivity; and (3) the most recent case of application. With the information obtained, three relationship matrices were elaborated with the possible combinations between the investigated tools for FDCT, CT and the integration between both.

All combinatorics without repetition between pairs of technologies were performed for the three search groups executed, according to Table 1. Table 2 identifies the total technologies per relationship matrix and the potential interoperability cases present in the AIC industry. The possible combinatorics are calculated according to Equation (1), where n is the total number of technologies to be chosen and r is the chosen technologies.

$$\frac{n!}{r!(n-r)!} = \binom{n}{r}, \tag{1}$$

Table 2. Details of combinatorics generated according to the relationship matrix.

Type of Connection	Relationship Matrix	Possible Combinations
Intra FDCT	FDCT ∪ FDCT	21
Intra CT	CT ∪ CT	253
Inter FDCT and CT	FDCT ∪ CT ∩ (FDCT ∪ FDCT) ∩ (CT ∪ CT)	161

The results are presented in three relationship matrices. In the case of intra-connections, they are represented by a square matrix of upper triangular type of dimension nxn, where n is the number of technologies identified, particularly 7 × 7 and 23 × 23. On the other hand, the interconnections are given by the dimensions between rows and columns, where the rows represent the CTs and the columns represent the FDCTs, with dimensions of 23 × 7. When there is a link between two technologies, the box contains the corresponding reference declaring the link. Otherwise, it is represented by an X.

For network analysis, the concepts of full inter- or intra-connections are employed, the former for connections between the same type of technology subcategory and the latter for cases of different subcategories, where only the interactions between FDCT and CT, and not the relationships between the same subcategories, are measured. To obtain global information, i.e., interactions between the inter-array connections of FDCT and CT and the intra-array connections, the concept of full is utilized.

In addition, a tool that allows for interconnecting the elements and facilitating the dissemination and understanding of the results is network analysis, which allows for visualizing the interconnected elements and thus performing an analysis of the existing relationships via graph theory; for this purpose, the free software Gephi 0.9.2 was employed. The metrics applied for network analysis are described in Table 3.

Table 3. Description of metrics used in network analysis. Adapted from [39].

Metric	Description
Degree centrality	Describes the number of connections. Centrality measures are essential metrics for analyzing the position of an actor in a network.
Betweenness centrality	Quantifies the frequency or number of times a node acts as a bridge along the shortest path between two other nodes.
Closeness centrality	Corresponds to the level of influence of the nodes based on the shortest routes from a node to its well-connected neighbors.
Modularity class	Measurement of the network structure. It was designed to measure the strength of the division of a network into modules and detects communities.

3.3. Stage 3: Validation of Digital Tools

Two rating systems were developed for criteria considered critical to enable the automation of a monitoring and control system, according to the previously mentioned subdivision according to their function, i.e., information capture and communication and collaboration. Scores are provided according to the qualification obtained, i.e., satisfactory performance, intermediate performance and poor performance, marked with white, gray and black, respectively. Likewise, a score was assigned with values of 2, 1 and 0. For each of the criteria, a brief explanation is presented, as shown in Tables 4 and 5.

Table 4. Criteria scoring system for FDCT. Adapted from [10].

Criteria	Good Performance (2)	Intermediate Performance (1)	Poor Performance (0)
Utility	General occasion solution	Solution for general occasions with some limitations	Can only be used on limited occasions
Time efficiency	Instantaneous recovery of information or takes less time than the traditional method	Recovery of the information takes the same amount of time as the traditional method	Recovery of information takes longer than the traditional method
Accuracy	Most or all of the data obtained is accurate	Accuracy of some data	Errors in all data
Automation level	Most or all of the steps in the process are automated	Only a few steps of the process are automated	None
Training	None	Needs training, learning facility	Needs specialized personnel
Equipment	Portable and easy-to-use equipment within easy reach	Medium-sized equipment, difficult to transport	Oversized equipment, not movable

Table 5. Criteria scoring system for CT. Adapted from [9].

Criteria	Good Performance (2)	Intermediate Performance (1)	Poor Performance (0)
Interoperability	Communicates with different software by standardized interfaces and processes	Communicates with some software by standardized interfaces and processes	Does not communicate with different software
Virtualization	Collects and monitors progress via electronic media	Collects and monitors progress by electronic and manual means	Manually collects and monitors progress
Decentralization	Allows for the delegation of actions across organizations at CSC level and allows relevant stakeholders to independently make decisions.	Partially allows for the delegation of actions and/or decisions at the CSC level.	Does not allow for the delegation of actions to organizations at the CSC level or the decision making of stakeholders.
Real-time capacity	Instantaneous recovery of information or takes less time than the traditional method	Recovery of the information takes the same amount of time as the traditional method	Recovery of information takes longer than the traditional method
Service oriented	Satisfies customer requirements, internal interests and CSC participants.	Partially satisfies customer requirements, internal interests and CSC participants.	Does not meet the requirements of the customer, internal interests and CSC participants.
Flexibility	Adapts to changing stakeholder requirements	Partially adapts to changing stakeholder requirements	Failure to adapt to changing stakeholder requirements

In the case of FDCT, six criteria are considered as follows: usability, time efficiency, accuracy, level of automation, training required and equipment [10]. On the other hand, for CTs, six criteria are considered as follows: interoperability, virtualization, decentralization, real-time capability, service orientation and flexibility [8].

The ratings were assigned based on the information obtained in the first and second stages of this research. Although each of the explored technologies has unique advantages and limitations, this rating is framed to the application for monitoring and control in the execution stage. It is worth mentioning that there are criteria that do not apply to certain technologies, in which case they are registered with the acronym NA.

Additionally, the ratings were grouped according to the frequency obtained for each of the criteria, so that the results have greater representativeness, thus avoiding the linearity of the criteria, since the choice of one technology over another depends on factors such as application, accuracy and the scale of the project. For this reason, the adopters determine which factor or factors they wish to prioritize.

These qualifications were then validated by means of an expert judgment made up of academics and researchers. The experts summoned had to meet the following criteria: (i) more than 10 years of experience in the field of monitoring and control of works and (ii) experience in the application or research in management models for monitoring and control of civil works via the remote adoption of digital tools. Thus, five experts were invited to participate in this research (Table 6).

Table 6. Characterization of expert judgment.

Profession (Grade)	Occupation	Work Area	Years of Experience
Civil Engineer, PhD	Researcher	BIM, lean and programming	Brazil
Civil Engineer, PhD	Academic Consultant	BIM, lean and construction management	Chile
Civil Engineer, PhD	Academic	Technology, construction management and robotics	Chile
Industrial Engineer, PhD	Researcher Consultant	Monitoring technologies and construction management	USA
Civil Engineer, PhD	Academic Consultant	Lean, construction management and monitoring	Chile

4. Results and Discussion

4.1. Cyber-Physical Systems

A cyber-physical system is made up of layers and components specialized to processing, communication, sensing and control functions. Multiple devices, hardware components, computational resources and sensors are all connected by communication protocols. Actuators, machines, robots and devices all have computing resources that are responsible for local decision making and onboard control. On the other hand, more centralized systems acquire and process data from a variety of sources. Higher-level computational processes may be responsible for autonomous or semiautonomous decision making at the system level, whereas control algorithms may be utilized to maintain the process-specific parameters of a device or machine [8].

Additionally, 3D data structures are significant in the architecture and construction domains; they are used to record both design intents and the as-built condition. Multiple input and monitoring devices, such as user interfaces and displays, may be included in a cyber-physical system [8]. Deferred information and inefficient communication among project stakeholders limit the efficiency of construction monitoring. Therefore, it is necessary to inquire about collaborative work procedures, where the collection, analysis and dissemination of information are standardized [29].

Based on these findings, the variation between the planned and executed schedule and budget of construction projects is due to the absence of a system that integrates the necessary tools to manage this type of project, from their planning and design to their execution, follow-up and control [40]. Emphasizing the last-mentioned stage, we examine the tools whose innovative use of technology allows for the monitoring of progress in projects, which are shown in Figure 4, classified according to their data capture (FDCT) or collaboration (CT) status, where the main key characteristics associated with Industry 4.0 can be distinguished.

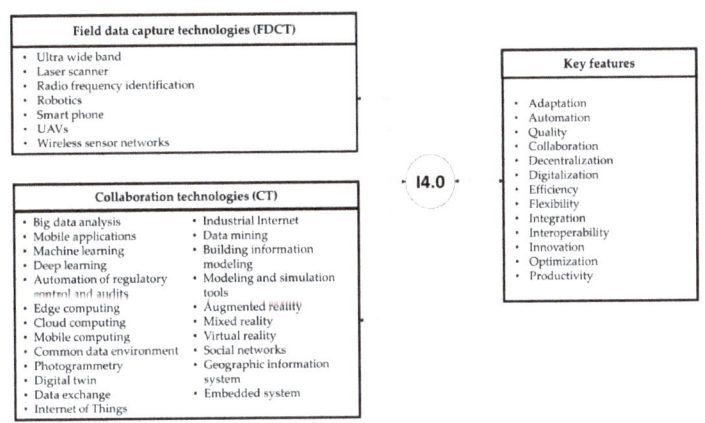

Figure 4. Key technologies and features of Industry 4.0. Adapted from [8].

4.1.1. Field Data Capture Technologies (FDCT)

Progress data have encouraged many researchers and practitioners to introduce various data acquisition technologies via digitization and an automated process of continuous and accurate monitoring of all activities. With these technologies, a better diagnosis and prognosis for the whole construction process is facilitated, with a direct impact on the improvement of production capacities [41]. Construction 4.0 proposes automating and digitizing design and construction processes, with a heavy focus on real-time data capture and technology integration in on-site construction. Some of these technologies have already been tried and used in construction, yielding some promising but limited outcomes; hence, Construction 4 is currently being implemented [41].

Technologies have proven their effectiveness in different functions, such as detection, counting, object identification, measurement of execution times and speeds, quality control, site conditions, location and tracking of elements [42]. To automate the tracking of construction progress, several methods can be used with different types of data acquisition; among the most used are imaging techniques and geospatial techniques [33].

These technologies are characterized according to their capture, visualization and/or geospatial pressure capabilities. Vision-based technologies, such as digital imaging, have enabled the development of civil engineering-related applications that aid in the design, construction and maintenance of construction projects [37]. Visual information from regions that are difficult to access can be easily collected; for example, daily photographs arranged in chronological order make it possible to track changes at a construction site. Therefore, there is the possibility to control the oldest state of the stored information, as all of them are collected in a database [43]. Studies using vision-based technologies for progress monitoring can be grouped into three categories: (i) studies on the generation of as-built 3D models integrated with BIM, (ii) studies using image processing and machine learning methods to monitor the construction progress and (iii) studies using unmanned aerial vehicles for autonomous data collection [44]. In recent years, the focus of researchers has been the use of unmanned aerial vehicles (UAVs) or drones that can be employed for photographic surveys to develop accurate three-dimensional models for intelligent monitoring of the construction progress of large-scale projects [45]. The use of unmanned aerial vehicles (UAVs) or drones has captured the attention of researchers and professionals [46]. UAVs can be employed in different uses depending on the types of sensors that are incorporated in the equipment [47]. Using UAVs with a camera sensor, UAVs allow for photographic surveys to develop accurate three-dimensional models for various applications, such as intelligent monitoring of the construction progress of large-scale projects [45,48], emergency assistance [49] and industrial operations safety monitoring [50–52].

It is common for several of these methods, such as laser scanning, photogrammetry and videogrammetry, to generate three-dimensional point clouds, which allow for the construction of as-built models that give the opportunity to identify the progress at a construction site from various angles and, thus, evaluate the current progress of a project. Among these methods, photogrammetry is the most economical and efficient technique for obtaining a 3D point cloud, which is necessary for progress documentation [53]. Additionally, data fusion using overlapping time frames or overlapping location information from multiple sources is possible by matching the same timestamps or locations. In this way, it is possible to validate the geometry of 3D models with geospatial techniques, such as GPS coordinates, and thus perform centimeter-level precision measurements, such as length, area, and volume measurements [43]. These accuracy measurements make it possible to evaluate the amount of work performed and efficiently and accurately identify construction deviations [19]. There are a considerable number of studies on outdoor progress monitoring; however, research on indoor progress monitoring is lacking. Researchers have used image processing methods or laser scanners but have not utilized photogrammetry [19]. Machine teams, robots and employees can collaborate on shared building activities thanks to device-to-device communication, which can connect many concurrent processes and enable new modalities of construction. These possibilities come with obstacles and necessitate the development of new enabling technologies, such as domain-specific hardware and software tools tailored to the needs and restrictions of the construction industry [54]. The use of the technologies could be conditioned to meteorological conditions [55]. In the case of the use of UAVs and image capture, recreational and semi-professional drones may not be prepared to withstand rain, for example. Additionally, light conditions on a sunny or cloudy day may affect the quality of the images, so it is necessary to consider weather conditions when planning flights [55–57].

4.1.2. Communication and Collaboration Technologies (CT)

Industry 4.0 generates a large amount of data. This information needs to be processed, analyzed and used efficiently, which demands solid technological management. By definition, this data management must occur in real time and needs human support. [58]. In this sense, the quality of the information generated, its efficiency, the format in which it is transferred, its applications and its subsequent uses are key elements for effective management [35] for correct processing to provide meaningful information [59]. The application of virtual collaboration in construction necessitates the use of various novel technologies and communications, collaboration software and visualization applications to create a better collaborative environment. The goal of planning and executing virtual collaboration is to provide a communication platform that has the potential to share essential visual information to support communication and knowledge sharing among planners, engineers and other team members [60].

It is widely accepted that to improve productivity and performance in construction, the industry requires an integrated collaborative approach to project execution. Collaborative environments provide tools that support the exchange of information among different applications, simultaneous access to data and the sharing of information sources across a network to enable collaboration among different users and support for all teams, addressing the challenges of fragmentation. The correct management of information is crucial to achieve these objectives [60]. Although there has been rapid growth in the development of collaborative tools and systems in recent years, especially in communication, visualization, information and knowledge management, the adoption and application of these tools has been slow and with mixed levels of success. Therefore, a well-defined methodology for collaborative work is needed [60].

A large amount of stored and processed data requires a robust storage system, which allows for access almost anywhere and by any stakeholder, inside or outside a network, enabling global data management, control and analysis. [61]. This system is particularly complex in an industry that is characterized by fragmentation and cascade planning of processes and performance stages of its disciplines and professionals. These barriers must be overcome, both inside and outside an organization, by actively involving suppliers in the logistics and innovation processes [62].

4.2. Interoperability: Integrated Use of FDCT and CT

The selection of a solution that works for each organization will depend on its role in CSCs and the investments being made. In choosing a particular solution, organizations should consider the compatibility among technologies [40] according to the actions carried out both individually and concurrently [63]. There are technologies with higher development; in this sense, Table 7 presents eight actions aligned with Construction 4.0 in relation to the monitoring and control of tasks during the construction phase. It is evident that the same technology can have several different actions, giving it a competitive advantage. Given the need to identify the unobstructed data exchange among different cyber-physical systems, a search for interoperability is carried out within the tools offered in the market. In this sense, Tables 8 and 9 represent the intra-connections for the cases of FDCT and CT, respectively, and Table 10 identifies the interconnections, which represent the existing combination between the two subdivisions of concurrent technologies.

Table 7. Actions associated with the technologies applied during the construction phase: monitoring and control. Construction 4.0. Adapted from [62].

Action	Description	Technologies
Automate	Total or partial execution of technical tasks without human intervention.	Automated regulation checking and audits (ARCA); big data analytics (BDA); data mining (DM); deep learning (DL); embedded system (ES); Internet of Things (IoT); industrial Internet (II); machine learning (ML); robotics (R)
Communicate	Transmit data, information or knowledge to a human.	Building information modeling (BIM); common data environment (CDE); cloud computing (CC); data sharing (DS); edge computing (EC); Internet of Things (IoT); industrial Internet (II); mixed reality (MR); mobile applications (MAA); mobile computing (MC); radio frequency identification (RFID); smartphone (SMP); ultrawide band (UWB); virtual reality (VR); wireless sensor network (WSN)
Locate	Track the positioning of humans and materials in space and time.	Augmented reality (AR); deep learning (DL); geographic information system (GIS); global positioning system (GPS); laser scanning (LS); mixed reality (MR); radio frequency identification (RFID); unmanned aerial vehicle (UAV); virtual reality (VR); wireless sensor networks (WSN)
Rebuild	This action translates an existing physical system into a digital model.	Building information model (BIM); geographic information system (GIS); laser scanning (LS); photogrammetry (PHT); unmanned aerial vehicle (UAV)
Simulate	To represent the behavior of a given process.	Building information model (BIM); digital twin (DT); geographic information system (GIS); machine learning (ML); simulations models and tools (SMT); virtual reality (VR)
Visualize	It allows for a visual digital model to be made available to interested parties.	Augmented reality (AR); building information model (BIM); digital twins (DT); mobile applications (MAA); mobile computing (MC); mixed reality (MR); smartphones (SMP); virtual reality (VR)
Transfer data	Longitudinal transferability of information of interest among stakeholders.	Building information model (BIM): industrial Internet (II); Internet of Things (IoT); mobile applications (MAA); mobile computing (MC); radio frequency identification (RFID); smartphones (SMP); wireless sensor network (WSN)

Table 8. Interoperability matrix among field data capture technologies—intra-FDCT.

FDCT	RFID	LS	UWB	WSN	UAV	R	SMP
RFID		[63]	[64]	[65]	[66]	[67]	X
LS			[68]	X	[69]	[70]	X
UWB				[71]	[72]	[72]	X
WSN					X	[73]	X
UAV						[74]	[75]
R							[76]
SMP							

Note: "[number]" is the reference number that evidence an integration between the two technologies. "X" implies that the authors do not find any refence between the two technologies.

Table 9. Interoperability matrix among communication technologies and intra-CT collaboration.

CT	AR	VR	DT	MR	MC	SMT	FTG	MAA	ARCA	BDA	DM	DL	ES	GIS	II	IoT	ML	BIM	CC	CDE	DS	EC	SN
SN	X	X	X	X	X	X	X	X	X	X	X	X	X	X	X	X	X	[77]	X	X	X	X	
EC	X	[78]	[78]	X	[79]	X	[80]	X	X	X	[81]	X	X	X	X	X	[82]	X	[83]	X	X		
DS	[84]	X	[85]	X	[86]	X	X	X	X	[87]	X	X	[88]	X	[89]	X	X	X	X	X			
CDE	X	X	X	X	X	X	X	[89]	X	X	X	X	X	X	X	X	X	[90]	X				
CC	X	X	[91]	X	X	X	[92]	X	X	X	X	X	[93]	X	[94]	[95]	[96]						
BIM	[97]	[98]	[99]	X	[100]	[101]	[72]	[98]	X	[102]	[103]	[104]	[105]	[106]	X	[107]	[108]						
ML	X	[11]	[109]	X	[31]	[110]	X	X	[111]	X	X	X	[112]	X	X								
IoT	[113]	[114]	[115]	[116]	X	X	[117]	X	X	X	X	X	X	[118]									
II	[113]	X	X	[119]	X	X	X	[120]	X	[121]	X	X	X	X									
GIS	[122]	X	[109]	[123]	[124]	X	[125]	X	X	X	X	X											
ES	X	X	X	X	X	X	[126]	X	X	[127]	X	X											
DL	[128]	X	[129]	[130]	X	X	[131]	X	X	X													
DM	[132]	X	X	X	X	[133]	[134]	X	X														
BDA	X	X	X	X	X	[135]	X	[86]	X														
ARCA	X	X	X	X	X	X	X																
MAA	[132]	X	X	[116]	[136]	X	[137]																
FTG	[138]	X	[139]	X	X																		
SMT	[140]	[31]	[141]	[142]	X																		
MC	[143]	X	X	X																			
MR	[144]	X	X																				
DT	[145]	[146]																					
VR	[114]																						
AR																							

Note: "[number]" is the reference number that evidence an integration between the two technologies. "X" implies that the authors do not find any refence between the two technologies.

Based on the links identified in the matrices in Tables 8–10 of the previous chapter, a network analysis was performed. The results are represented by three types of connections: (1) inter-edge, (2) intra-edge and (3) full. Each node represents the different technologies, either FDCT or CT. The connections, which are represented by edges, symbolize the existing interoperability. It is worth mentioning that the nodes that do not have edges are those that do not manage to connect with other technologies, such as the automation of regulatory control and audits (ARCA), for which there is no reported evidence that this technology is linked with other technologies, which is inferred because it is a recent technology. Although it has the potential to be utilized for monitoring and control, it still fails to integrate with existing technologies, at least in the AECO industry. Table 11 shows the metrics analyzed with a description of their interpretation and how they are graphically represented in the diagrams generated.

Table 10. Integration matrix between field data capture and communication and collaboration technologies—inter-FDCT and -CT.

	RFID	LS	UWB	WSN	UAV	R	SMP
AR	X	X	X	[147]	[148]	[149]	[150]
ARCA	X	X	X	X	X	X	X
BDA	[151]	[152]	X	X	X	[153]	X
BIM	[154]	[155]	X	[156]	[157]	[158]	[159]
CC	[160]	X	X	X	[161]	X	X
CDE	X	X	X	X	X	X	x
DL	X	[162]	X	X	[163]	[164]	X
DM	[90]	X	X	X	X	X	[165]
DS	X	X	X	X	X	X	X
DT	X	[166]	X	[156]	X	X	[167]
EC	X	X	X	X	[168]	[83]	X
ES	[169]	X	X	[170]	X	[171]	[172]
FTG	X	[173]	[174]	X	[92]	[175]	[176]
GIS	[174]	[106]	[177]	[178]	[179]	[180]	[181]
II	[118]	X	X	[118]	[182]	[183]	[184]
IoT	X	X	X	[118]	[182]	[185]	[186]
MAA	X	X	X	[187]	[137]	X	X
MC	X	X	X	[187]	X	[188]	[189]
ML	[190]	[191]	X	X	[168]	[192]	[193]
MR	X	X	X	X	[194]	X	X
SMT	[133]	[195]	X	[196]	[197]	X	X
SN	X	X	X	X	X	X	X
VR	[198]	[199]	X	[200]	[201]	[202]	[150]

Note: "[number]" is the reference number that evidence an integration between the two technologies. "X" implies that the authors do not find any refence between the two technologies.

Table 11. Interpretation and graphical representation of network analysis.

Metric	Interpretation	Graphic Representation
Degree centrality	Indicates technologies with the highest number of relationships, hence, the most influential within the groups.	Node size: the larger is the node size, the higher is the degree centrality
Betweenness degree	Technologies that occupy an intermediary position. Technologies that are more intermediary have a great potential to be combined with other technologies. They have high interoperability.	Node color intensity: the darker the color, the greater the degree of intermediation.
Closeness centrality	It is based on the idea that nodes with a short distance to other nodes can collaborate, which demonstrates that technologies allow for interoperability given their functional capacity and despite having few connections. It represents an excellent position to monitor the flow of information at the network level.	Node color intensity: the darker the color, the greater its closeness centrality.
Modularity class	It arranges technologies as groups or communities according to the existing interoperability among digital tools.	Colors of communities: each color of the nodes represents a community.

For intra-FCCT connections, specifically in field data capture technologies, Figure 5 presents the metrics of modularity and centrality of the intermediary, (a) and (b), respectively.

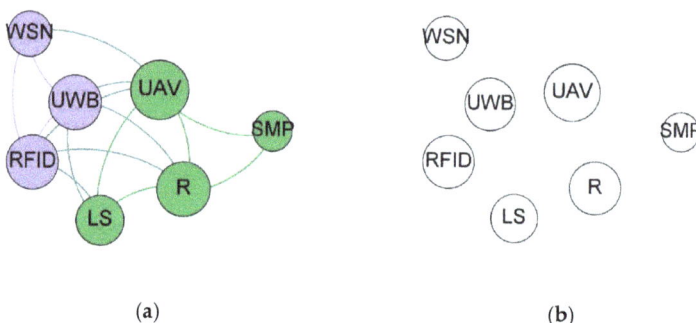

Figure 5. Intra-FDCT network analysis. (**a**) Modularity: capture technologies form two communities. (**b**) Betweenness: UAV stands out subtly; the balance of betweenness between capture technologies is maintained, as represented by the pale green color.

The purple group is characterized by technologies that are highly automated, i.e., the man–machine relationship for acquiring information is almost nil. On the other hand, it is natural that the analysis of betweenness maintains the balance, since each technology has its capabilities, and none is better than another. Thus, each technology is selected according to the function it satisfies. In contrast, for the intra-connections' analysis of the CTs, Figure 6 shows that BIM is the most powerful intermediary or the intermediary with the highest interoperability among the TC tools, i.e., it can exchange a large amount of data and knowledge. The technologies that achieve this to a lesser extent but still significantly are augmented reality (AR), photogrammetry (PHT), the Internet of Things (IoT) and digital twins (DT).

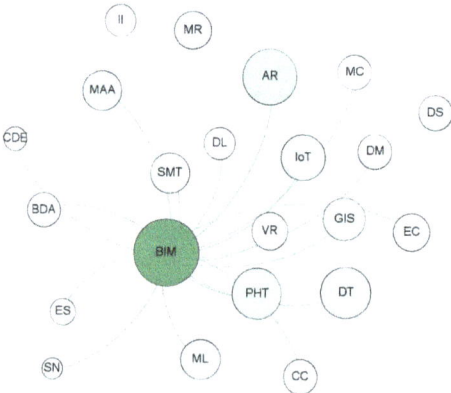

Figure 6. Analysis of intra-connections networks CT—betweenness.

Figure 7 shows the interconnections, with FDCT technologies in red and CT technologies in green. Given the magnitude of the nodes, FDCT technologies take the leading role, and drones (UAVs) again become an influential technology within the group, which is related to technologies that have different functions, mainly reconstruction, simulation and visualization. Regarding the role of intermediaries, in general, FDCTs stand out over TCs, in this case, R, WSN, SMP and RFID. It is worth mentioning that BIM, under this arrangement, does not have a major role because BIM is characterized by being a communicative and collaborative technology (refer to Figure 8).

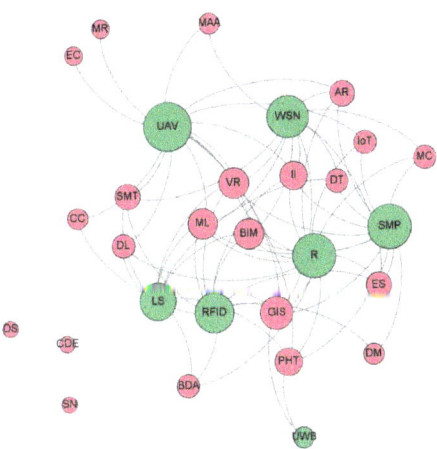

Figure 7. Interconnection network analysis—category.

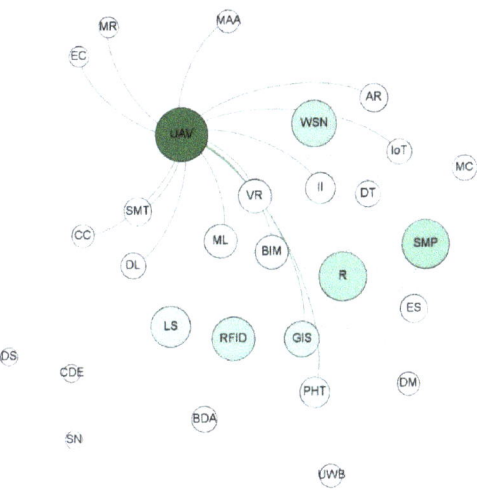

Figure 8. Interconnection network analysis—betweenness.

Figure 9 shows the full connections, where it is evident that BIM, at the moment of relating with the totality of the interactions, undoubtedly becomes the major intermediary. BIM has great potential to be employed with other technologies, whether FDCT or CT. BIM is able to connect with 80% of the technologies that have the potential to be utilized for monitoring and control in construction projects according to the evidence found in the literature.

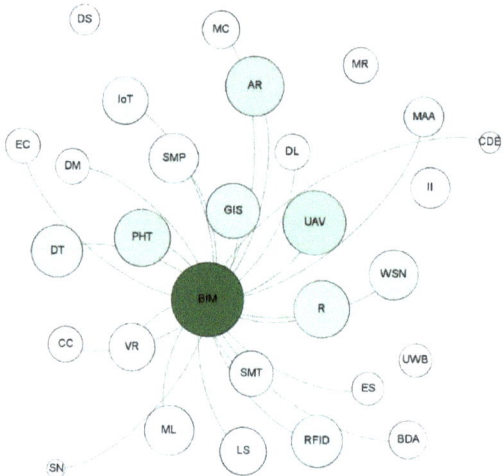

Figure 9. Full network analysis—betweenness.

In general, three communities can be distinguished in Figure 10, where the leaders are building information modeling (BIM), unmanned aerial vehicles (UAVs) and photogrammetry (PHT). The first group, led by BIM, is in charge of being the intermediary with the technologies that present less interoperability. The second group, led by UAVs, presents mostly CT-type technologies. For the orange community, CT technologies are mainly grouped and influential for data transfer, reconstruction and simulations.

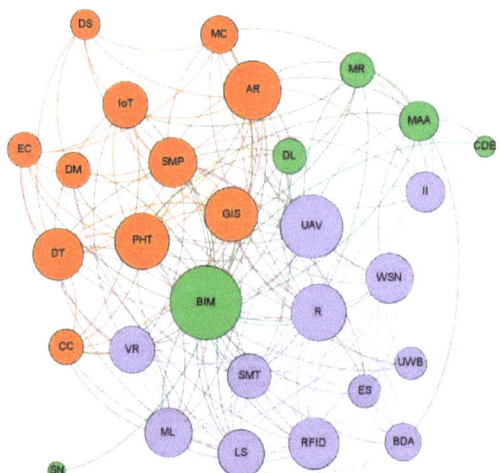

Figure 10. Full network analysis—modularity.

This triad formed by BIM, UAVs and PHTs does not coincide; the construction industry has been widely adopting these technologies, and there is evidence of success in their application. The benefits of their use extend to not only monitoring and control but also all phases of the construction process, including operation. In addition, the models generated can also be integrated with surveys of other technologies, such as laser scanners, which provide more accurate information.

Regarding the groups or communities formed in Figure 6, it can be seen that each group is formed by FDCT and CT technologies, which shows the need to have a source of data capture for the subsequent communication and collaboration that they provide. It is noteworthy that the group led by UAV is not connected to PHT or GIS.

Figure 11 shows the abovementioned BIM, which is in charge of housing the information that is extracted and analyzed by means of the other technologies. BIM becomes a valuable platform due to its high interoperability, which allows for the integration of the tools, allowing those involved in the construction process to unify the information collected in a single platform.

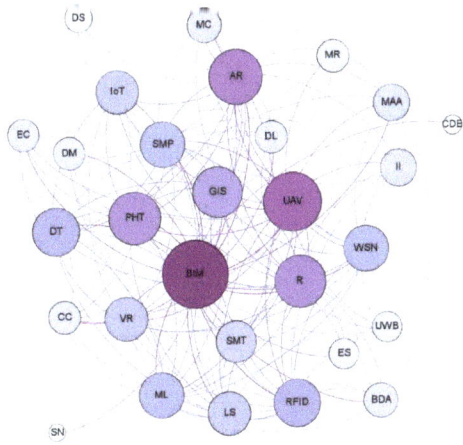

Figure 11. Full network analysis—closeness.

4.3. Validation of Digital Tools

To validate the performance of the 30 technologies that present the potential use for monitoring and control within the AIC industry, the judgment of five experts was used to consider the evaluations determined by the author. Each expert gave their judgment on each of the six criteria for the FDCT and CT groups, the ratings of which are shown in Tables 12 and 14, respectively.

In the frequency tables (refer to Tables 12 and 14), it is possible to note the level of performance achieved by each technology according to the criteria of each group, excluding the "not applicable" cases (Table 12).

Note that, in general (Table 13), technologies have a general solution or certain limitations; only smartphones have limited usefulness. For time efficiency, 100% of the technologies do not take more time than the traditional method, and 57% of the technologies meet the objective of obtaining as-built data in less time compared to the traditional method, where those that would take more time are matched by the accuracy of the information obtained, with the exception of smartphones. The accuracy of the data is 86% reliable and provides accurate information about the current state of progress. Automation is achieved in 86% of the technologies, where smartphones are again the exception.

The obstacles to the implementation of technologies, as usual, are centered on the required training and equipment. It is essential to have qualified personnel with the necessary knowledge to perform the methods. Although these technologies have become more prevalent in recent years, as they are complex techniques, their manipulation is not trivial. UAVs have become the tool that has been able to close this gap and are increasingly implemented with greater frequency due to their versatility, ease of learning and portable equipment.

Table 12. Performance matrix of digital tools for in situ data acquisition.

FDCT	Utility	Time Efficiency	Accuracy	Level of Automation	Training Required	Equipment
	UTI	EFI	ACC	AUT	TRA	EQUI
Ultrawide band (UWB)	2	2	2	2	0	1
Laser scanner (LS)	2	2	2	2	0	1
Robotics ®	1	2	2	2	1	1
Radio frequency identification (RFID)	1	1	2	2	0	1
Smartphones (SMT)	0	1	1	0	1	2
Unmanned aerial vehicle (UAV)	2	1	2	1	1	2
Wireless sensor networks (WSN)	1	2	2	2	1	2

Table 13. Frequencies obtained according to the criteria for FDCT.

Criteria	Frequency of Performance		
	Good (2)	Intermediate (1)	Poor (0)
Utility	LS; UWB; UAV	RFID; WSN; R	SMT
Time efficiency	LS; UWB; WSN; R	RFID; UAV; SMT	
Accuracy	RFID; LS; UWB; WSN; UAV; R	SMT	
Level of automation	LS; UWB; WSN; R	RFID; UAV	SMT
Training required		WSN; UAV; SMT; R	RFID; LS; UWB
Equipment	WSN; UAV; SMT	RFID; LS; UWB; R	

For the results in Table 14, the interoperability of the technologies was measured based on the information obtained in the previous chapter, considering the full relationships, that is, the combination between the intra-connections relationship and interconnections relationship of the FDCT and CT technologies. Some technologies received a score of 0 because no document was identified where these technologies interact with others. A score of 1 was assigned to technologies that presented a betweenness less than or equal to 5. Technologies with a score of 1 or 2 presented the highest values: in decreasing order, building information modeling, photogrammetry, digital twins, mobile applications and the Internet of Things.

All of the technologies collect information via electronic means; however, 39% still use the traditional method as a complementary method. In terms of decentralization, there is a distinct trend towards partial delegation of actions and/or decisions at the CSC level. On the other hand, 80% of communicative and collaborative technologies present an instantaneous retrieval of information or take less time than the traditional method.

There is a balance towards meeting the needs of stakeholders within the capabilities offered by each of the technologies. However, there is a medium adaptation to changing requirements, especially for visualization or simulation technologies such as AR, VR, DT, MR and SMT, since the changes generated in the field will not always be represented in these technologies (Table 15).

Table 14. Communication and collaboration technology tool performance matrix.

TC	Interoperability	Virtualization	Decentralization	Real-Time Capacity	Service-Oriented	Flexibility
	INT	VIR	DES	RTC	SO	FLE
Big data analysis (BDA)	1	2	NA	2	2	NA
Mobile applications (MAA)	2	1	1	1	1	1
Machine learning (ML)	1	2	2	1	1	1
Deep learning (DL)	1	2	1	2	1	1
Automated regulation checking and audits (ARCA)	0	2	1	2	1	1
Edge computing (EC)	1	2	NA	2	1	NA
Cloud computing (CC)	1	1	2	2	2	2
Mobile computing (MC)	1	1	NA	1	2	1
Common data environment (CDE)	1	1	2	2	2	2
Photogrammetry (PHT)	2	2	NA	2	2	2
Digital twins (DT)	2	2	1	1	2	1
Data sharing (DS)	1	1	2	2	2	2
Internet of Things (IoT)	2	2	1	2	2	2
Industrial Internet (II)	1	2	NA	2	2	NA
Data mining (DM)	1	1	NA	2	1	NA
Building information model (BIM)	2	1	2	2	2	2
Simulations models and tools (SMT)	2	1	1	NA	2	1
Augmented reality (AR)	1	2	1	2	2	1
Mixed reality (MR)	1	2	1	2	1	1
Virtual reality (VR)	1	2	1	2	2	1
Social networks (SN)	0	1	1	1	1	2
Geographic information system (GIS)	1	2	NA	2	2	1
Embedded system (ES)	1	2	NA	2	1	1

Table 15. Frequencies obtained according to the criteria for CT.

Criteria	Frequency of Performance		
	Good (2)	Intermediate (1)	Poor (0)
Interoperability	AR; BIM; DT; GIS; IoT; PHT	BDA; CC; CDE; DL; DM; DS; EC; ES; II; MAA; MC; ML; MR; SMT; VR	ARCA; SN
Virtualization	AR; ARCA; BDA; DL; DT; EC; ES; GIS; II; IoT; ML; MR; PHT; VR	BIM; CC; CDE; DM; DS; MAA; MC; SMT; SN	
Decentralization	BIM; CC; CDE; DS; ML	AR; ARCA; DL; DT; IoT; MAA; MR; SMT; SN; VR	
Real-time capacity	AR; ARCA; BDA; BIM; CC; CDE; DL; DM; DS; EC; ES; GIS; II; IoT; MR; PHT; VR	DT; MC; MAA; ML; SN	
Service orientation	BDA; BIM; CC; CDE; DS; GIS; II; IoT; PHT; SMT	AR; ARCA; DL; DM; DT; EC; ES; MAA; MC; ML; MR; SN; VR	
Flexibility	BIM; CC; CDE; DS; IoT; PHT; SN	AR; ARCA; DL; DT; ES; GIS; MAA; MC; ML; MR; SMT; VR	

5. Conclusions

The use of technologies for monitoring and control in construction projects involves challenges and opportunities for the AIC industry. With the current progress, it has been possible to facilitate data acquisition in the field due to the key features offered by their implementation, including automation, digitization, integration and interoperability, which are supported by the Industry 4.0 revolution.

While there is a wide range of technologies on the market that promise different functions, their use and the interoperability among them are still limited. The contribution of this research is to show an analysis of the integration and interoperability of digital tools for monitoring and control of construction projects during the execution phase. The most widely used technologies in recent decades, such as BIM, unmanned aerial vehicles and photogrammetry, are technologies that present a greater compatibility, which allows for a reduction in the fragmentation of information and automation to a certain extent, since human support is still required. As evidenced, BIM is a tool with a great capacity to store data and it allows for the dissemination of this information to stakeholders. BIM marks a new paradigm, and the interoperability and integration it has with different technologies allows us to improve the efficiency of information management in construction due to the unification of information. In summary, BIM was characterized by generating connections with emerging technologies in the industry and, therefore, its potential is evidently superior to that of other technologies.

A limitation that emerged from the research was the limited evidence found in the literature. Although technological advances have increased, the AEC industry has slowly adopted these innovative techniques, even more so for the function and stage of focus of this research. Implementing this type of technique in monitoring and control is still considered experimental in nature and, therefore, requires a significant investment from companies (whether public or private). Only a minority of companies dare to innovate and leave behind the traditional method. Another limitation of the study was that a set of tools was not applied in a real case of professional practice with the objective of analyzing and evaluating the benefits and obstacles of implementing this type of technologies associated with Construction 4.0; therefore, it is recommended to carry out studies in real cases and evaluate the impact of their implementation from a quantitative and qualitative perspective.

On the other hand, it is difficult to point out which technology or technologies will lead the monitoring and control process, since the technologies investigated in this paper will not necessarily be the technologies used in the medium term. In addition, given the dynamic and changing nature of construction, it is currently difficult to standardize the monitoring and control process. Defining which technologies and the sequence of the process is complex, since there are different factors and variables that must be controlled according to the particularity of each project. In addition, it is necessary to individually or jointly evaluate the benefits and limitations of the technologies according to a given context. It is also a challenge to promote a culture of implementation and adaptation of these innovative methods.

From a practical point of view, this research seeks to contribute to the discussion and promotion of other studies on the integration and interoperability of these tools studied. It is interesting to study the advances made by the industry, analyze how these technologies are implemented and evaluate the added value in the projects where they are implemented. Researchers are also encouraged to experiment with finding interoperability, which was not possible in the current literature. In this way, researchers can document them and provide more information to contribute to a greater supply of techniques to achieve full automation of the process.

As a future line of work, to complement what has been pointed out in this document, it is possible to work on new criteria considered critical to allow for the automation of a monitoring and control system. Adding new criteria, weighting them, adding multiplying factors, etc. would make it possible to determine with greater certainty and rigor which are more relevant when choosing one tool over another.

Additionally, we encourage developing an incremental methodology of adoption of a group of technologies, where the levels of adoption of automation are defined and where the technologies are subject to the functionality according to the objective to be achieved due to the flexibility provided by the technologies. Thus, they can manage the construction process of any building by (1) allowing for the development of integrated solutions via a shift towards the flexibility of processes and autonomy and communication among devices connected by multiple parallel processes and (2) allowing for new modes of networked and collective construction in common construction tasks.

Author Contributions: Conceptualization, R.F.H. and L.D.-V.; methodology, R.F.H.; software, L.D.-V.; validation, F.M.-L.R. and E.A.; formal analysis, L.D.-V.; writing—original draft preparation, L.D.-V.; writing—review and editing, R.F.H., E.A. and F.M.-L.R.; visualization, L.D.-V. and F.M.-L.R.; supervision, R.F.H. and E.A. All authors have read and agreed to the published version of the manuscript.

Funding: The APC was paid by Pontificia Universidad Católica de Valparaíso. This research was funded by CONICYT grant number CONICYT-PCHA/InternationalDoctorate/2019-72200306 for funding the graduate research of Muñoz-La Rivera.

Institutional Review Board Statement: Not applicable.

Informed Consent Statement: Not applicable.

Data Availability Statement: Not applicable.

Acknowledgments: The authors wish to thank the TIMS space (Technology, Innovation, Management, and Innovation) of the School of Civil Engineering of the Pontificia Universidad Católica de Valparaíso (Chile), where part of the research was carried out.

Conflicts of Interest: The authors declare no conflict of interest.

References

1. Restrepo, D. La Importancia del Sector de la Construcción en Materia Económica—LA.Network. Available online: https://la.network/la-importancia-del-sector-de-la-construccion-en-materia-economica/ (accessed on 27 October 2021).
2. Alarcón, L.F. ¿Sabías que la Productividad de la Construcción es más baja que la de toda la Economía? Available online: https://www.claseejecutiva.com.ec/blog/articulos/sabias-que-la-productividad-de-la-construccion-es-mas-baja-que-la-de-toda-la-economia/ (accessed on 21 October 2021).
3. CICA; CCHC; FIIC. *Situación Mundial COVID-19 Industria de la Construcción y General*; Informe Internacional: Santiago, Chile, 2020.
4. Sanchéz, A. Digitalización y Sostenibilidad, los Principales Desafíos para la Reactivación de las Empresas tras la Crisis del COVID-19. Available online: https://www.eurochile.cl/es/noticias/cooperacion-empresarial/digitalizacion-y-sostenibilidad-los-principales-desafios-para-la-reactivacion-de-las-empresas-tras-la-crisis-del-covid-19/ (accessed on 27 October 2021).
5. Griffor, E.R.; Greer, C.; Wollman, D.A.; Burns, M.J. Framework for Cyber-Physical Systems: Volume 1, Overview NIST Special Publication 1500-201 Framework for Cyber-Physical Systems: Volume 1. Overview. *Nist* **2017**, *1*, 79.
6. Muñoz-La Rivera, F.; Mora-Serrano, J.; Valero, I.; Oñate, E. Methodological-Technological Framework for Construction 4.0. *Arch. Comput. Methods Eng.* **2021**, *28*, 689–711. [CrossRef]
7. Chen, J.; Zhou, J. Revisiting Industry 4.0 with a Case Study. In Proceedings of the IEEE 2018 International Congress on Cybermatics: 2018 IEEE Conferences on Internet of Things, Green Computing and Communications, Cyber, Physical and Social Computing, Smart Data, Blockchain, Computer and Information Technology, iThings/Gree, Halifax, NS, Canada, 30 July–3 August 2018; pp. 1–6.
8. Sawhney, A.; Riley, M.; Irizarry, J. *Construction 4.0: An Innovation Platform for the Built Environment*, 2020th ed.; Sawhney, A., Riley, M., Irizarry, J., Eds.; Routledge: New York, NY, USA, 2020; Volume 53, ISBN 9788578110796.
9. Braun, A.; Tuttas, S.; Borrmann, A.; Stilla, U. A Concept for Automated Construction Progress Monitoring Using BIM-based Geometric Constraints and Photogrammetric Point Clouds. *J. Inf. Technol. Constr.* **2015**, *20*, 68–79.
10. Kopsida, M.; Brilakis, I.; Vela, P. A Review of Automated Construction Progress and Inspection Methods. In Proceedings of the 32nd CIB W78 Conference on Construction IT, Eindhoven, The Netherlands, 27–29 October 2015; pp. 1–12.
11. Dib, H.; Adamo-Villani, N.; Issa, R.A. Gis-Based Integrated Information Model to Improve Building Construction Management: Design and Initial evaluation. *Proc. CONVR* **2014**, 769–781.
12. Li, Y.; Liu, C. Applications of Multirotor Drone Technologies in Construction Management. *Int. J. Constr. Manag.* **2019**, *19*, 401–412. [CrossRef]
13. García, J.; Salazar, P. Métodos de Administración y Evaluación de Riesgos. Bachelor's Thesis, Universidad de Chile, Santiago, Chile, 2005.
14. Trejo, N. Estudio de Impacto del uso de la Metodología BIM en la Planifiación y Control de Proyectos de Ingeniería y Construcción. Bachelor's Thesis, Universidad de Chile, Santiago, Chile, 2018.
15. Jrade, A.; Lessard, J. An Integrated BIM System to Track the Time and Cost of Construction Projects: A Case Study. *J. Constr. Eng.* **2015**, *2015*, 579486. [CrossRef]
16. GMC Ingeniería Planificación y Control de la Edificación y Obra Civil. Available online: https://www.gmcingenieria.com/servicios/planificacion-y-control-de-la-edificacion-y-obra-civil/ (accessed on 21 May 2020).
17. *Dirección de Planeamiento MOP Gestión y Monitoreo de Planes de Obras Públicas Implementación, Metas e Indicadores*; MOP Chile: Santiago de Chile, Chile, 2012.
18. Villavicencio, A.; Gabriela, M.; Fajardo, M.; De Fátima, A. Uso de Tecnologías de Adquisición de datos para Optimizar los Tiempos de Monitoreo del Progreso de la Construcción en Edificios Residenciales. Bachelor's Thesis, Universidad Peruana de Ciencias Aplicadas, Lima, Peru, 2019.
19. Mahami, H.; Nasirzadeh, F.; Ahmadabadian, A.H.; Nahavandi, S. Automated progress controlling and monitoring using daily site images and building information modelling. *Buildings* **2019**, *9*, 70. [CrossRef]
20. Project Management Institute. *Guía de los Fundamentos para la Dirección de Proyectos (Guía del PMBOOK)*; Quinta Edition; Project Management Institute, Inc.: Newtown Square, PA, USA, 2013; ISBN 0428630790.
21. CORFO. PMG Informe final fase 3, Hoja de Ruta PyCS 2025. In *Programa Estratégico Nac. Product. y Sustentabilidad en la Construcción 2016*; CORFO Chile: Santiago de Chile, Chile, 2016; p. 172.
22. Navon, R.; Goldschmidt, E. Can Labor Inputs be Measured and Controlled Automatically? *J. Constr. Eng. Manag.* **2003**, *129*, 437–445. [CrossRef]
23. Muñoz Velasco, M. Aplicación de un Modelo de Planificaciòn Financiera en Pequeñas y Medianas Empresas Constructoras en Chile (AVG). Bachelor's Thesis, Universidad del Bío-Bío, Concepción, Chile, 2016.
24. Petrov, I.; Hakimov, A. Digital Technologies in Construction Monitoring and Construction Control. In *Proceedings of the IOP Conference Series: Materials Science and Engineering*; IOP Publishing: St. Petersburg, Russia, 2019; Volume 497, pp. 1–9.
25. Alvarenga, T.W.; Neves Da Silva, E.; Brasil de Brito Mello, L. BIM and Lean Construction: The Evolution Obstacle in the Brazilian Civil Construction Industry. *Technol. Appl. Sci. Res.* **2017**, *7*, 1904–1908. [CrossRef]
26. de, M. Nascimento, D.L.; Sotelino, E.D.; Caiado, R.G.G.; Ivson, P.; Faria, P.S. Sinergia entre Princípios do Lean Thinking e Funcionalidades de BIM na Interdisciplinaridade de Gestão em Plantas Industriais. *J. Lean Syst.* **2017**, *2*, 80–105.

27. Saad, I.M.H. The project reporter: Multimedia Progress Reporting for Construction Projects. In *Construction Congress VI: Building Together for a Better Tomorrow in an Increasingly Complex World*; ASCE: Orlando, FL, USA, 2000; Volume 278, pp. 1165–1176. [CrossRef]
28. Xu, Q.; Chong, H.Y.; Liao, P.C. Collaborative Information Integration for Construction Safety Monitoring. *Autom. Constr.* **2019**, *102*, 120–134. [CrossRef]
29. Alizadehsalehi, S.; Yitmen, I. A Concept for Automated Construction Progress Monitoring: Technologies Adoption for Benchmarking Project Performance Control. *Arab. J. Sci. Eng.* **2019**, *44*, 4993–5008. [CrossRef]
30. Ahmed Memon, Z.; Abd Majid, M.Z.; Mustaffar, M. Digitalizing Construction Monitoring (DCM): An Overview of Malaysian Construction Industry and Proposing Prototype Software. In Proceedings of the 6th Asia-Pacific Structural Engineering and Construction Conference (APSEC 2006), Kuala Lumpur, Malaysia, 5–6 September 2006.
31. Kapogiannis, G.; Mlilo, A. Digital Construction Strategies and BIM in Railway Tunnelling Engineering. In *Tunnel Engineering-Selected Topics*; IntechOpen: London, UK, 2019. [CrossRef]
32. Yang, L.R.; O'Connor, J.T.; Chen, J.H. Assessment of Automation and Integration Technology's Impacts on Project Stakeholder Success. *Autom. Constr.* **2007**, *16*, 725–733. [CrossRef]
33. Pour Rahimian, F.; Seyedzadeh, S.; Oliver, S.; Rodriguez, S.; Dawood, N. On-demand Monitoring of Construction Projects through a Game-like Hybrid Application of BIM and Machine Learning. *Autom. Constr.* **2020**, *110*, 103012. [CrossRef]
34. Navon, R.; Shpatnitsky, Y. A Model for Automated Monitoring of Road Construction. *Constr. Manag. Econ.* **2005**, *23*, 941–951. [CrossRef]
35. Memon, Z.A.; Majid, M.Z.A.; Mustaffar, M. An Automatic Project Progress Monitoring Model by Integrating Auto CAD and Digital Photos. In Proceedings of the International Conference on Computing in Civil Engineering, Cancun, Mexico, 12–15 July 2005; pp. 1–13.
36. Robuffo, F. An Innovative Approach for Automated Jobsite Work Progress Assessment. Ph.D. Thesis, Università Politecnbica delle Marche, Ancona, Italy, 2013.
37. Brilakis, I.; Soibelman, L.; Shinagawa, Y. Material-Based Construction Site Image Retrieval. *J. Comput. Civ. Eng.* **2005**, *19*, 341–355. [CrossRef]
38. Kapogiannis, G.; Sherratt, F. Impact of Integrated Collaborative Technologies to form a Collaborative Culture in Construction Projects. *Built Environ. Proj. Asset Manag.* **2018**, *8*, 24–38. [CrossRef]
39. Kuz, A.; Falco, M.; Giandini, R. Análisis de redes sociales: Un caso práctico. *Comput. Sist.* **2016**, *20*, 89–106. [CrossRef]
40. Cárdenas, C.; Zapata, P.; Lozano, N. Building Information Modeling 5D and Earned Value Management methodologies integration through a computational tool. *Rev. Ing. Constr.* **2018**, *33*, 263–278. [CrossRef]
41. García de Soto, B.; Agustí-Juan, I.; Hunhevicz, J.; Joss, S.; Graser, K.; Habert, G.; Adey, B.T. Productivity of digital fabrication in construction: Cost and time analysis of a robotically built wall. *Autom. Constr.* **2018**, *92*, 297–311. [CrossRef]
42. Alias, C.; Jawale, M.; Goudz, A.; Noche, B. Applying Novel Future-internet-based Supply Chain Control Towers to the Transport and Logistics Domain. In Proceedings of the ASME 2014 12th Biennial Conference on Engineering Systems Design and Analysis, ESDA 2014, Copenhagen, Denmark, 25–27 June 2014; Volume 3.
43. Ersoz, A.B.; Pekcan, O.; Tokdemir, O.B. Lean Project Management using Unmanned Aerial Vehicles. *Tamap J. Eng.* **2019**, *2018*, 65. [CrossRef]
44. Asadi, K.; Ramshankar, H.; Pullagurla, H.; Bhandare, A.; Shanbhag, S.; Mehta, P.; Kundu, S.; Han, K.; Lobaton, E.; Wu, T. Vision-based Integrated Mobile Robotic System for Real-time Applications in Construction. *Autom. Constr.* **2018**, *96*, 470–482. [CrossRef]
45. Coetzee, G.L. Smart Construction Monitoring of Dams with UAVS—Neckartal Dam Water Project Phase 1. In Proceedings of the Smart Dams and Reservoirs—Proceedings of the 20th Biennial Conference of the British Dam Society, Swansea, UK, 13–15 September 2018; pp. 445–456.
46. Zohdi, T.I. Multiple UAVs for Mapping: A review of Basic Modeling, Simulation, and Applications. *Annu. Rev. Environ. Resour.* **2018**, *43*, 523–543. [CrossRef]
47. Cummings, A.R.; Mckee, A.; Kulkarni, K.; Markandey, N. The Rise of UAVs. *Photogramm. Eng. Remote Sens.* **2017**, *83*, 317–325. [CrossRef]
48. Jacob-Loyola, N.; Muñoz-La Rivera, F.; Herrera, R.F.; Atencio, E. Unmanned Aerial Vehicles (Uavs) for Physical Progress Monitoring of Construction. *Sensors* **2021**, *21*, 4227. [CrossRef]
49. Vallejo, D.; Castro-Schez, J.J.; Glez-Morcillo, C.; Albusac, J. Multi-agent architecture for information retrieval and intelligent monitoring by UAVs in Known Environments Affected by Catastrophes. *Eng. Appl. Artif. Intell.* **2020**, *87*, 103243. [CrossRef]
50. Ko, Y.D.; Song, B.D. Application of UAVs for Tourism Security and Safety. *Asia Pac. J. Mark. Logist.* **2021**, *33*, 1829–1843. [CrossRef]
51. Liu, D.; Chen, J.; Hu, D.; Zhang, Z. Dynamic BIM-augmented UAV Safety Inspection for Water Diversion Project. *Comput. Ind.* **2019**, *108*, 163–177. [CrossRef]
52. Jeelani, I.; Gheisari, M. Safety Challenges of UAV Integration in Construction: Conceptual Analysis and Future Research Roadmap. *Saf. Sci.* **2021**, *144*, 105473. [CrossRef]
53. Wehbe, R.; Shahrour, I. Use of BIM and Smart Monitoring for buildings' Indoor Comfort Control. In Proceedings of the MATEC Web of Conferences, Lille, France, 8–10 October 2019; Volume 295, p. 02010.

54. Spengler, A.J.; Alias, C.; Magallanes, E.G.C.; Malkwitz, A. Benefits of Real-Time Monitoring and Process Mining in a Digitized Construction Supply Chain. *Mobilität Zeiten Veränderung* **2019**, 411–435. [CrossRef]
55. Prosser-Contreras, M.; Atencio, E.; La Rivera, F.M.; Herrera, R.F. Use of Unmanned Aerial Vehicles (Uavs) and Photogrammetry to Obtain the International Roughness Index (iri) on Roads. *Appl. Sci.* **2020**, *10*, 8788. [CrossRef]
56. Romero-Chambi, E.; Villarroel-Quezada, S.; Atencio, E.; Rivera, F.M. La Analysis of Optimal Flight Parameters of Unmanned Aerial Vehicles (UAVs) for Detecting Potholes in Pavements. *Appl. Sci.* **2020**, *10*, 4157. [CrossRef]
57. Jofré-Briceño, C.; Muñoz-La Rivera, F.; Atencio, E.; Herrera, R.F. Implementation of Facility Management for Port Infrastructure through the Use of UAVs, Photogrammetry and BIM. *Sensors* **2021**, *21*, 6686. [CrossRef] [PubMed]
58. Mohamed, M. Challenges and Benefits of Industry 4.0: An overview. *Int. J. Supply Oper. Manag.* **2018**, *5*, 256–265.
59. Fard, M.G.; Peña-Mora, F. Application of Visualization Techniques for Construction Progress Monitoring. In *Proceedings of the Computing in Civil Engineering*; American Society of Civil Engineers: Reston, VA, USA, 2007; Volume 40937, pp. 216–223.
60. Sepehr, A.; Ibrahim, Y. Impact of the Virtual Collaboration on Project Progress Monitoring in the Construction Industry. In Proceedings of the AEI 2017: Resilience of the Integrated Building, Oklahoma City, OK, USA, 11–13 April 2017; pp. 964–982.
61. Da Costa, M.B.; Dos Santos, L.M.A.L.; Schaefer, J.L.; Baierle, I.C.; Nara, E.O.B. Industry 4.0 technologies basic network identification. In *Proceedings of the Scientometrics*; Springer International Publishing: Berlin/Heidelberg, Germany, 2019; Volume 121, pp. 977–994.
62. Perrier, N.; Bled, A.; Bourgault, M.; Cousin, N.; Danjou, C.; Pellerin, R.; Roland, T. Construction 4.0: A survey of research trends. *J. Inf. Technol. Constr.* **2020**, *25*, 416–437. [CrossRef]
63. Valero, E.; Adán, A.; Bosché, F. Semantic 3D Reconstruction of Furnished Interiors Using Laser Scanning and RFID Technology. *J. Comput. Civ. Eng.* **2016**, *30*, 04015053. [CrossRef]
64. Aryan, A. Evaluation of Ultra-Wideband Sensing Technology for Position Location in Indoor Construction Environments. Master's Thesis, University of Waterloo, Waterloo, ON, Canada, 2011.
65. Xiong, Z.; Song, Z.; Scalera, A.; Ferrera, E.; Sottile, F.; Brizzi, P.; Tomasi, R.; Spirito, M.A. Hybrid WSN and RFID indoor positioning and tracking system. *EURASIP J. Embed. Syst.* **2013**, *2013*, 6. [CrossRef]
66. Won, D.; Chi, S.; Park, M.W. UAV-RFID Integration for Construction Resource Localization. *KSCE J. Civ. Eng.* **2020**, *24*, 1683–1695. [CrossRef]
67. Oliveira, L.F.P.; Silva, M.F.; Moreira, A.P. Agricultural robotics: A state of the art survey. In Proceedings of the Robots in Human Life—Proceedings of the 23rd International Conference on Climbing and Walking Robots and the Support Technologies for Mobile Machines, CLAWAR 2020, Moscow, Russia, 24–26 August 2020; pp. 279–286.
68. Masiero, A.; Fissore, F.; Guarnieri, A.; Pirotti, F.; Visintini, D.; Vettore, A. Performance Evaluation of Two Indoor Mapping Systems: Low-Cost UWB-aided photogrammetry and backpack laser scanning. *Appl. Sci.* **2018**, *8*, 416. [CrossRef]
69. Said, K.O.; Onifade, M.; Githiria, J.M.; Abdulsalam, J.; Bodunrin, M.O.; Genc, B.; Johnson, O.; Akande, J.M. On the Application of Drones: A progress report in mining operations. *Int. J. Min. Reclam. Environ.* **2021**, *35*, 235–267. [CrossRef]
70. Kim, P.; Chen, J.; Kim, J.; Cho, Y.K. SLAM-driven Intelligent Autonomous Mobile Robot Navigation for Construction Applications. In *Workshop of the European Group for Intelligent Computing in Engineering*; Springer: Cham, Switzerland, 2018; pp. 254–269. [CrossRef]
71. Courtay, A.; Le Gentil, M.; Berder, O.; Scalart, P.; Fontaine, S.; Carer, A. Anchor Selection Algorithm for Mobile Indoor Positioning using WSN with UWB Radio. In Proceedings of the SAS 2019—2019 IEEE Sensors Applications Symposium, Sophia Antipolis, France, 11–13 March 2019; pp. 1–4.
72. Lakas, A.; Belkhouche, B.; Benkraouda, O.; Shuaib, A.; Alasmawi, H.J. A Framework for a Cooperative UAV-UGV System for Path Discovery and Planning. In Proceedings of the 2018 13th International Conference on Innovations in Information Technology, IIT 2018, Al Ain, United Arab Emirates, 18–19 November 2018; pp. 42–46.
73. Ahmed, S.; Shakev, N.; Milusheva, L.; Topalov, A. Neural Net tracking Control of a Mobile Platform in Robotized Wireless Sensor Networks. In Proceedings of the 2015 IEEE International Workshop of Electronics, Control, Measurement, Signals and their Application to Mechatronics, ECMSM 2015, Liberec, Czech Republic, 22–24 June 2015.
74. Bulgakov, A.G.; Pakhomova, E.G. Coordination of Construction Manipulation Robotic System Using UAV. In Proceedings of the IOP Conference Series: Materials Science and Engineering, Brasov, Romania, 7–10 October 2020; Volume 789.
75. Mishra, D.; Natalizio, E. A Survey on Cellular-connected UAVs: Design Challenges, Enabling 5G/B5G Innovations, and Experimental Advancements. *Comput. Netw.* **2020**, *182*, 107451. [CrossRef]
76. Hamledari, H.; McCabe, B.; Davari, S. Automated Computer Vision-based Detection of Components of Under-construction Indoor Partitions. *Autom. Constr.* **2017**, *74*, 78–94. [CrossRef]
77. Turk, Ž.; Klinc, R. A social–product–process framework for construction. *Build. Res. Inf.* **2020**, *48*, 747–762. [CrossRef]
78. Zhang, J.; Deng, C.; Zheng, P.; Xu, X.; Ma, Z. Development of an Edge Computing-based Cyber-physical Machine Tool. *Robot. Comput. Integr. Manuf.* **2021**, *67*, 102042. [CrossRef]
79. Sodhro, A.H.; Pirbhulal, S.; De Albuquerque, V.H.C. Artificial Intelligence-Driven Mechanism for Edge Computing-Based Industrial Applications. *IEEE Trans. Ind. Inform.* **2019**, *15*, 4235–4243. [CrossRef]
80. Sacco, A.; Flocco, M.; Esposito, F.; Marchetto, G. An Architecture for Adaptive Task Planning in Support of IoT-based Machine Learning Applications for Disaster Scenarios. *Comput. Commun.* **2020**, *160*, 769–778. [CrossRef]

81. Kireev, V.S.; Bochkaryov, P.V.; Guseva, A.I.; Kuznetsov, I.A.; Filippov, S.A. Monitoring System for the Housing and Utility Services Based on the Digital Technologies IIoT, Big Data, Data Mining, Edge and Cloud Computing. In *Proceedings of the Communications in Computer and Information Science*; Springer: Istanbul, Turkey, 2019; Volume 1054, pp. 193–205.
82. Kochovski, P.; Stankovski, V. Supporting Smart Construction with Dependable Edge Computing Infrastructures and Applications. *Autom. Constr.* **2018**, *85*, 182–192. [CrossRef]
83. Lee, C.M.; Kuo, W.L.; Tung, T.J.; Huang, B.K.; Hsu, S.H.; Hsieh, S.H. Government Open Data and Sensing Data Integration Framework for Smart Construction Site Management. In Proceedings of the 36th International Symposium on Automation and Robotics in Construction, ISARC 2019, Banff, AB, Canada, 21–24 May 2019; pp. 1261–1267.
84. Wang, X.; Yew, A.W.W.; Ong, S.K.; Nee, A.Y.C. Enhancing Smart Shop Floor Management withUbiquitous Augmented Reality. *Int. J. Prod. Res.* **2020**, *58*, 2352–2367. [CrossRef]
85. Tang, S.; Wang, R.; Zhao, X.; Nie, X. Building Cloud Services for Monitoring Offshore Equipment and Operators. In Proceedings of the Annual Offshore Technology Conference, Houston, TX, USA, 30 April–3 May 2018; Volume 2, pp. 852–864.
86. Gomez, J.A.; Talavera, J.; Tobon, L.E.; Culman, M.A.; Quiroz, L.A.; Aranda, J.M.; Garreta, L.E. A Case Study on Monitoring and Geolocation of Noise in Urban Environments Using the Internet of Things. In Proceedings of the Second International Conference on Internet of things, Data and Cloud Computing, Cambridge, UK, 22–23 March 2017.
87. Sawamura, M.; Iwamoto, S.; Kashihara, K. First application of CIM to tunnel construction in Japan. In Proceedings of the ISRM International Symposium—8th Asian Rock Mechanics Symposium, ARMS 2014, Sapporo, Japan, 14–16 October 2014; pp. 1054–1063.
88. Zhitong, S.; Liang, G.; Shaozhi, L.; Sifang, Y. Research and appilcation of intergrated 2D&3D spatial geographical information sharing and monitoring platform—case study on the Pazhou new district of Guangzhou. *J. Geomatics* **2019**, *44*, 101–103. [CrossRef]
89. Panteli, C.; Kylili, A.; Fokaides, P.A. Building information modelling applications in smart buildings: From design to commissioning and beyond A critical review. *J. Clean. Prod.* **2020**, *265*, 121766. [CrossRef]
90. Stransky, M. Functions of Common Data Environment Supporting Procurement of Subcontractors. *Eng. Rural Dev.* **2020**, *19*, 793–799. [CrossRef]
91. Liu, C.; Le Roux, L.; Körner, C.; Tabaste, O.; Lacan, F.; Bigot, S. Digital Twin-enabled Collaborative Data Management for Metal Additive Manufacturing Systems. *J. Manuf. Syst.* **2020**. [CrossRef]
92. Mahami, H.; Nasirzadeh, F.; Hosseininaveh Ahmadabadian, A.; Esmaeili, F.; Nahavandi, S. Imaging network design to improve the automated construction progress monitoring process. *Constr. Innov.* **2019**, *19*, 386–404. [CrossRef]
93. Kirti, K.C.; Singla, A. Architecture for Garbage Monitoring System using Integrated Technologies with Short Literature Survey. In Proceedings of the ICRITO 2020—IEEE 8th International Conference on Reliability, Infocom Technologies and Optimization (Trends and Future Directions), Noida, India, 4–5 June 2020; pp. 27–32.
94. Štefanič, M.; Stankovski, V. A Review of Technologies and Applications for Smart Construction. *Proc. Inst. Civ. Eng. Civ. Eng.* **2018**, *172*, 83–87. [CrossRef]
95. Mansouri, S.; Castronovo, F.; Akhavian, R. Analysis of the Synergistic Effect of Data Analytics and Technology Trends in the AEC/FM Industry. *J. Constr. Eng. Manag.* **2020**, *146*, 04019113. [CrossRef]
96. Arif, F.; Khan, W.A. A Real-Time Productivity Tracking Framework Using Survey-Cloud-BIM Integration. *Arab. J. Sci. Eng.* **2020**, *45*, 8699–8710. [CrossRef]
97. Elshafey, A.; Saar, C.C.; Aminudin, E.B.; Gheisari, M.; Usmani, A. Technology Acceptance Model for Augmented Reality and Building Information Modeling Integration in the Construction Industry. *J. Inf. Technol. Constr.* **2020**, *25*, 161–172. [CrossRef]
98. Vacanas, B.; Salem, O.; John Samuel, I.; Sid, S.H.; Dewberry, R. BIM and VR/AR technologies: From project development to lifecycle asset management. *Proc. Int. Struct. Eng. Constr.* **2020**, 1–6. [CrossRef]
99. Xie, X.; Lu, Q.; Rodenas-Herraiz, D.; Parlikad, A.K.; Schooling, J.M. Visualised Inspection System For Monitoring Environmental Anomalies During Daily Operation and Maintenance. *Eng. Constr. Archit. Manag.* **2020**, *27*, 1835–1852. [CrossRef]
100. Ratajczak, J.; Schimanski, C.P.; Marcher, C.; Riedl, M.; Matt, D.T. Mobile Application for Collaborative Scheduling and Monitoring of Construction Works According to Lean Construction Methods. In *Proceedings of the Lecture Notes in Computer Science (Including Subseries Lecture Notes in Artificial Intelligence and Lecture Notes in Bioinformatics)*; Springer: Mallorca, Spain, 2017; Volume 10451, pp. 207–214.
101. Santos, M.C.F.; Costa, D.B.; Ferreira, E.D.A.M. Conceptual Framework for Integrating Cost Estimating and Scheduling with BIM. In *Lecture Notes in Civil Engineering*; Springer: Berlin/Heidelberg, Germany, 2021; Volume 98, pp. 613–625.
102. Talamo, C.; Atta, N. FM Services Procurement and Management: Scenarios of Innovation. In *Springer Tracts in Civil Engineering*; Springer: Berlin/Heidelberg, Germany, 2019; pp. 201–242.
103. Kouhestani, S.; Nik-Bakht, M. IFC-based Process Mining for Design Authoring. *Autom. Constr.* **2020**, *112*, 103069. [CrossRef]
104. Angah, O.; Chen, A.Y. Removal of Occluding Construction Workers in Job site Image Data Using U-Net based Context Encoders. *Autom. Constr.* **2020**, *119*, 103332. [CrossRef]
105. Wang, Z.; Zhang, Q.; Yang, B.; Wu, T.; Lei, K.; Zhang, B.; Fang, T. Vision-Based Framework for Automatic Progress Monitoring of Precast Walls by Using Surveillance Videos during the Construction Phase. *J. Comput. Civ. Eng.* **2021**, *35*, 04020056. [CrossRef]
106. Soilán, M.; Justo, A.; Sánchez-Rodríguez, A.; Riveiro, B. 3D point cloud to BIM: Semi-automated framework to define IFC alignment entities from MLS-acquired LiDAR data of highway roads. *Remote Sens.* **2020**, *12*, 2301. [CrossRef]

107. Ding, K.; Shi, H.; Hui, J.; Liu, Y.; Zhu, B.; Zhang, F.; Cao, W. Smart steel bridge construction enabled by BIM and Internet of Things in industry 4.0: A framework. In *Proceedings of the ICNSC 2018—15th IEEE International Conference on Networking, Sensing and Control*; Institute of Electrical and Electronics Engineers Inc.: Zhuhai, China, 2018; pp. 1–5.
108. Braun, A.; Tuttas, S.; Borrmann, A.; Stilla, U. Improving progress monitoring by fusing point clouds, semantic data and computer vision. *Autom. Constr.* **2020**, *116*, 103210. [CrossRef]
109. Hakdaoui, S.; Emran, A.; Oumghar, F. Mobile mapping, machine learning and digital twin for road infrastructure monitoring and maintenance: Case study of mohammed VI bridge in Morocco. In Proceedings of the Proceedings—2020 IEEE International Conference of Moroccan Geomatics, MORGEO 2020, Casablanca, Morocco, 11–13 May 2020.
110. Braun, A.; Borrmann, A. Combining inverse photogrammetry and BIM for automated labeling of construction site images for machine learning. *Autom. Constr.* **2019**, *106*, 102879. [CrossRef]
111. You, Z.; Wu, C.; Zheng, L.; Feng, L. An Informatization Scheme for Construction and Demolition Waste Supervision and Management in China. *Sustainability* **2020**, *12*, 1672. [CrossRef]
112. Tsao, Y.C.; Hsu, P.H. 3D scene reconstruction from multi-view stereo images using machine learning. In Proceedings of the 40th Asian Conference on Remote Sensing, ACRS 2019: Progress of Remote Sensing Technology for Smart Future, Daejeon, Korea, 14–18 October 2020; pp. 1–10.
113. Khalil, M.; Bergs, C.; Papadopoulos, T.; Wuchner, R.; Bletzinger, K.U.; Heizmann, M. IIoT-based fatigue life indication using augmented reality. In Proceedings of the 2019 IEEE 17th International Conference on Industrial Informatics (INDIN), Helsinki-Espoo, Finland, 22–25 July 2019; pp. 746–751. [CrossRef]
114. Davila Delgado, J.M.; Oyedele, L.; Demian, P.; Beach, T. A research agenda for augmented and virtual reality in architecture, engineering and construction. *Adv. Eng. Inform.* **2020**, *45*, 101122. [CrossRef]
115. Lu, Y.; Liu, C.; Wang, K.I.K.; Huang, H.; Xu, X. Digital Twin-driven smart manufacturing: Connotation, reference model, applications and research issues. *Robot. Comput. Integr. Manuf.* **2020**, *61*, 101837. [CrossRef]
116. Stark, E.; Kučera, E.; Haffner, O.; Drahoš, P.; Leskovský, R. Using augmented reality and internet of things for control and monitoring of mechatronic devices. *Electronics* **2020**, *9*, 1272. [CrossRef]
117. Ventrella, A.V.; Esposito, F.; Sacco, A.; Flocco, M.; Marchetto, G.; Gururajan, S. APRON: An Architecture for Adaptive Task Planning of Internet of Things in Challenged Edge Networks. In Proceedings of the 2019 IEEE 8th International Conference on Cloud Networking, CloudNet 2019, Coimbra, Portugal, 4–6 November 2019.
118. Chen, W. Intelligent manufacturing production line data monitoring system for industrial internet of things. *Comput. Commun.* **2020**, *151*, 31–41. [CrossRef]
119. Hoppenstedt, B.; Kammerer, K.; Reichert, M.; Spiliopoulou, M.; Pryss, R. Convolutional Neural Networks for Image Recognition in Mixed Reality Using Voice Command Labeling. In *International Conference on Augmented Reality, Virtual Reality and Computer Graphics*; Springer: Cham, Switzerland, 2019; pp. 63–70. [CrossRef]
120. Liu, Y.; Kashef, M.; Lee, K.B.; Benmohamed, L.; Candell, R. Wireless Network Design for Emerging IIoT Applications: Reference Framework and Use Cases. *Proc. IEEE* **2019**, *107*, 1166–1192. [CrossRef]
121. Trziszka, M. Internet of Things in the Enterprise as a Production Process Control System. In *Proceedings of the Advances in Intelligent Systems and Computing*; Springer: San Diego, CA, USA, 2020; Volume 1216, pp. 56–62.
122. Fernandez, F.; Sanchez, A.; Velez, J.F.; Moreno, B. The Augmented Space of a Smart City. In Proceedings of the International Conference on Systems, Signals, and Image Processing, Niteroi, Brazil, 1–3 July 2020; Volume 2020, pp. 465–470.
123. Nurminen, A.; Kruijff, E.; Veas, E. HYDROSYS—A mixed reality platform for on-site visualization of environmental data. In *International Symposium on Web and Wireless Geographical Information Systems*; Springer: Berlin/Heidelberg, Germany, 2011; Volume 6574, pp. 159–175. [CrossRef]
124. Maia, A.; Rodrigues, A.; Lemos, R.; Capitão, R.; Fortes, C.J. Developing a responsive web platform for the systematic monitoring of coastal structures. *Commun. Comput. Inf. Sci.* **2019**, *936*, 176–197. [CrossRef]
125. Bognot, J.R.; Candido, C.G.; Blanco, A.C.; Montelibano, J.R.Y. Building construction progress monitoring using unmanned aerial system (UAS), low-cost photogrammetry, and geographic information system (GIS). *ISPRS Ann. Photogramm. Remote Sens. Spat. Inf. Sci.* **2018**, *4*, 41–47. [CrossRef]
126. Anghel, A.; Vasile, G.; Boudon, R.; d'Urso, G.; Girard, A.; Boldo, D.; Bost, V. Combining spaceborne SAR images with 3D point clouds for infrastructure monitoring applications. *ISPRS J. Photogramm. Remote Sens.* **2016**, *111*, 45–61. [CrossRef]
127. Wazid, M.; Das, A.K.; Hussain, R.; Succi, G.; Rodrigues, J.J.P.C. Authentication in cloud-driven IoT-based big data environment: Survey and outlook. *J. Syst. Archit.* **2019**, *97*, 185–196. [CrossRef]
128. Wang, S.; Zargar, S.A.; Xu, C.; Yuan, F.G. An efficient augmented reality (AR) system for enhanced visual inspection. In *Proceedings of the Structural Health Monitoring 2019: Enabling Intelligent Life-Cycle Health Management for Industry Internet of Things (IIOT)—Proceedings of the 12th International Workshop on Structural Health Monitoring*; DEStech Publications Inc.: Stanford, CA, USA, 2019; Volume 1, pp. 1543–1550.
129. Wang, Q.; Jiao, W.; Zhang, Y. Deep learning-empowered digital twin for visualized weld joint growth monitoring and penetration control. *J. Manuf. Syst.* **2020**, *57*, 429–439. [CrossRef]
130. Mudassar, R.; Zailin, G.; Jabir, M.; Lei, Y.; Hao, W. Digital twin-based smart manufacturing system for project-based organizations: A conceptual framework. In *Proceedings of the International Conference on Computers and Industrial Engineering*; CIE: Beijing, China, 2019; Volume 2019.

131. Saovana, N.; Yabuki, N.; Fukuda, T. Development of an unwanted-feature removal system for Structure from Motion of repetitive infrastructure piers using deep learning. *Adv. Eng. Inform.* **2020**, *46*, 101169. [CrossRef]
132. Boonbrahm, P.; Kaewrat, C.; Boonbrahm, S. Effective collaborative design of large virtual 3D model using multiple AR markers. *Procedia Manuf.* **2020**, *42*, 387–392. [CrossRef]
133. Altaf, M.S.; Bouferguene, A.; Liu, H.; Al-Hussein, M.; Yu, H. Integrated production planning and control system for a panelized home prefabrication facility using simulation and RFID. *Autom. Constr.* **2018**, *85*, 369–383. [CrossRef]
134. van der Heijden, R.; Tai, A.; Fagerström, G. Volatile data mining: A proof of concept for performance evaluation of the built environment using drones. *Simul. Ser.* **2017**, *49*, 225–232. [CrossRef]
135. Brulé, M.R. Big data in E&P: Real-time adaptive analytics and data-flow architecture. In Proceedings of the Society of Petroleum Engineers—SPE Digital Energy Conference and Exhibition 2013, The Woodlands, TX, USA, 5–7 March 2013; pp. 305–311.
136. Xia, Z.W.; Chen, Y.; Wan, G.C.; Tong, M.S. Design of interconnected mobile application for visualized information system of monitoring risks. In Proceedings of the 2017 Progress in Electromagnetics Research Symposium-Fall (PIERS-FALL), Singapore, 19–22 November 2017; pp. 2123–2126. [CrossRef]
137. Battulwar, R.; Winkelmaier, G.; Valencia, J.; Naghadehi, M.Z.; Peik, B.; Abbasi, B.; Parvin, B.; Sattarvand, J. A practical methodology for generating high-resolution 3D models of open-pit slopes using UAVs: Flight path planning and optimization. *Remote Sens.* **2020**, *12*, 2283. [CrossRef]
138. Machado, R.L.; Vilela, C. Conceptual framework for integrating bim and augmented reality in construction management. *J. Civ. Eng. Manag.* **2020**, *26*, 83–94. [CrossRef]
139. Pawlewitz, J.; Mankel, A.; Jacquin, S.; Basile, N. The Digital Twin in a Brownfield Environment: How to Manage Dark Data. In Proceedings of the Offshore Technology Conference, Houston, TX, USA, 4–7 May 2020.
140. Revetria, R.; Tonelli, F.; Damiani, L.; Demartini, M.; Bisio, F.; Peruzzo, N. A real-time mechanical structures monitoring system based on digital Twin, IOT and augmented reality. *Simul. Ser.* **2019**, *51*, 1–10. [CrossRef]
141. Ruppert, T.; Abonyi, J. Integration of real-time locating systems into digital twins. *J. Ind. Inf. Integr.* **2020**, *20*, 100174. [CrossRef]
142. Berman, I.; Zereik, E.; Kapitonov, A.; Bonsignorio, F.; Khassanov, A.; Oripova, A.; Lonshakov, S.; Bulatov, V. Trustable Environmental Monitoring by Means of Sensors Networks on Swarming Autonomous Marine Vessels and Distributed Ledger Technology. *Front. Robot. AI* **2020**, *7*, 70. [CrossRef] [PubMed]
143. Erra, U.; Capece, N. Engineering an advanced geo-location augmented reality framework for smart mobile devices. *J. Ambient Intell. Humaniz. Comput.* **2019**, *10*, 255–265. [CrossRef]
144. Pavlov, D.; Sosnovsky, I.; Dimitrov, V.; Melentyev, V.; Korzun, D. Case Study of Using Virtual and Augmented Reality in Industrial System Monitoring. In Proceedings of the Conference of Open Innovation Association, FRUCT, Yaroslavl, Russia, 20–24 April 2020; Volume 2020, pp. 367–375.
145. Auer, M.E.; Ram B, K. (Eds.) *Cyber-Physical Systems and Digital Twins*; Lecture Notes in Networks and Systems; Springer International Publishing: Cham, Switzerland, 2020; Volume 80, ISBN 978-3-030-23161-3.
146. Pérez, L.; Rodríguez-Jiménez, S.; Rodríguez, N.; Usamentiaga, R.; García, D.F. Digital twin and virtual reality based methodology for multi-robot manufacturing cell commissioning. *Appl. Sci.* **2020**, *10*, 3633. [CrossRef]
147. Angrisani, L.; Arpaia, P.; Esposito, A.; Moccaldi, N. A Wearable Brain-Computer Interface Instrument for Augmented Reality-Based Inspection in Industry 4.0. *IEEE Trans. Instrum. Meas.* **2020**, *69*, 1530–1539. [CrossRef]
148. Unal, M.; Bostanci, E.; Sertalp, E. Distant augmented reality: Bringing a new dimension to user experience using drones. *Digit. Appl. Archaeol. Cult. Herit.* **2020**, *17*, e00140. [CrossRef]
149. Dodevska, Z.A.; Kvrgić, V.M.; Mihić, M.M.; Delibašić, B.V. The concept and application of simplified robotic models. *Serb. J. Electr. Eng.* **2019**, *16*, 419–437. [CrossRef]
150. Soedji, B.; Lacoche, J.; Villain, E. Creating AR Applications for the IOT: A New Pipeline. In Proceedings of the 26th ACM Symposium on Virtual Reality Software and Technology, Virtual, 1–4 November 2020; pp. 1–2. [CrossRef]
151. Zhong, R.Y.; Xu, C.; Chen, C.; Huang, G.Q. Big Data Analytics for Physical Internet-based intelligent manufacturing shop floors. *Int. J. Prod. Res.* **2017**, *55*, 2610–2621. [CrossRef]
152. Anderson, S.; Barvik, S.; Rabitoy, C. Innovative digital inspection methods. In Proceedings of the Annual Offshore Technology Conference, Houston, TX, USA, 6–9 May 2019; Volume 2019, pp. 6–9.
153. Lim, S.; Chung, S.; Chi, S. Developing a pattern model of damage types on bridge elements using big data analytics. In Proceedings of the ISARC 2017—34th International Symposium on Automation and Robotics in Construction, Taipei, Taiwan, 28 June–1 July 2017; pp. 849–855.
154. Chen, Q.; Adey, B.T.; Haas, C.; Hall, D.M. Using look-ahead plans to improve material flow processes on construction projects when using BIM and RFID technologies. *Constr. Innov.* **2020**, *20*, 471–508. [CrossRef]
155. Bosché, F.; Ahmed, M.; Turkan, Y.; Haas, C.T.; Haas, R. The value of integrating Scan-to-BIM and Scan-vs-BIM techniques for construction monitoring using laser scanning and BIM: The case of cylindrical MEP components. *Autom. Constr.* **2015**, *49*, 201–213. [CrossRef]
156. Lin, Y.-C.; Cheung, W.-F. Developing WSN/BIM-Based Environmental Monitoring Management System for Parking Garages in Smart Cities. *J. Manag. Eng.* **2020**, *36*, 04020012. [CrossRef]

157. Gheisar, M.; Irizarry, J. A User-centered Approach to Investigate Unmanned Aerial System (UAS) Requirements for a Department of Transportation Applications. In Proceedings of the 2015 Conference on Autonomous and Robotic Construction of Infrastructure, Ames, IA, USA, 2–3 June 2015; Volume 6, pp. 118–131.
158. Pathak, K.; Bandara, J.M.S.; Agrawal, R. *Recent Trends in Civil Engineering*, 1st ed.; Springer: Berlin/Heidelberg, Germany, 2021; ISBN 9789811551949.
159. Zaher, M.; Greenwood, D.; Marzouk, M. Mobile augmented reality applications for construction projects. *Constr. Innov.* **2018**, *18*, 152–166. [CrossRef]
160. Ko, H.S.; Azambuja, M.; Felix Lee, H. Cloud-based Materials Tracking System Prototype Integrated with Radio Frequency Identification Tagging Technology. *Autom. Constr.* **2016**, *63*, 144–154. [CrossRef]
161. Zhang, Y.; Qin, W.; Zheng, D.; Zhou, C. *Geoinformatics in Sustainable Ecosystem and Society*; Xie, Y., Li, Y., Yang, J., Xu, J., Deng, Y., Eds.; Communications in Computer and Information Science; Springer: Singapore, 2020; Volume 1228, ISBN 978-981-15-6105-4.
162. Yoon, S.; Ju, S.; Park, S.; Heo, J. A Framework Development for Mapping and Detecting Changes in Repeatedly Collected Massive Point Clouds. In Proceedings of the 36th International Symposium on Automation and Robotics in Construction, ISARC 2019, Banff, AB, Canada, 21–24 May 2019; pp. 603–609.
163. Liao, X.; Sahran, S.; Abdul Shukor, S. An experimental study of vehicle detection on aerial imagery using deep learning-based detection approaches. *J. Phys. Conf. Ser.* **2020**, *1550*, 032005. [CrossRef]
164. Abedi, M.; Jazizadeh, F. Deep-Learning for Occupancy Detection Using Doppler Radar and Infrared Thermal Array Sensors. In Proceedings of the 36th International Symposium on Automation and Robotics in Construction, ISARC 2019, Banff, AB, Canada, 21–24 May 2019; pp. 1098–1105.
165. Cheng, J.C.P.; Deng, Y. Automated Cycle Time Measurement and Analysis of Excavator's Loading Operation Using Smart Phone-Embedded IMU Sensors. *Comput. Civ. Eng.* **2015**, *2015*, 667–674.
166. Sánchez-Rodríguez, A.; Esser, S.; Abualdenien, J.; Borrmann, A.; Riveiro, B. From point cloud to IFC: A masonry arch bridge case study. In Proceedings of the EG-ICE 2020 Workshop on Intelligent Computing in Engineering, Berlin, Germany, 30 June–3 July 2020; pp. 422–431.
167. Naets, F.; Geysen, J.; Desmet, W. An Approach for Combined Vertical Vehicle Model and Road Profile Identification from Heterogeneous Fleet Data. In Proceedings of the 2019 IEEE International Conference on Connected Vehicles and Expo (ICCVE), Graz, Austria, 4–8 November 2019; pp. 1–5.
168. Ullah, Z.; Al-Turjman, F.; Moatasim, U.; Mostarda, L.; Gagliardi, R. UAVs joint optimization problems and machine learning to improve the 5G and Beyond communication. *Comput. Netw.* **2020**, *182*, 107478. [CrossRef]
169. Raza, S.; Faheem, M.; Guenes, M. Industrial wireless sensor and actuator networks in industry 4.0: Exploring requirements, protocols, and challenges—A MAC survey. *Int. J. Commun. Syst.* **2019**, *32*, e4074. [CrossRef]
170. Dzemydienė, D.; Radzevičius, V. An Approach for Networking of Wireless Sensors and Embedded Systems Applied for Monitoring of Environment Data. In *Studies in Computational Intelligence*; Springer: Berlin/Heidelberg, Germany, 2020; Volume 869, pp. 61–82.
171. Oyetoke, O.O. A practical application of ARM cortex-M3 processor core in embedded system engineering. *Int. J. Intell. Syst. Appl.* **2017**, *9*, 70–88. [CrossRef]
172. Hsiao, S.-J.; Lian, K.-Y.; Sung, W.-T. Employing Cross-Platform Smart Home Control System with IOT Technology Based. In Proceedings of the 2016 International Symposium on Computer, Consumer and Control (IS3C), Xi'an, China, 4–6 July 2016; pp. 264–267.
173. Chen, J.; Kira, Z.; Cho, Y.K. Deep Learning Approach to Point Cloud Scene Understanding for Automated Scan to 3D Reconstruction. *J. Comput. Civ. Eng.* **2019**, *33*, 04019027. [CrossRef]
174. Moselhi, O.; Bardareh, H.; Zhu, Z. Automated data acquisition in construction with remote sensing technologies. *Appl. Sci.* **2020**, *10*, 2846. [CrossRef]
175. Siu, M.F.F.; Lu, M.; AbouRizk, S. Combining photogrammetry and robotic total stations to obtain dimensional measurements of temporary facilities in construction field. *Vis. Eng.* **2013**, *1*, 4. [CrossRef]
176. Wojnarowski, A.E.; Leonteva, A.B.; Tyurin, S.V.; Tikhonov, S.G.; Artemeva, O.V. Photogrammetric technology for remote high-precision 3D monitoring of cracks and deformation joints of buildings and constructions. *Int. Arch. Photogramm. Remote Sens. Spat. Inf. Sci. ISPRS Arch.* **2019**, *42*, 95–101. [CrossRef]
177. Yu, Z.; He, Z. Research on the Substation Intelligent Protection System for Operations Personnel Based on the UWB Positioning Technology. In *Proceedings of the MATEC Web of Conferences*; Li, J.Y., Liu, T.Y., Deng, T., Tian, M., Eds.; EDP Sciences: Les Ulis, France, 2015; Volume 22, p. 02001.
178. Fadiya, O.; Georgakis, P.; Chinyio, E.; Nwagboso, C. Development of an ICT-based logistics framework for the construction industry. In Proceedings of the Association of Researchers in Construction Management, ARCOM 2010—26th Annual Conference, Leeds, UK, 6–8 September 2010; pp. 63–72.
179. Kochetkova, L.I. Pipeline monitoring with unmanned aerial vehicles. *J. Phys. Conf. Ser.* **2018**, *1015*, 042021. [CrossRef]
180. Moretti, N.; Dejaco, M.C.; Maltese, S.; Re Cecconi, F. An information management framework for optimised urban facility management. In Proceedings of the ISARC 2018—35th International Symposium on Automation and Robotics in Construction and International AEC/FM Hackathon: The Future of Building Things, Berlin, Germany, 20–25 July 2018.

181. Rezendez, A.; Marros, R.J.; Farrag, K.; Acharya, S. Reducing Excavation Damage in the Natural Gas Industry Using Real-Time GIS and Sensors. In *Proceedings of the Volume 3: Operations, Monitoring, and Maintenance; Materials and Joining*; American Society of Mechanical Engineers: Calgary, AB, Canada, 2018.
182. Salhaoui, M.; Guerrero-González, A.; Arioua, M.; Ortiz, F.J.; El Oualkadi, A.; Torregrosa, C.L. Smart industrial iot monitoring and control system based on UAV and cloud computing applied to a concrete plant. *Sensors* **2019**, *19*, 3316. [CrossRef] [PubMed]
183. Elloumi, M.; Dhaou, R.; Escrig, B.; Idoudi, H.; Saidane, L.A.; Fer, A. Traffic Monitoring on City Roads Using UAVs. In Proceedings of the 18th International Conference on Ad-Hoc Networks and Wireless, ADHOC-NOW 2019, Luxembourg, 1–3 October 2019; Volume 11803, pp. 588–600.
184. Odelius, J.; Famurewa, S.M.; Forslöf, L.; Casselgren, J.; Konttaniemi, H. Industrial internet applications for efficient road winter maintenance. *J. Qual. Maint. Eng.* **2017**, *23*, 355–367. [CrossRef]
185. Akinlolu, M.; Haupt, T.C.; Edwards, D.J.; Simpeh, F. A bibliometric review of the status and emerging research trends in construction safety management technologies. *Int. J. Constr. Manag.* **2020**, 1–13. [CrossRef]
186. Jin, R.; Zhang, H.; Liu, D.; Yan, X. IoT-based detecting, locating and alarming of unauthorized intrusion on construction sites. *Autom. Constr.* **2020**, *118*, 103278. [CrossRef]
187. Wen, W.; Zhao, S.; Shang, C.; Chang, C.Y. EAPC: Energy-aware path construction for data collection using mobile sink in wireless sensor networks. *IEEE Sens. J.* **2018**, *18*, 890–901. [CrossRef]
188. Hussin, Z. Fast-converging indoor mapping for wireless indoor localization. In Proceedings of the 2014 IEEE International Conference on Pervasive Computing and Communication Workshops (PERCOM WORKSHOPS), Budapest, Hungary, 24–28 March 2014; pp. 171–173.
189. Kim, C.; Park, T.; Lim, H.; Kim, H. On-site construction management using mobile computing technology. *Autom. Constr.* **2013**, *35*, 415–423. [CrossRef]
190. Soltani, M.M.; Motamedi, A.; Hammad, A. Enhancing Cluster-based RFID Tag Localization using artificial neural networks and virtual reference tags. *Autom. Constr.* **2015**, *54*, 93–105. [CrossRef]
191. Merkle, D.; Schmitt, A.; Reiterer, A. Sensor evaluation for crack detection in concrete bridges. *Int. Arch. Photogramm. Remote Sens. Spat. Inf. Sci.* **2020**, *43*, 1107–1114. [CrossRef]
192. Menolotto, M.; Komaris, D.-S.; Tedesco, S.; O'Flynn, B.; Walsh, M. Motion Capture Technology in Industrial Applications: A Systematic Review. *Sensors* **2020**, *20*, 5687. [CrossRef] [PubMed]
193. Yang, Z.; Yuan, Y.; Zhang, M.; Zhao, X.; Tian, B. Assessment of Construction Workers' Labor Intensity Based on Wearable Smartphone System. *J. Constr. Eng. Manag.* **2019**, *145*, 04019039. [CrossRef]
194. Lopez-Peña, F.; Deibe, A.; Orjales, F. On the initiation phase of a mixed reality simulator for air pollution monitoring by autonomous UAVs. In Proceedings of the 2017 IEEE 9th International Conference on Intelligent Data Acquisition and Advanced Computing Systems: Technology and Applications, IDAACS 2017, Bucharest, Romania, 21–23 September 2017; Volume 1, pp. 1–7.
195. Wu, X.; Lu, M.; Mao, S.; Shen, X. As-built modeling and visual simulation of tunnels using real-time TBM positioning data. In Proceedings of the 2013 Winter Simulation Conference—Simulation: Making Decisions in a Complex World, WSC 2013, Washington, DC, USA, 8–11 December 2013; pp. 3066–3073.
196. Jayaram, A.; Deb, S. EA-MAC: A QoS Aware Emergency Adaptive MAC Protocol for Intelligent Scheduling of Packets in Smart Emergency Monitoring Applications. *J. Circuits Syst. Comput.* **2020**, *29*, 2050205. [CrossRef]
197. Nagaty, A.; Thibault, C.; Seto, M.; Trentini, M.; Li, H. Construction, modelling, and control of an autonomous unmanned aerial vehicle for target localization. *Can. Aeronaut. Space J.* **2015**, *61*, 23–35. [CrossRef]
198. Li, C.Z.; Xue, F.; Li, X.; Hong, J.; Shen, G.Q. An Internet of Things-enabled BIM platform for on-site assembly services in prefabricated construction. *Autom. Constr.* **2018**, *89*, 146–161. [CrossRef]
199. Dinis, F.M.; Sanhudo, L.; Martins, J.P.; Ramos, N.M.M. Improving project communication in the architecture, engineering and construction industry: Coupling virtual reality and laser scanning. *J. Build. Eng.* **2020**, *30*, 101287. [CrossRef]
200. Salmeri, A.; Licciardi, C.A.; Lamorte, L.; Valla, M.; Giannantonio, R.; Sgroi, M. An Architecture to Combine Context Awareness and Body Sensor Networks for Health Care Applications. *Proc. Natl. Acad. Sci. USA* **2009**, *104*, 90–97.
201. Sakib, M.N.; Chaspari, T.; Ahn, C.R.; Behzadan, A.H. An experimental study of wearable technology and immersive virtual reality for drone operator training. In Proceedings of the EG-ICE 2020 Workshop on Intelligent Computing in Engineering, Berlin, Germany, 1–4 July 2020; pp. 154–163.
202. Trabucco, D. Robotics in construction: The next 50 years. In Proceedings of the 50 Forward 50 Back: The Recent History and Essential Future of Sustainable Cities—CTBUH 10th World Congress, Chicago, IL, USA, 28 October–2 November 2019; pp. 269–274.

Article

Towards the Integration and Automation of the Design Process for Domestic Drinking-Water and Sewerage Systems with BIM

Edison Atencio [1,2,*], Pablo Araya [1], Francisco Oyarce [1], Rodrigo F. Herrera [1], Felipe Muñoz-La Rivera [1,3,4] and Fidel Lozano-Galant [2]

1 School of Civil Engineering, Pontificia Universidad Católica de Valparaíso, Av. Brasil 2147, Valparaíso 2340000, Chile
2 Department of Civil Engineering, Universidad de Castilla-La Mancha, Av. Camilo Jose Cela s/n, 13071 Ciudad Real, Spain
3 School of Civil Engineering, Universitat Politecnica de Catalunya, 08034 Barcelona, Spain
4 International Center for Numerical Methods in Engineering (CIMNE), 08034 Barcelona, Spain
* Correspondence: edison.atencio@pucv.cl

Abstract: The use of building information modelling (BIM) in construction projects is expanding, and its usability throughout building lifecycles, from planning and construction to operation and maintenance, is gaining increasing proof. In the design of domestic drinking-water and sewerage systems (DDWSSs), BIM focuses on coordinating disciplines and their design. Despite studies promoting BIM environments for DDWSSs that take into account the regulatory frameworks of corresponding countries, these efforts do not include the use of parametric tools that enhance the efficiency of the design process. Therefore, engineers still use conventional 2D design, which requires many rounds of iteration, and manual work is also generally still used. In this research, we developed and validated an intuitive methodology for solving a specific DDWSS problem, using a design science research method (DSRM) as an applied science approach. This was addressed by developing an artefact and validating it through two case studies. The obtained solution combines BIM models and parametric tools to automate the manual activities of the traditional design method. This article aims to bring abstract BIM concepts into practice and encourage researchers and engineers to adopt BIM for DDWSSs.

Keywords: building information modelling (BIM); automatisation; facilities design; domestic plumbing and sanitation

Citation: Atencio, E.; Araya, P.; Oyarce, F.; Herrera, R.F.; Muñoz-La Rivera, F.; Lozano-Galant, F. Towards the Integration and Automation of the Design Process for Domestic Drinking-Water and Sewerage Systems with BIM. *Appl. Sci.* **2022**, *12*, 9063. https://doi.org/10.3390/app12189063

Academic Editors: Mariusz Szóstak, Marek Sawicki and Jarosław Konior

Received: 30 June 2022
Accepted: 6 September 2022
Published: 9 September 2022

Publisher's Note: MDPI stays neutral with regard to jurisdictional claims in published maps and institutional affiliations.

Copyright: © 2022 by the authors. Licensee MDPI, Basel, Switzerland. This article is an open access article distributed under the terms and conditions of the Creative Commons Attribution (CC BY) license (https://creativecommons.org/licenses/by/4.0/).

1. Introduction

Mechanical, electrical and plumbing (MEP) design and coordination tools have an important role in the design process [1]. However, the use of building information modelling (BIM) in MEP disciplines has traditionally focused on coordinating specialities and not on the actual design of each of the specialities [2]. The design and capacity of pipe network systems—as well as domestic drinking-water and sewerage systems (DDWSSs)—and their critical points/connections inside a building may affect the building components, such as architectural spaces and structural systems. Service systems, such as heating, ventilation and air conditioning (HVAC), can also be affected [1].

Previous studies promoted the design of drinking water facilities in BIM environments in order to take into account the regulatory frameworks of corresponding countries [3]. However, these efforts did not include the use of parametric tools that increase the efficiency of the design process. The MEP tool from one of the most common BIM software packages [4], REVIT, is based on the International Plumbing Code (IPC), but it cannot be edited and it does not take into account different country regulations, preventing its use as a drinking water design tool

The use of BIM for plumbing systems has also been explored in the context of asset management in the building operation stages, even using 4.0 technologies; e.g., augmented reality and sensors [5]. However, these uses are mainly focused on visualisation-based analysis and not the design of plumbing systems.

This research developed and validated an intuitive methodology for solving a specific problem in DDWSS design, using a design science research method (DSRM) as an applied science approach. This problem is addressed by developing a methodology as a workflow. To validate the proposed methodology in the case of the Chilean standard for DDWSS design, two case studies (one simple and one complex) were developed and compared with the traditional DDWSS design methodology. These case studies were validated in terms of the effort required, the complexity, the real automation level and the quality of the results. The obtained solution combines BIM models and parametric tools to automate the manual activities from the traditional design method.

This work aims to bring abstract BIM concepts into practice and encourage researchers and MEP engineers to adopt BIM for DDWSSs and to automate manual tasks. This issue is relevant in light of the findings of the most recent, extensive BIM report developed in Chile [6], which reveals that the MEP speciality is the least advanced in BIM in terms of adherence to BIM standards, motivation, satisfaction and perceived value.

The remainder of this paper is organised as follows. In Section 2, the research methodology is described. Section 3 discusses the background of this paper through a literature review. Section 4 describes the proposed workflow for DDWSS design. The case studies are developed in Section 5, and their results are provided and discussed in Section 6. Finally, the conclusions of this paper are described in Section 7.

2. Research Methodology

This research is based on a design science research method (DSRM) suitable for engineering innovation research projects [7]. The DSRM is structured in five stages: (1) identification of observed problems and motivations, (2) definition of the potential solution, (3) design and development, (4) demonstration and (5) evaluation. Figure 1 shows a summary of these stages, along with their objectives and the tasks and tools required.

The first stage focused on understanding the challenges in DDWSS design project approaches. Moreover, the current research and designing standards were collected; these include the Regulation on Domestic Installations of Drinking Water and Sewerage (Reglamento de Instalaciones Domiciliarias de Alcantarillado y Agua Potable, RIDAA [8]), a Chilean standard. Additionally, complementary standards are applied for verification, determination of materials, identification of the ranges of allowable values and the presentation of deliverables. For this paper, the following Chilean standards (https://www.inn.cl/nch-aprobadas, accessed on 8 September 2022) were used: NCh 398, NCh 399, NCh 951, NCh 1635, NCh 2038, NCh 2485, NCh 2592 and NCh 2836.

The literature review was used a tool in this first stage to identify research articles and documented projects related to key concepts: automation using BIM, parametric design automation and plumbing design in BIM environments. The Google Scholar, Scopus and Web of Science platforms were used to search for relevant literature published between 2000 and 2022.

In the second stage, we developed automatised methods for optimising time, costs, human resources and the overall development of DDWSS design projects. This goal was achieved by creating algorithms and automatic processes. The parameterisation of regulatory calculation guidelines in a BIM environment reduces the amount of manual work required during the design stage of DDWSS projects.

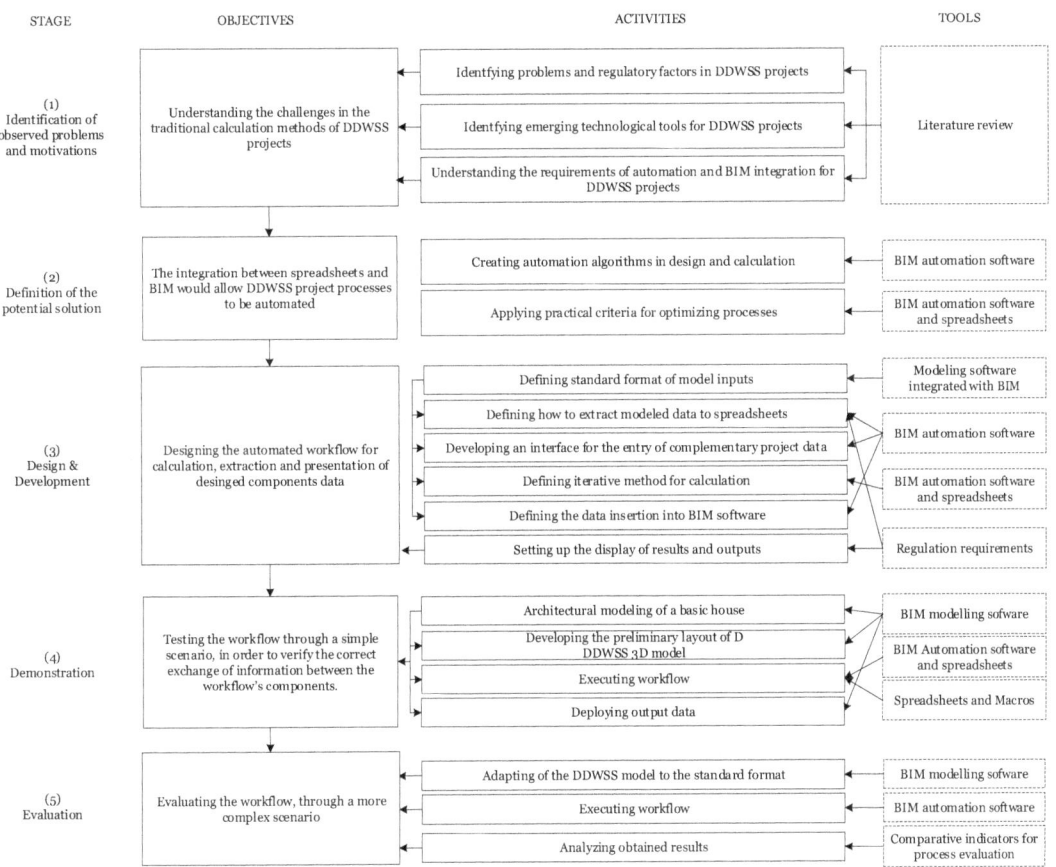

Figure 1. Research methodology workflow.

In the third stage, we identified a sequence of activities based on the literature review and determined the different functionalities of the selected software for the development of the expected solution:

1. Defining the format for the input data of the project so that it corresponds to parameters obtained both directly and indirectly from the three-dimensional model;
2. Defining the way to extract these data into spreadsheets, if required;
3. Defining the procedure for transferring the complementary project information that corresponds to the indirect input data of the model—i.e., project characteristics that cannot be obtained from the model and must be entered by the user;
4. Determining the iterative method for the calculation and verification of the resulting parameters as a convergence process;
5. Defining the method for the insertion of data into the BIM software, which are then returned to the model and modify the initial values of the project—for instance, pipe diameters;
6. Configuring deliverables corresponding to floor plans and tabulations with results in a standard plan, as defined by the RIDAA.

In the fourth stage, the described workflow was implemented for a simple case study with basic characteristics involving a residential building with a distributed system on a single floor. This process involved checking the functionality of the algorithm and the data management for the design parameters.

Finally, in the fifth stage, a more complex case study was introduced to evaluate the behaviour of the programmed algorithms. This case study corresponded to a known and previously solved real-world project, which allowed the results obtained to be compared with values resulting from traditional calculation. The comparison of the results was conducted by calculating the relative errors with the known values of the project.

3. Background

This section is organised into as follows: firstly, digitalisation processes in the building sector are reviewed. Interoperability problems among disciplines are also analysed. Then, the traditional design method meeting the requirements of DDWSSs is analysed. Finally, the application of BIM for the design of DDWSSs is presented.

3.1. Digitalisation of the Construction Sector

MEP systems provide particular, basic services that help to expand and complement the economic efficiency, utility and durability of buildings and create comfortable environments throughout the entire lifecycles of buildings [9]. These services commonly include heating, ventilation and air conditioning (HVAC), along with electric, electronic, plumbing and anti-power systems [10]. For housing projects, the functionality and efficiency of these facilities can have a huge impact on a project's sustainability and success [11]. As building designs and requirements evolve into more complete and complex installations [12], progress in computational technologies and their continuous integration into the construction industry have proven effective in dealing with these new challenges [13].

BIM represents a work methodology that combines information and communication technologies, improving project management and facilitating the application of international standards throughout buildings' lifecycles [12]. BIM encompasses digital modelling, simulation, coordination, optimisation and automatic drawing generated by computers; it facilitates information storage and exchange between different users, enhancing project understanding and outcome development [14]. A major challenge of the BIM methodology is the accurate exchange of information between the software of different disciplines (interoperability) [15]. In addition, working with BIM requires collaborative work methodologies that focus on information exchange, coordination and collaboration among the stakeholders [16].

Solving problems of interoperability between the software for different disciplines is one of the major challenges in implementing the BIM methodology. In fact, this kind of problem can also appear within the software of a single company. Table 1 provides a selection of the many studies in the literature dealing with the interoperability problems in Autodesk® software. This table includes the aim of the study, the different software used for the modelling (modelling software) and the software used to deliver the information (visual programming tool). These visual programming language software packages are based on textual programming languages; they visually represent actions composed of linear string sequences of code. The analysis in Table 1 shows that interoperability problems can be found in different disciplines (such as facility management, heritage, coordination and MEP). Regarding the software analysis, the table shows that Revit is the most common modelling software, and, for the delivery of the information, visual programming language software packages (Dynamo or Grasshopper) are commonly used.

Table 1. Examples of studies using Autodesk software that deal with interoperability problems.

Reference	Aim	Modelling Software	Visual Programming Language Software
[17]	Facility management	Revit, Solibri	Dynamo
[18]	State of the art review	Revit, Robot, Tricalc	Dynamo, Grasshoper

Table 1. Cont.

Reference	Aim	Modelling Software	Visual Programming Language Software
[19]	Heritage	Revit	Dynamo
[20]	State of the art review	Revit, Rhino, Archicad	Dynamo, Grasshoper
[21]	Coordination	Revit, Robot	-
[22]	Interoperability study	Revit, gbXML, IFC	EnergyPlus engine
[23]	State of the art review	Revit, Archicad, IFC	Dynamo, Grasshoper
[15]	State of the art review	Revit, Rhino, Archicad	Dynamo, Grasshoper
[19]	State of the art review	Revit, Solibri, Archicad	Dynamo, Grasshoper
[24]	MEP	Revit	Dynamo

3.2. Traditional Design Method for DDWSSs

The spatial and mechanical design of sanitary plumbing and drainage systems and components has not undergone substantial change in recent years [25]. CAD and 2D layouts are still suggested by standard regulations for the calculation methodology and required as the official presentation format [26]. At present, calculation methods must be guided by the specific codes and standards of the relevant country. The design, installation and specifications of plumbing systems in buildings are covered by guidelines [27]. Although standards specify equations, limits, considerations and minimum criteria, different hydraulic formulations have their respective justifications. In this context, RIDAA is currently the official code/regulation in Chilean practice.

The conventional method for DDWSS design involves laborious work based on permanent user interaction with different professional software packages [6,28]. Data manipulation and association comprise a highly manual process. These methods generally involve the following steps:

- The design process starts with interpreting two-dimensional models (2D/CAD) and visualising and locating relevant data according to the potential input parameters presented in the project floor plan, such as types of plumbing fixtures, locations, quantities, rooms, equipment and other operational function requirements;
- Specialist sub-contractors use this information to develop their system routing, connecting elements of all building systems in compliance with architectural and structural designs [29]. Then, spreadsheets are commonly used to discretise data about piping systems, adding conditions and general project specifications that are not incorporated in the data extracted from the layout [6,28];
- These spreadsheets, as shown in RIDAA's content, represent the official calculation and presentation formats. The system, material and geometric parameters of pipe paths are adjusted until an optimal solution is reached: the process is repeated until design parameters reach acceptable values according to the code ranges for pressure/flow rates and slopes in drinking water networks and drainage networks, respectively [26];
- Finally, contractors summarise and compile results, incorporating them into a two-dimensional model of the floor plans, elevations and isometrics, according to the symbology defined by the standards. This symbology is established for the water supply and sanitary treatment companies in the area, which are responsible for providing and maintaining connections with the public network.

Urbanisation through residential projects involves integration of several engineering specialisms. Contractors develop their models individually, following project requirements for optimal functionality. Therefore, they generally do not consider other systems from an interrelated viewpoint [29]. Coordination is needed to detect and eliminate spatial and functional interferences among systems; drawings for multiple layouts must be overlaid and compared [30]. In two-dimensional design, visualisation is limited, the data-sharing process easily comes into conflict with the work of other professionals and work lacks efficiency [31]. This repeatedly generates rework and coordination meetings, as re-routing systems and the relocation of elements are required to avoid clashes with a significantly increasing number of iterations [29]. Moreover, these issues are time-consuming and expensive [12]. When such changes occur later in the construction stage, unscheduled delays, and even accidents, can occur. These situations force engineers to respond to and resolve these situations quickly and may cause confusion and problems for technicians and/or malfunctions in installations [32].

3.3. BIM as a DDWSS Design Tool

BIM modelling utilises a virtual platform to design and develop building projects. The Autodesk® Revit® platform integrates operational and functional characteristics, providing strong tools for the management of relevant and necessary information when the BIM methodology is used. MEP design tools have been developed and represent a small part of Autodesk's priority of developing BIM solutions. At this stage, these efforts are intended to provide all the tools needed by MEP engineers, and Autodesk® is focused on evolving into a single-source BIM solution for the AEC community.

In DDWSS design projects, BIM modelling tools generate a database with element, component and technical information in the form of either geometric (diameter, elevation, length, slope) or mechanical parameters (materials, roughness, flow rate, flow velocity, pressure). These elements can be modified as a project requires, allowing parameters to be modified in conformity with geometrical or model connections, which makes the modelling process more efficient and intuitive [33]. However, these tools lack the potential for customisation, and adaptation to local design guidelines can still be achieved through traditional approaches [27,34].

Parametric modelling combines 3D modelling with external data, enhancing information storage within various project elements as designers generate the 3D model, set instance parameters and place the model in the 3D interface. This information can be used for different projects after saving the initial configurations for further usage [35,36]. Among the parametric programming tools, Dynamo can be highlighted. This tool allows 3D elements to be correlated and enables spatial control with geometric and parametric information. The above allows custom data management and visualisation; all changes appear as analytical data in real time [37,38].

4. Proposed Workflow for DDWSS Design

When proposing a sequence of semi-automated activities for DDWSS design, design standards, technical recommendations, software characteristics, the scope of use, work templates, interoperability between software packages and interface usability must be considered in terms of the effort required. Nevertheless, not all the tasks in a process may be suitable for automation. Complexity levels and the manual work required to perform a task impact the level of effort required. Therefore, as a rule, if automating a task involves more work than performing it manually, it can be considered to either not be automatable or to be a high-cost automation task [39].

The workflow and algorithms developed in this research allow the automation of the development of MEP design in building projects. The workflow provides simple steps for the automation of the design process. Furthermore, the algorithms developed in Dynamo allow the recognition of a BIM project in the Revit environment, identifying its elements and constraints, enabling its calculation with different standards in Excel and updating the design information for the BIM environment.

Figure 2 shows the proposed workflow, including the automation and documentation processes. It consists of four stages: (I) inputs, (II) data processing, (III) parameter management and (IV) output documentation.

Figure 2 shows the steps for automating the design process for DDWSSs. The formulations and calculation guidelines are based on the RIDAA. These stages are easily distinguishable in order to intuitively guide users through the automation process in a simple way. This workflow represents only a guide for the development; the details of the individual sections are given below. Later, in Section 5, the detailed workflows and associated codes for the automation process for both the design and verification of the required MEP elements are presented.

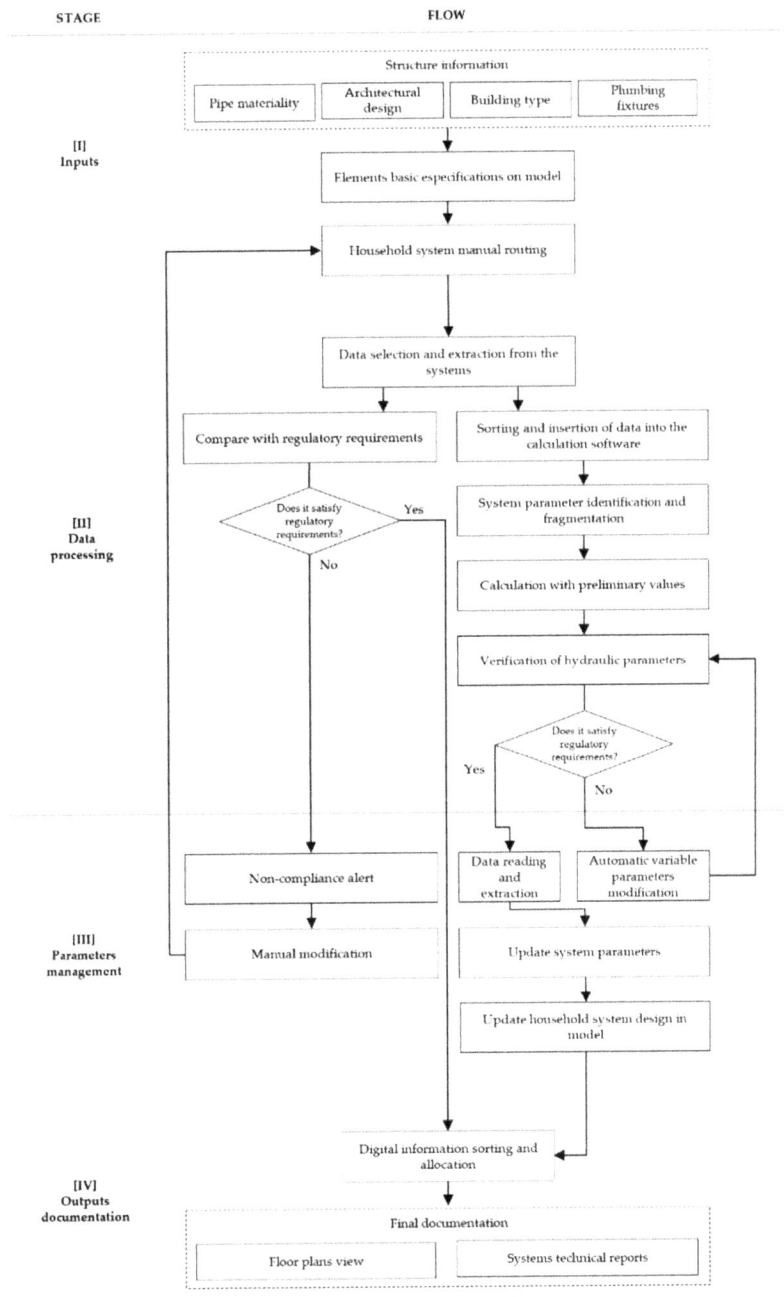

Figure 2. Workflow for drinking-water system (white boxes) and sewerage (grey boxes) design automation.

4.1. Inputs for Automation

For conceptual design, the preliminary information on the structure is delivered by an architectural design that provides the basic conditions for DDWSS operation, such as the type of building, the location, the number of occupants or users, the estimated consumption and the existing public distribution and collection networks [26]. It is then necessary to define the type, location and quantity of artefacts used in the building in order to connect them and create a system through a manual routing process. The purpose of the manual routing process is to facilitate the modeller's freedom and establish criteria for the design of the network. Within the routing process, the materials and preliminary diameters of the pipes must be assigned, as well as any mechanical systems that may be required [40].

As a preliminary step for the generation of BIM project models, the different elements used for the calculations must be configured in Autodesk® Revit®, especially those belonging to MEP systems. These elements are included by default with predefined parametric values. Autodesk® Revit® contains a database of default families and family types, the configuration of which is predefined according to International Plumbing Code (IPC) specifications. It is necessary to modify and integrate objects' families according to the regulations relating to their characteristics, dimensions and mechanical parameters applicable in the country where the automation process will be applied. This also applies to pipe families, for which—although they have a predefined database—it is necessary to add new family types. In the case of materials, it is necessary to add those that are available in the market and those that are increasingly used to establish the commercial diameters for the modelling.

Regarding the system's artefacts, for each sanitary fixture, the corresponding acronym, installed flow rates, height and inlet and outlet diameters must be defined. Each pipe section drawn will be subject to both geometrical and mechanical constraints, so it is important to consider the performance of the connections of the elements in the model when building the piping system. It is thus necessary to undertake detailed modelling of each component of each system. All the fittings and connectors must be described and positioned consistently in order to avoid any deviations or differences in the data readings provided by the model. Moreover, the modeller must correctly select the type of system to which each traced pipe belongs; i.e., the cold-water, hot-water or drainage systems.

In sewerage systems, it is important to indicate the intended use of the sanitary fixtures according to the type of building and the number of people to be served in order to correctly define the wastewater flow that returns to the public system. Water flow sewage systems are gravity-fed; therefore, certain considerations must be taken into account when laying out the pipes so as not to generate points at which solid waste can be deposited and obstruct the continuous discharge of the flow. According to the regulations, a minimal pipe slope must be provided to allow for self-cleaning, the creation of geometries with closed singularities should be avoided and manholes should be placed correctly.

4.2. Data Processing and Parameter Management

In this part of the study, the parameters of the sanitary devices present in the model were identified and extracted, then ordered and tabulated in such a way that the pipe sections can be automatically generated. In the case of drinking water systems, the creation of tables for export is intended for the iterative calculation of their parameters, which are then returned to the model, replacing the existing values. In the case of sewerage systems, data extraction is carried out to verify the slope, waste load and diameter parameters, which are compared with those from the regulations used. If the regulatory values are not satisfied, an alert is shown indicating regulatory non-compliance; this means that the parameters in the model must be manually modified and, subsequently, re-verified until they comply with the permitted ranges.

In addition, depending on the regulations used when carrying out automation, it may be necessary to make practical changes, replacing the designer's criteria when establishing ranges of values for certain parameters that may be subject to special external conditions or simply in order to maximise safety.

4.3. Output Documentation

The presentation of a project includes various documents and outputs for each of the systems to be designed, commonly involving the use of floor plans, detail drawings and calculation tables. However, the features and capabilities of BIM tools allow all these documents to be stored in a single model, enabling the stored information to be updated in real-time and, thus, reducing the amount of repeated work required if changes must be made to the project [36]. Two types of templates were thus created to export the BIM models within Autodesk® Revit®. In the case of sanitary plumbing networks, the plan views of each floor of the structure were shown separately, with separate floors for the cold-water and hot-water systems. On the other hand, for the drainage networks, the template showed the floor plan view of each floor and the floor plan view of the manholes, together with information on the hydraulic equivalent units (HEUs, a quantification of the wastewater discharge generated by each fixture with reference to the RIDAA) and the slope corresponding to each pipe (Figure 3).

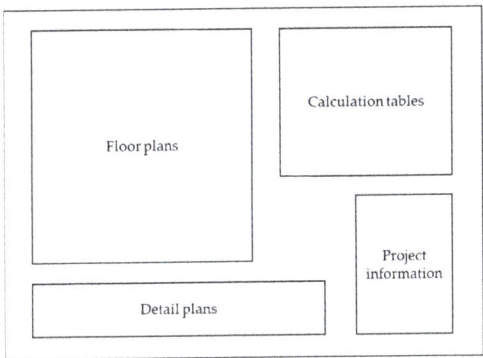

Figure 3. Reference format for the exported plans.

5. Case Study

This section describes the methods used for the case studies and the results obtained from the automation processes. Section 5.1 shows the results associated with drinking-water systems, including the hot- and cold-water networks. Section 5.2 shows the results of the automation for sewerage systems.

5.1. Drinking-Water System Calculation

A series of tables with numerical information associated with the parameters involved in the regulations for sanitary plumbing networks were created as a starting point for the calculation processes [26]. These values are not subject to variation since the formulations and design methods have not undergone significant modifications, so they can be considered as fixed input data for any type of project.

The acronyms established for the denomination of sanitary fixtures were based on the RIDAA and are shown in Table 2.

In some cases, values were adapted to optimise automatic calculation processes and avoid erroneous criteria related to network sizing, consumption and the calculation of pressure losses. The values provided by the RIDAA are organised according to the range of endowments in single-family housing or "dwelling house"-type structures, which can fluctuate between 80 and 450 L/inhab/day, according to the regulation. In our case, the minimum starting value was rounded to 250 L/inhab/day, taking as a reference the regulatory minimum of 211.6 L/inhab/day. These values are related to the Chilean average residential drinking water consumption and take into consideration the fact that any housing unit will have at least one bathroom (including a toilet, bathtub and sink), a kitchen and a washing machine [39]. This value rises as the number of bathrooms in the network increases, as shown in Table 3. For practical purposes, the term "medium" is used for bathrooms that do not have a bathtub or rain bath; in other words, those which contain only a sink, a toilet and possibly a bidet.

Table 2. Acronyms used for the design of artefacts based on the Chilean RIDAA standard.

Plumbing Fixture	Abbreviation Used	Plumbing Fixture	Abbreviation Used
Trough	BE	Glasswasher	LC
Bidet	BI	Handwasher	LO
Bathtub	BO	Dishwasher	LP
Rain shower	BOLL	Laundry	LV
Shower with perforated pipe	BP	Dishwashing machine	LVV
Heater	CAL	Washing machine	MLV
Wet net	GRH	Urinal	UR
Yard tap 13 (mm)	LLJ13	Urinal with perforated pipe	URP
Yard tap 19 (mm)	LLJ19	Toilet	WC

Table 3. Adapted supplies for automatic calculation.

Building Type	Subtype	Endowment
Social housing	Not applicable	70 Lt/inhab/day
Single-family house	With one bathroom, a kitchen and a washing machine	250 Lt/inhab/day
	With two and a half bathrooms, a kitchen and a washing machine	300 Lt/inhab/day
	With two bathrooms, a kitchen and a washing machine	350 Lt/inhab/day
	With three bathrooms, a kitchen and a washing machine	400 Lt/inhab/day
	With more than three bathrooms, a kitchen and a washing machine	450 Lt/inhab/day
Single-start apartment building	Not applicable	450 Lt/inhab/day
Apartment building with an independent meter or sub-meter	Not applicable	Endowment varies by department and according to the subtypes applied for single-family dwellings
Commercial or office premises	Occupation per employee	150 Lt/inhab/day
	Occupancy by surface area	10 Lt/m^2/day
Bar, restaurant, fountain and similar	Not applicable	40 Lt/m^2/day

For the water-meter calculation, the regulation indicates that the maximum admissible head loss is 5 mca (meters of water column). This value is calculated according to the maximum probable flow rate, which is obtained from the fixtures' consumption rates, and the maximum daily consumption rates, obtained according to the building's supply and occupancy. However, since the networks studied are directly connected to the public network (without storage tanks or elevation mechanisms), they are limited to only the maximum probable consumption. The tabulated maximum daily consumption rates (Table 3) only represent reference values [26]. For this reason, the actual value of the maximum daily consumption should be used only when the resulting loss in the meter is less than 5 mca. Otherwise, the maximum daily consumption rate corresponding to the meter diameter that complies with the probable maximum flow rate, as indicated in Table 4, can be used to calculate the losses.

Table 4. Table of meter capacities [26].

Diameter (mm)	Maximum Daily Consumption (m³/day)	Probable Maximum Consumption (L/min)
13	3	50
19	5	80
25	7	117
38	20	333
50	30	500

In the case in which a network has sub-meters with diameters that are not consecutive values in relation to the general meter diameter, according to the order tabulated in Table 4, the final diameter of the sub-meter should be considered as the value immediately below the diameter of the general meter of the system. This is for construction purposes because, if diameters differ by more than two values, the connection of the meter to the sub-meter segment would lead to the incorporation of an unnecessary number of diameter reduction accessories, making the sub-meter inefficient in terms of construction.

The material type of the pipe is defined according to the minimum acceptable nominal pressure for water conduction, which must be greater than or equal to 10 kgf/cm^2 [26]. In addition, the usable diameter is limited to a minimum, imposed by the RIDAA, of 13 mm for copper or 16 mm for plastic materials. In addition, diameters that are not commercially available are not considered, as is the case for 15 mm copper pipes.

Based on the parameters required as input data for the calculation of sanitary plumbing networks, as well as the considerations previously mentioned, a database spreadsheet was generated. The contents of this spreadsheet are summarised in Table 5.

Table 5. Contents of the database for the calculation of sanitary plumbing water networks.

Type of Data	Purpose	Adaptations
Drinking water supply ranges	Maximum daily consumption calculation	Interpolated values of endowments in single-family dwellings are calculated according to the number of existing bathrooms
Accessory loss coefficients	Singular loss calculation	No adaptations
Pipe materials and diameters	Determination of frictional losses and flow velocity	Limitations imposed by the regulations and the commercial diameters available in Chile
Consumption according to the type of sanitary appliance	Calculation of installed and probable maximum flow rates	No adaptations
Common artefact elevations	Calculation of elevation losses	An average elevation is assumed based on architecture and depending on the type of artefact

As an adaptation for the automatic iteration processes, and in order to obtain results within the normative margins, a safety criterion can be established regarding the flow velocities in the pipes for the conduction of drinking water. The maximum flow velocity must be 2.5 m/s for external or main distribution pipes and 2.0 m/s for the internal pipes of the network [26]. As a result, the restriction limiting velocity can be modified to 2.0 m/s in all the pipes of the system, regardless of their conditions, except for the section located immediately downstream of the house connection, where the restriction can be maintained at 2.5 m/s. These values are selected under the assumption that this section will be the one with the largest diameter in the entire network. This definition has the purposes of giving continuity to the changes in diameter between the different sections and protecting the system from possible damage due to water hammer effects.

Applying the workflow described in Figure 2, different series of codes were analysed and tested in order to extract the relevant information from the model. The optimal and lightest way to transfer the data corresponding to each object in the project model in an organised and accurate way was sought. The process obtained after applying the workflow is shown in Figure 4.

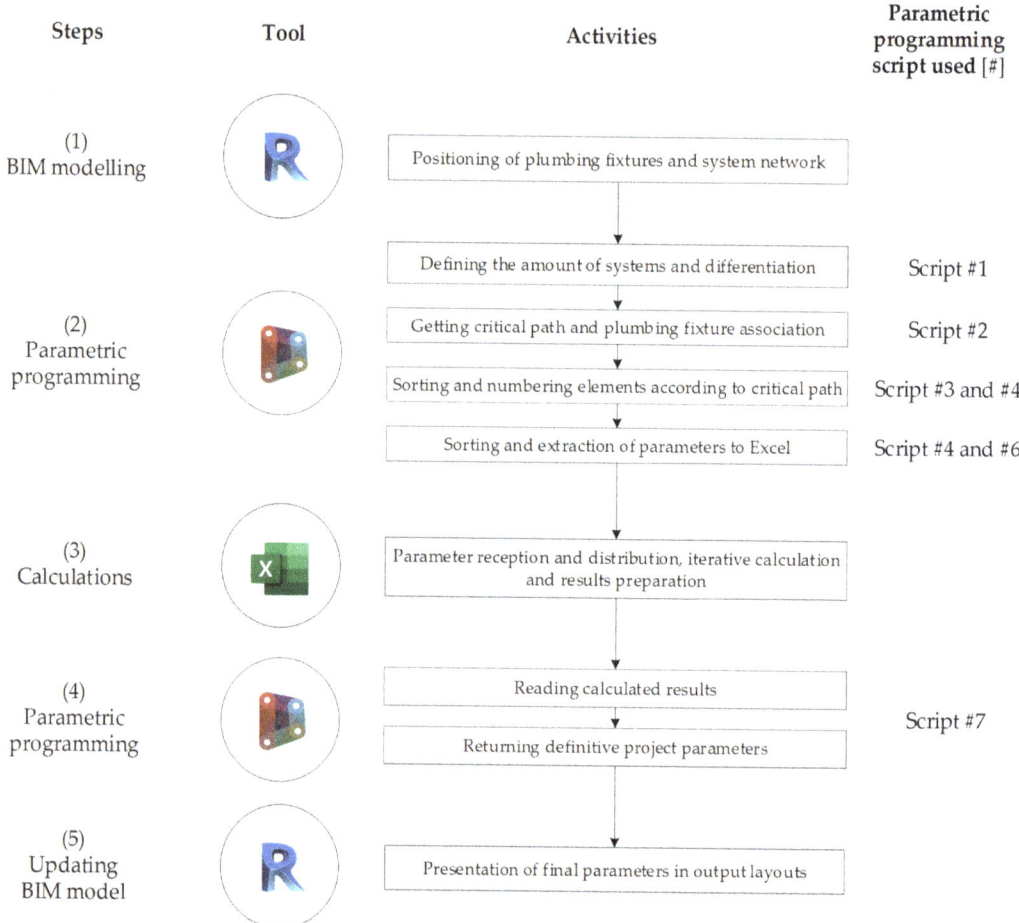

Figure 4. Activity sequence for automatic calculation of sanitary plumbing systems organised into five steps: (1) BIM modelling, (2) parametric programming, (3) calculations, (4) parametric programming and (5) updating BIM model. All the scripts are presented as Appendix A.

Figure 4 shows the process, which contains five steps, for the automation of the calculation of sanitary plumbing systems. Each step is composed of activities. The activities, supported by parametric programming with Dymamo, are related to the number of scripts used, which contain the programming code. These scripts are available in Appendix A to allow for the replication of the process and are as follows. (1) BIM modelling: this step is carried out with the software Revit and enables the location of the sanitary plumbing facility and the system's network. (2) Parametric programming: the Dynamo parametric tool is used to identify the different systems and their quantities, as well as to obtain the critical path of the systems. All systems' components are sorted according to the critical path and exported to a spreadsheet. (3) Calculations: The information obtained by Dynamo is organised in tables and exported to a database in the software Excel. In this software, iterative calculations are carried out to achieve compliance with the design standards until the final design parameters are obtained. (4) Parametric programming: the Dynamo tool is used to transfer the information in Excel to Revit. (5) Updating the BIM model: the final MEP model, along with the required data, is updated in Revit with the information from Dynamo.

5.1.1. Drinking-Water System Calculation—Case Study 1

The designed house has the basic appliances for a structure of this type distributed on a single floor. It is assigned an occupancy of five habitants and a garden area of 20 m^2. As mentioned in Section 4.1, the input parameters incorporated by the user correspond to the preliminary layout of the drinking water and sewerage networks and the global characteristics of the project. Figure 5 shows the 3D model created with Autodesk® Revit®, where the fixtures and cold-water, hot-water and drainage networks are identified. Table 6 details the global project data in accordance with the requirements of Chilean regulations.

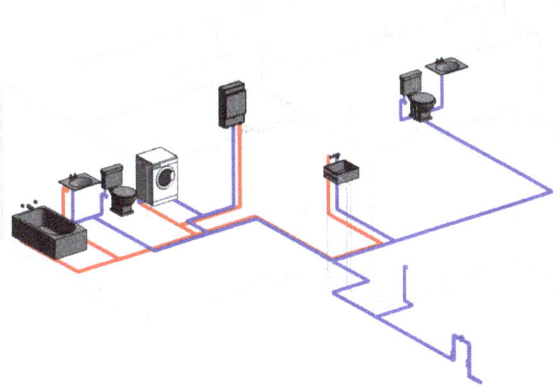

Figure 5. Sanitary plumbing network layout in a single-family house.

Table 6. Global data for the single-family house case study.

Building Type	Subtype	Number of Occupants	Garden or Lawn Surface	Recirculating Pool Volume	Pool Volume without Recirculation	Firefighting Network
Single-family house	House with one and a half bathrooms	Five residents	20 m^2	0	0	0

The second step of the process is the insertion of the preliminary information for the project into the project information chart, as shown in Table 6 and Figure 6.

Once the information for the architectural model has been complemented with the manual routing and the general information for the project, it is necessary to ensure the correct modelling of the connections between pipes and fixtures so that the data will not present errors resulting from manual routing. Then, script 1 and script 2 from Dynamo are executed, which extract and order the information, identify the fixed and variable parameters for the design, group them and create nodes that determine the pipe sections and their connections in order to determine of the critical route of the system. This information assigns names to the systems and nodes (through scripts 3 and 4). Figure 7 shows the result of the identification of the nodes and systems within the structure.

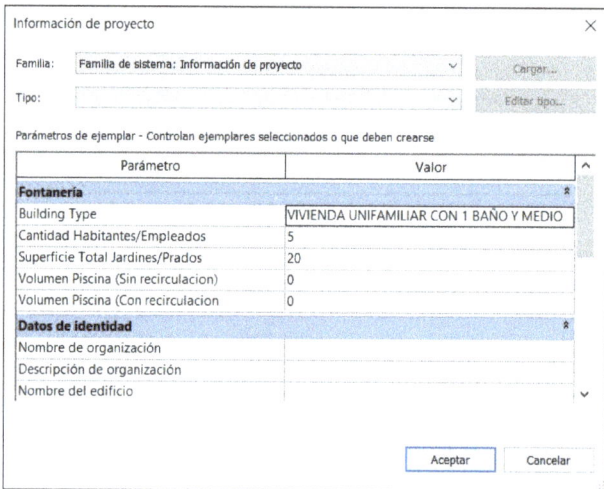

Figure 6. Project information chart for single-family house.

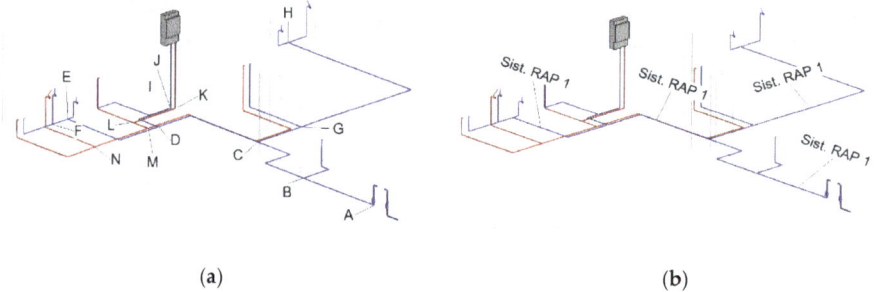

Figure 7. Results after the application of scripts 3 and 4: (**a**) node identification and (**b**) sanitary plumbing system recognition.

The information generated and extracted from the model is inserted into the spreadsheet. Each system present in the house (cold water and hot water) is differentiated and the information is sorted according to the entities derived from the 3D model. Figure 8 shows the distribution of the information in the parameter insertion spreadsheet.

The information is then processed using automatic iterative formulations in MS Excel®, modifying the diameter of each pipe section until a value that complies with the minimum standards of the regulation and the previously adopted criteria is obtained (Table 4).

As a result, the data corresponding to the tables for the total flow rates of the fixtures contained in the house (Table 7) and for the calculation of the meter (Table 8) are obtained.

The diameter of and losses in the drinking water-meter can be calculated similarly. According to Table 9, the actual maximum daily consumption is not high enough for the pressure losses to be less than 5 mca. Therefore, the consumption corresponding to the 19 mm meter, obtained from Table 4 with the probable maximum flow, is automatically assigned, resulting in a pressure loss that complies with the regulations.

Table 9 shows the losses in the cold-water network of the house, summarising the results subject to restrictions and improving the general visualisation of the values.

Table 10 summarises the losses corresponding to the hot-water network of the house.

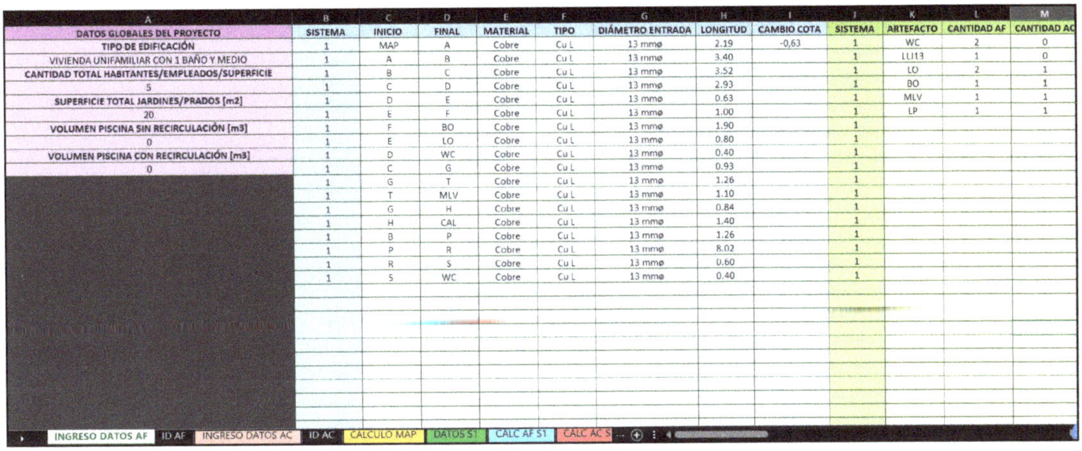

A	B	C	D	E	F	G	H	I	J	K	L	M
DATOS GLOBALES DEL PROYECTO	SISTEMA	INICIO	FINAL	MATERIAL	TIPO	DIÁMETRO ENTRADA	LONGITUD	CAMBIO COTA	SISTEMA	ARTEFACTO	CANTIDAD AF	CANTIDAD AC
TIPO DE EDIFICACIÓN	1	MAP	A	Cobre	Cu L	13 mmø	2.19	-0,63	1	WC	2	0
VIVIENDA UNIFAMILIAR CON 1 BAÑO Y MEDIO	1	A	B	Cobre	Cu L	13 mmø	3.40		1	LLJ13	1	0
CANTIDAD TOTAL HABITANTES/EMPLEADOS/SUPERFICIE	1	B	C	Cobre	Cu L	13 mmø	3.52		1	LO	2	1
5	1	C	D	Cobre	Cu L	13 mmø	2.93		1	BO	1	1
SUPERFICIE TOTAL JARDINES/PRADOS [m2]	1	D	E	Cobre	Cu L	13 mmø	0.63		1	MLV	1	1
20	1	E	F	Cobre	Cu L	13 mmø	1.00		1	LP	1	1
VOLUMEN PISCINA SIN RECIRCULACIÓN [m3]	1	F	BO	Cobre	Cu L	13 mmø	1.90		1			
0	1	E	LO	Cobre	Cu L	13 mmø	0.80		1			
VOLUMEN PISCINA CON RECIRCULACIÓN [m3]	1	D	WC	Cobre	Cu L	13 mmø	0.40		1			
0	1	C	G	Cobre	Cu L	13 mmø	0.93		1			
	1	G	T	Cobre	Cu L	13 mmø	1.26		1			
	1	T	MLV	Cobre	Cu L	13 mmø	1.10		1			
	1	G	H	Cobre	Cu L	13 mmø	0.84		1			
	1	H	CAL	Cobre	Cu L	13 mmø	1.40		1			
	1	B	P	Cobre	Cu L	13 mmø	1.26		1			
	1	P	R	Cobre	Cu L	13 mmø	8.02		1			
	1	R	S	Cobre	Cu L	13 mmø	0.60		1			
	1	S	WC	Cobre	Cu L	13 mmø	0.40		1			

INGRESO DATOS AF | ID AF | INGRESO DATOS AC | ID AC | CALCULO MAP | DATOS S1 | CALC AF S1 | CALC AC S

Figure 8. Insertion of the information for the pipes in an MS Excel® spreadsheet using scripts 5 and 6.

Table 7. System flow rates table for a single-family house.

Plumbing Fixture	Acronym	Plumbing Fixture Supply (Lt/min)	Cold Water Quantity	Hot Water Quantity	Supply (Lt/min)
Toilet	WC	10	2	0	20
Yard tap	LLJ13	20	1	0	20
Handwasher	LO	8	2	1	24
Bathtub	BO	15	1	1	30
Washing machine	MLV	15	1	1	30
Dishwasher	LP	12	1	1	24
		Total installed flow (L/min)			148

Table 8. Water-meter calculation report for a single-family house.

Water-Meter Calculation Report	
System installed flow rate	148.00 Lt/min
Maximum probable flow rate	54.43 Lt/min
Meter diameter	19.00 mm
Maximum daily consumption	5 m^3/day
Meter head loss	4.27 mca

Table 9. Loss summary table for cold-water system in a single-family house.

Segment		Material	Type	Length (m)	Diameter (mm)	Speed (m/s)	Total Loss (mca)	Final Pressure (mca)
MAP	A	Copper	Cu L	2.19	25	1.70	1.92	12.08
A	B	Copper	Cu L	3.40	25	1.54	0.69	11.40
B	C	Copper	Cu L	3.52	25	1.28	0.42	10.98
C	D	Copper	Cu L	2.93	19	1.03	0.86	10.12
D	E	Copper	Cu L	0.63	19	0.80	0.08	10.04
E	F	Copper	Cu L	1.00	19	0.60	0.04	9.99
F	BO	Copper	Cu L	1.90	13	1.25	3.17	6.83
E	LO	Copper	Cu L	0.80	13	0.81	1.27	8.76
D	WC	Copper	Cu L	0.40	13	0.94	0.77	9.35
C	G	Copper	Cu L	0.93	19	1.65	0.47	10.50
G	T	Copper	Cu L	1.26	19	0.60	0.06	10.44

Table 9. Cont.

Segment		Material	Type	Length (m)	Diameter (mm)	Speed (m/s)	Total Loss (mca)	Final Pressure (mca)
T	MLV	Copper	Cu L	1.10	13	1.25	2.23	8.21
G	H	Copper	Cu L	0.84	19	1.37	0.17	10.33
H	CAL	Copper	Cu L	1.40	19	1.37	2.72	7.61
B	P	Copper	Cu L	1.26	19	0.97	0.15	11.25
P	R	Copper	Cu L	8.02	19	0.68	0.35	10.90
R	S	Copper	Cu L	0.60	19	0.45	0.03	10.87
S	WC	Copper	Cu L	0.40	13	0.94	0.75	10.12
R	LO	Copper	Cu L	0.80	13	0.81	1.30	9.60
P	Q	Copper	Cu L	1.64	19	0.51	0.19	11.06
Q	LP	Copper	Cu L	0.80	13	1.07	1.59	9.46
A	LLJ13	Copper	Cu L	0.60	13	1.52	2.29	9.80

Table 10. Table summarising hot-water-system losses in a single-family house.

Segment		Material	Type	Length (m)	Diameter (mm)	Speed (m/s)	Total Loss (mca)	Final Pressure (mca)
CAL	I	Copper	Cu L	1.40	19	1.37	−0.37	7.98
I	J	Copper	Cu L	1.05	19	1.37	0.23	7.75
J	O	Copper	Cu L	1.30	19	0.60	0.06	7.69
O	MLV	Copper	Cu L	1.10	13	1.25	2.23	5.46
J	K	Copper	Cu L	0.88	19	1.07	0.16	7.59
K	N	Copper	Cu L	6.31	19	0.51	0.20	7.39
N	LP	Copper	Cu L	0.80	13	1.07	1.59	5.79
K	L	Copper	Cu L	3.61	19	0.80	0.26	7.33
L	LO	Copper	Cu L	0.80	13	0.81	1.27	6.06
L	M	Copper	Cu L	1.00	19	0.60	0.06	7.27
M	BO	Copper	Cu L	1.90	13	1.25	3.17	4.10

As shown in Table 7, the developed algorithm allows the fixtures present in the system to be identified and assigns the acronyms used for their abbreviation in the tabulations. It interprets the connections of the appliances to the hot- and cold-water systems in order, finally, to calculate the values of the final and total flow rates installed in the house. As shown in Table 9, the flow velocity values comply with the maximum of 2.0 m/s, even in the first section. Likewise, all the final pressures of the system comply with the required minimum of 4 mca. The diameters show a downward variation but maintain a value of 13 mm in the sections connected to the fixtures. In addition, the calculation table is ordered starting with the critical path from the meter to the most unfavourable device, which in this case is the rain bath (BOLL). Since there is only one floor in the house, this is the point of highest elevation in the entire system. In the hot-water system, the flow rates and final pressures satisfy the regulatory restrictions. As shown in Table 10, the initial section (CAL-I) starts at the heating device (CAL), so the initial pressure is not equal to the 14 mca coming from the meter but to the pressure resulting from the section that ends at the heater, indicated in Table 9 (H-CAL).

Based on the results obtained for the single-family-house case, it is possible to verify the correct performance of the algorithms in terms of their effective recognition of the devices and maintenance of the desirable diameters (in the case of copper pipes, 13 mm) in the sections they are present in. In addition, algorithms demonstrated the ability to perform iterative calculations of diameters progressively until the minimum pressures are met, and the calculation criteria were correctly followed along the critical path.

5.1.2. Drinking-Water System Calculation—Case Study 2

The second case study represents a real project, previously designed and with values calculated using the traditional methods described in Section 3.1 and approved by the DDWSS local administration. The building is structured over two floors and contains six retail stores inside the building and a garden area of 20 m^2. Each store has a water submeter fed from a general water meter. In addition, it has a 200 L/min wet fire suppression network. This project belongs to the category "commercial or office premises", so its occupancy is measured in terms of its surface area. Figure 9 shows the 3D model of the project, which is based on interpretation of the CAD drawings. Table 11 summarises the global data for the project.

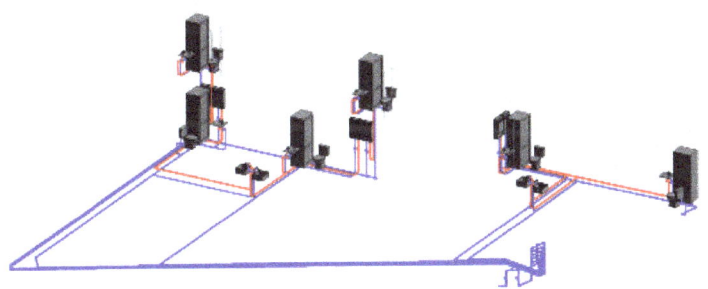

Figure 9. Sanitary plumbing network layout for retail stores and offices.

Table 11. Global data for the retail store building case study.

Building Type	Subtype	Number of Occupants	Garden or Lawn Surface	Recirculating Pool Volume	Pool Volume without Recirculation	Firefighting Network
Commercial or office premises	Occupancy by surface area	321.5 m^2	20 m^2	0	0	200 Lt/Min

The architecture was designed with two different retail store patterns: those on the second floor have one less fixture than those on the first floor. Therefore, two different cases were considered for the study, choosing the stores on each floor that are farthest from the drinking water meter, as these are the most unfavourable in terms of friction losses due to the greater length of their pipes. Retail store type 1 has a toilet, sink, dishwasher and rain bath, while retail store type 2 has a toilet, sink and dishwasher. Figure 10 shows the systems for each of the types of stores used for the study.

(a)   (b)

Figure 10. Sanitary plumbing water systems for the retail stores under study: (**a**) type 1 (first floor); (**b**) type 2 (second floor).

As in the case of the single-family house, the information corresponding to the general characteristics of the building is incorporated into a chart (Figure 11).

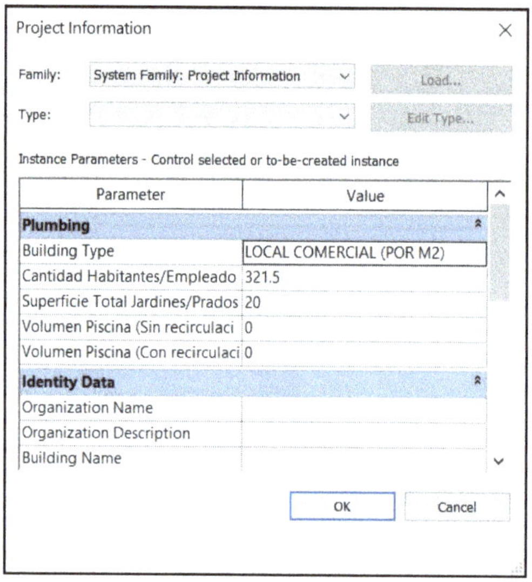

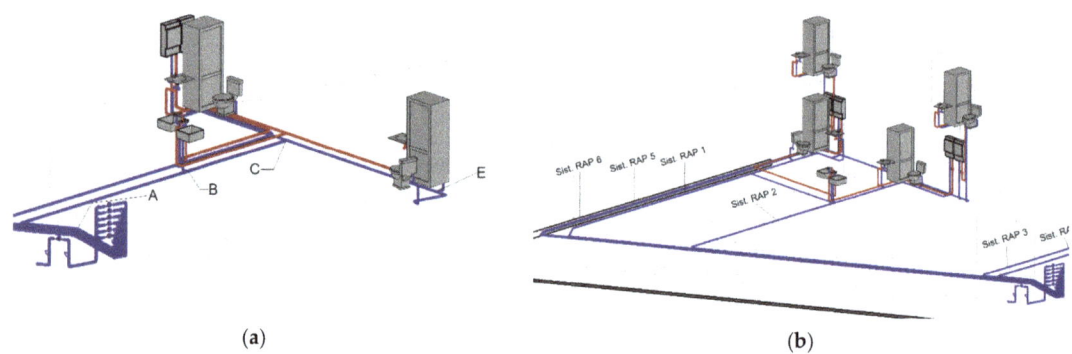

Figure 11. Project information chart for a retail building.

With the information from the architectural model, scripts 1 and 2 are executed in Dynamo to obtain the nodes and sections of each system present in the structure, assign names to them (scripts 3 and 4) and determine the critical path. Figure 12 shows the results of the identification of the nodes and systems within the structure.

Figure 12. Results of the application of scripts 3 and 4: (**a**) identification of nodes with the assignment of letters; (**b**) recognition of the drinking-water network system.

As in the first case, the information is inserted into the MS Excel® spreadsheet. The results for the second case study are presented as follows. The general table of installed loads in the system is shown in Table 12, and the calculation memory of the drinking water meter that feeds the entire system is shown in Table 13.

Table 12. Total installed loads chart for a commercial building.

Plumbing Fixture	Acronym	Plumbing Fixture Flow Rate (Lt/min)	Cold Water Quantity	Hot Water Quantity	Final Flow Rate (Lt/min)
Toilet	WC	10	6	0	60
Handwasher	LO	8	6	6	96
Rain shower	BOLL	10	6	6	120
Dishwasher	LP	12	4	4	96
Wet net					200
Total installed flow (Lt/min)					572

Table 13. General meter calculation report for a commercial building.

Meter Calculation Report	
System installed flow rate	572.00 Lt/min
Maximum probable flow rate	138.13 Lt/min
Meter diameter	38.00 mm
Maximum daily consumption	20.00 m^3/day
Meter head loss	1.72 mca

As mentioned above, calculations for the networks were undertaken for two types of retail stores, which are representative of the configurations found in the project. Thus, we can begin by showing the results for the type 1 store. Table 14 shows the flow rate chart.

Table 14. Flow rate table for type 1 retail store.

Plumbing Fixture	Acronym	Plumbing Fixture Flow Rate (Lt/min)	Cold Water Quantity	Hot Water Quantity	Final Flow (Lt/min)
Toilet	WC	10	1	0	10
Handwasher	LO	8	1	1	16
Rain shower	BOLL	10	1	1	20
Dishwasher	LP	12	1	1	24
Total installed flow (Lt/min)					70

As for the previous building, the contributions of the fixtures to the system were identified and quantified, and the total installed flow was calculated.

Once the system flow rates were determined, the calculation for the water sub-meter associated with the retail store was performed, yielding the results shown in Table 15.

Table 15. Sub-meter calculation report for type 1 retail store.

Sub-Meter Calculation Report	
System installed flow rate	70.00 Lt/min
Maximum probable flow rate	32.49 Lt/min
Sub-meter diameter	25.00 mm
Maximum daily consumption	7.00 m^3/day
Sub-meter head loss	0.78 mca

Table 16 shows a summary of the losses chart for the type 1 retail store.

Table 16. Loss summary table for cold potable water system for type 1 retail store.

Segment		Material	Type	Length (m)	Diameter (mm)	Speed (m/s)	Total Loss (mca)	Final Pressure (mca)
MAP	RAP	Copper	Cu L	3.30	25	1.02	0.76	13.24

Table 16. Cont.

Segment		Material	Type	Length (m)	Diameter (mm)	Speed (m/s)	Total Loss (mca)	Final Pressure (mca)
RAP	A	Copper	Cu L	2.56	25	1.02	0.21	13.03
A	B	Copper	Cu L	28.39	25	1.02	1.76	11.27
B	C	Copper	Cu L	2.20	19	1.52	0.60	10.68
C	D	Copper	Cu L	0.99	19	1.34	0.29	10.39
D	BOLL	Copper	Cu L	1.43	13	0.94	2.15	8.24
D	E	Copper	Cu L	0.85	19	1.14	0.26	10.13
E	CAL	Copper	Cu L	1.72	19	0.97	1.59	8.54
E	LO	Copper	Cu L	1.46	13	0.81	1.02	9.11
C	WC	Copper	Cu L	0.67	13	0.94	0.57	10.11
B	LP	Copper	Cu L	5.51	13	1.07	1.67	9.61

Table 17 shows the summary of hot water network losses.

Table 17. Loss summary table for hot potable water system, retail store type 1.

Segment		Material	Type	Length (m)	Diameter (mm)	Speed (m/s)	Total Loss (mca)	Final Pressure (mca)
CAL	F	Copper	Cu L	1.19	19	0.97	2.04	6.49
F	G	Copper	Cu L	0.28	19	0.97	0.15	6.35
G	LO	Copper	Cu L	0.86	13	0.81	0.97	5.37
G	H	Copper	Cu L	1.33	19	0.78	0.21	6.14
H	BOLL	Copper	Cu L	1.18	13	0.94	2.12	4.02
H	LP	Copper	Cu L	8.19	13	1.07	2.08	4.06

The flow rate table corresponding to the type 2 store in the retail store project is shown in Table 18.

Table 18. Total installed loads chart for type 2 retail store.

Plumbing Fixture	Acronym	Plumbing Fixture Flow Rate (Lt/min)	Cold Water Quantity	Hot Water Quantity	Total Flow Rate (Lt/min)
Toilet	WC	10	1	0	10
Handwasher	LO	8	1	1	16
Rain shower	BOLL	10	1	1	20
		Total installed flow (Lt/min)			46

As previously, the resulting values for the installed flow rates were correctly obtained, counting the fixtures with cold and hot water. In this case, a lower value for the total flow is evident due to the lower number of fixtures in this system. The result for the corresponding sub-meter is shown in Table 19. The table of losses for the type 2 retail store is shown in Table 20.

Table 19. Sub-meter calculation report for type 2 retail store.

R.A.P. Calculation Report	
System installed flow rate	46.00 Lt/min
Maximum probable flow rate	24.33 Lt/min
Sub-meter diameter	25.00 mm
Maximum daily consumption	7.00 m^3/day
Sub-meter head loss	0.78 mca

Table 20. Losses in the cold potable water system for type 2 retail store.

Segment		Material	Type	Length (m)	Diameter (mm)	Speed (m/s)	Total Loss (mca)	Final Pressure (mca)
MAP	RAP	Copper	Cu L	11.93	25	0.76	0.73	13.60
RAP	A	Copper	Cu L	2.06	25	0.76	0.11	13.49
A	B	Copper	Cu L	41.92	25	0.76	1.54	11.95
B	C	Copper	Cu L	0.43	19	0.92	3.32	8.63
C	D	Copper	Cu L	3.16	19	0.92	0.25	8.39
D	E	Copper	Cu L	0.33	19	0.92	0.10	8.29
E	F	Copper	Cu L	0.15	19	0.68	0.05	8.24
F	BOLL	Copper	Cu L	1.22	13	0.94	2.13	6.11
F	LO	Copper	Cu L	1.56	13	0.81	1.03	7.21
E	WC	Copper	Cu L	1.29	13	0.94	0.67	7.61
B	CAL	Copper	Cu L	1.46	19	0.68	1.50	10.46

As a final result, the loss table for the hot-water network of the store under analysis is shown in Table 21.

Table 21. Summary table for hot potable water system losses for type 2 commercial premises.

Segment		Material	Type	Length (m)	Diameter (mm)	Speed (m/s)	Total Loss (mca)	Final Pressure (mca)
CAL	G	Copper	Cu L	3.16	19	0.68	3.94	6.51
G	H	Copper	Cu L	0.76	19	0.68	0.09	6.42
H	BOLL	Copper	Cu L	1.13	13	0.94	2.12	4.30
H	LO	Copper	Cu L	1.43	13	0.81	1.02	5.40

The results of the automatic export of the results are presented in Figure 13. The obtained data are compiled in a standard plan, displaying views of the structure together with the tables corresponding to the installed flow rates, head losses and meter and sub-meter calculations. This plan is based on the technical format imposed by the RIDAA, whereby the detailed drawings of the meter niche, sub-meters, starter and boiler device are included, resulting in a format ready to be presented in an application for sanitary service feasibility evaluation.

5.2. Sewage System Calculation

The sewerage network implementation methodology starts with a similar approach as for drinking water networks. Modelling considerations for sanitary drainage must cover all implications of gravitational flow. As defined in traditional methods and by code requirements, system routing must be constructed to ensure continuous operation and discharge of waste flow. The modelling process starts by assigning all plumbing fixtures a discharge route and assigning pipe segments throughout the construction plan, from the fixture outlets to the public collector, assuming initial slopes match the minimum values allowed by regulations. Manhole positioning must then be considered when routing pipes, as they must be placed on the outer part of the building and ensure continuity for every discharge.

Three-dimensional model geometry data information and parameters are extracted from plumbing fixtures and piping elements. This model is obtained by generating a system grouping and filtering according to sanitary classifications, at which point the relevant data are stored and distributed for further analysis. All drainage system piping elements are taken downstream from each fixture discharge sequentially, with the exception of manholes, as defined by the flow direction. This system is described in terms of the HEUs that every pipe must discharge in accordance with [26]. The application of the workflow described in Figure 2 for a sewerage system is detailed in Figure 14.

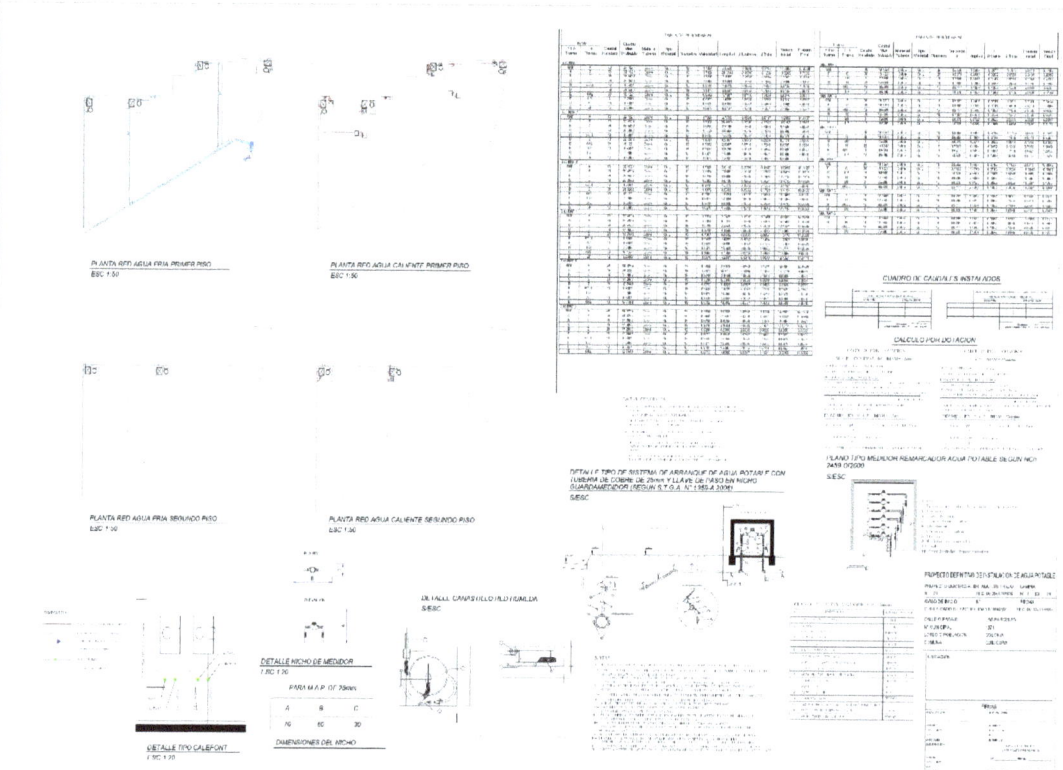

Figure 13. Presentation of the final results in standard layout format.

Figure 14 shows the sewerage system process verification. This process is composed of three steps. Each step is also composed of activities. The activities, supported by Dynamo, are related to the number of scripts used, which contain the programming code. These scripts are available in Appendix A to allow for the replication of the process and are as follows. (1) BIM modelling: this step, undertaken in Revit, involves the sewerage system layout and fixture description. (2) Parametric programming: Dynamo software is used to calculate the discharge slopes and pipe diameters in a first design. This design is divided into different pipes and sections in order to carry out the verification according to the design standard. (3) Verification: after the second step, two messages may be displayed. On the one hand, if issues (such as non-compliance with minimum and maximum slopes) are identified, a message is presented to the user requesting manual correction. On the other hand, if there are no issues to correct, a message stating "no issues" is displayed.

The objects or pipes assigned to each plumbing fixture extracted in the previous nodes are separated into sub-lists, in which each waste flow value is assigned according to sub-list indexes. The pipe positioning determines whether a single element is included in several different-sub lists and received discharges or waste flow from multiple fixtures. This allows the detection and determination, using summary functions, of the total discharge units for each pipe according to their repeated display in sub-lists and their HEU values. The obtained values are loaded into the element parameters of each pipeline. This is achieved by indicating the element instance, the parameter name and the new values. In addition, tagging pipe elements and calculating HEU values also provides relevant code-referenced information. This allows differentiation between pipes that contain streams from a single or multiple artefacts, called "secondary pipelines" and "primary pipelines", respectively (RIDAA). This procedure can also be applied in order to identify ventilation system elements (Figure 13, script 8), where the flow value must be equal to zero because they do not have any relationship with any fixture upstream of their components, as shown in Figure 15.

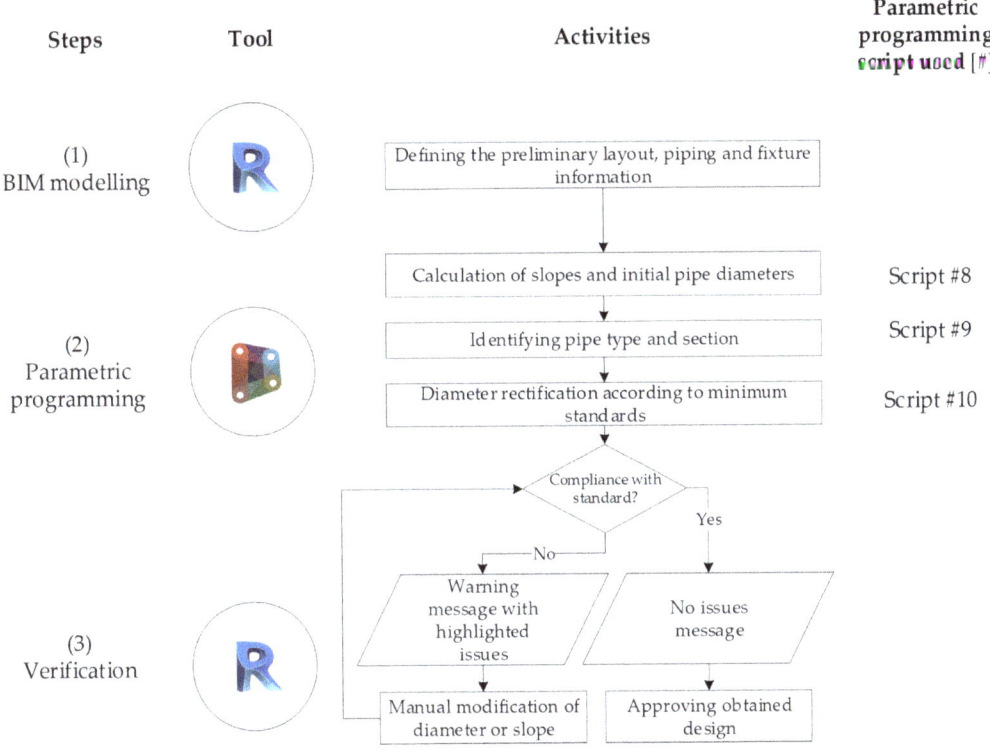

Figure 14. Processes for sewerage system verification organised into three steps: (1) BIM modelling, (2) parametric programming and (3) verification. All the scripts are presented in the Appendix A.

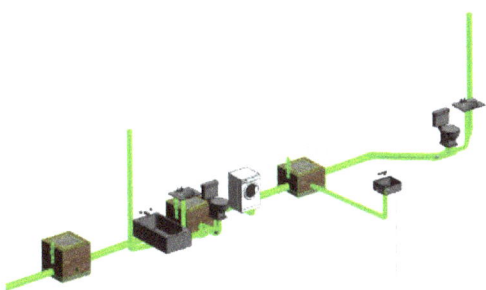

Figure 15. Household drainage network layout in a single-family house.

Regular design codes for house sewerage systems describe a verification-based methodology for network calculation. This involves establishing limits or variation ranges for design parameters, such as diameters, slopes, fitting angles, elevation and HEUs. This approach enables alignment for validation tests through database integration directly into the Dynamo workflow, without requiring external software connections or data transfer. Each verification table is referenced and explained in code documents for sewerage calculations, including its application form and limitations. Tables are changed into Dynamo nodes as a list, following the exact order shown in the code documents. These tables are organised after collecting all element parameters and associating and grouping geometry parameters, flow values and slopes into separated lists. They are then linked to verification tables through logic algorithms. The first verification ensures no diameter reduction between any plumbing fixture and its downstream discharges. Workflow criteria assume that the minimum discharge diameter of every fixture is uploaded as an element parameter, has been previously read and is stored in sub-lists. Piping elements from each fixture are uploaded and sorted according to the downstream flow into a list connected to a node that reads internal element diameters. The information obtained is compared with the minimum fixture discharge diameter and loaded into element parameters according to system routing and conditions (Figure 13, script 9).

The second verification requires all the information previously calculated and uploaded as element parameters. Where geometric pipe parameters are matched with their corresponding axes within verification tables, diameters and slopes are identified and compared in the tables' rows and columns to obtain the maximum HEU value for each pipe. These tabulated values are also distinguished in terms of primary and secondary pipelines, indicating whether it is necessary to modify segment parameters. Such modifications must be performed by editing parameters manually.

However, it is difficult to easily modify lengths and slopes simultaneously since adjacent elements share position parameters, so modifying them may cause disconnections or invalidity in the internal mechanics of the systems. Therefore, minimum diameters can be changed automatically. If pipe slopes do not satisfy specified limit values, the algorithm highlights the pipe in a striking colour, so that users know which section presents normative problems (as shown in Figure 16). This algorithm thus allows model configurations to be manually re-designed and adjusted until the whole system complies correctly when the verification code is executed again (Figure 13, script 10).

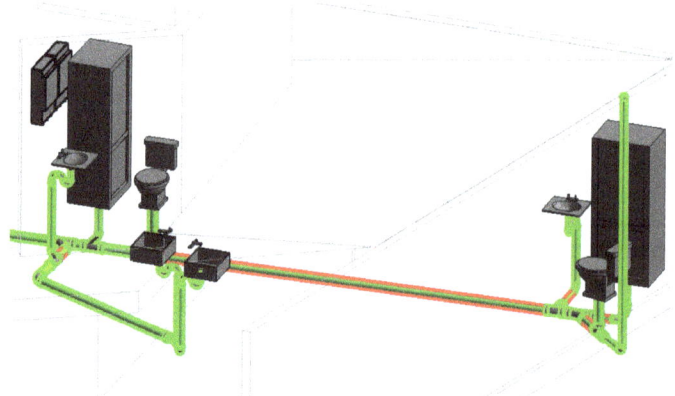

Figure 16. Example of an alert for pipelines that do not comply with the established minimum.

A check is also made for pipes that exceed the tabulated slopes or the established maximum of 15%, as well as whether they have sufficient slopes to withstand the discharge conditions. Once these verifications are finished and the sizing design is developed, the final documentation needs to be generated to assign floor plans and isometric views to corresponding sheets and implement the code presentation format (Figure 17).

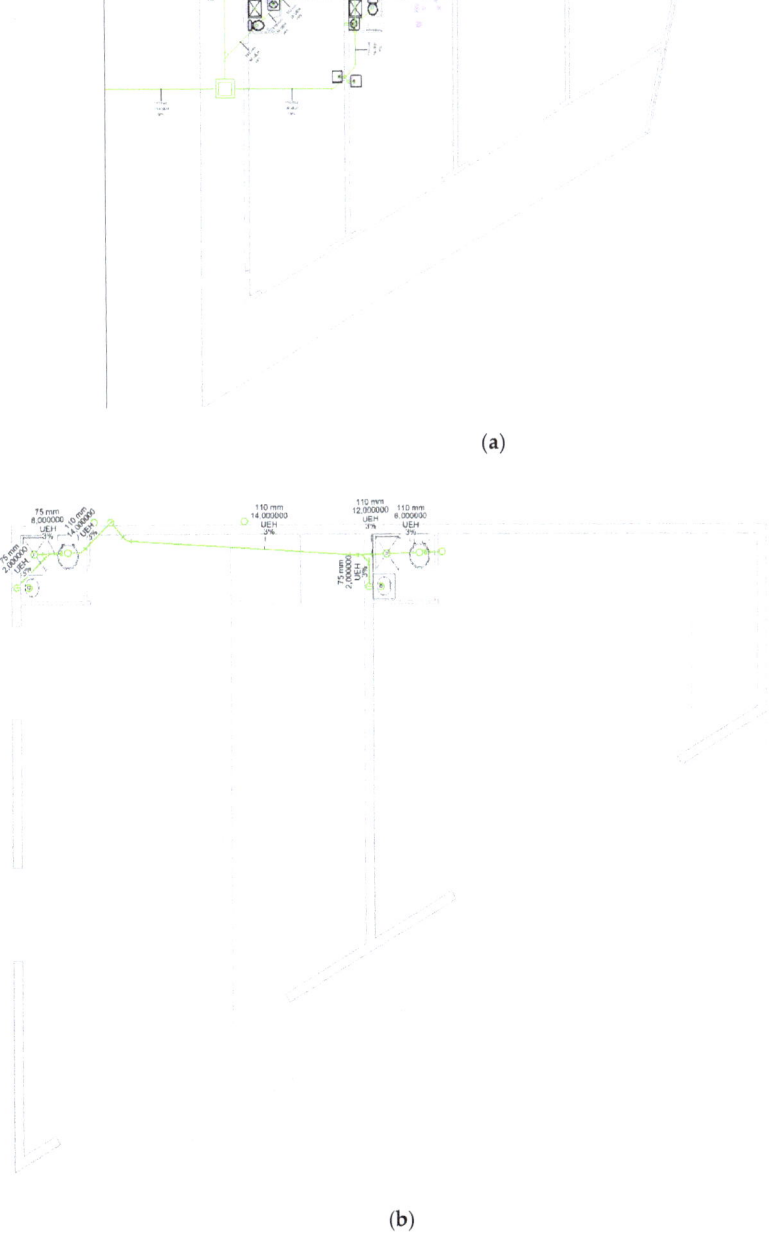

Figure 17. Sewerage system floor plan, including calculated HEUs for each segment on the first story (**a**) and the second story (**b**).

6. Results and Discussion

The developed workflow shows that several design activities can be automated. The interaction with DDWSS components makes it possible to simplify the algorithm's development through data management and calculations with visual programming. As sanitary calculation requires flow analysis, Dynamo provides specified nodes that work especially well with plumbing facilities, correctly organising and detecting every system element and evaluating its properties [41]. Node customisation allows specific data storage for verifications, modifications and calculations. The case studies prove that different approaches can be developed to satisfy different user requirements, meaning that the proposed approach is a generalisable solution for sanitary design. The Dynamo-friendly user interface makes programming comfortable and intuitive; node-based tasks make smooth workflows easier to structure and connect between different actions.

The ability to manipulate data between different software packages helps designers integrate specific calculations and more complex analyses. DDWSS design requires several iterations and developing an optimised design requires a large-scale mechanical design that modelling software cannot handle.

For the drinking-water system design in case study 1, different pipeline systems were read or extracted upstream from each fixture. Elements were divided into sub-lists, and each one was labelled according to a given numeration stored in the sub-meter properties and element parameters. Each element was then associated with each system uploading a mark type, allowing differentiation of the elements that feed each room domain (Figure 4, script 1). Pipe sections were then differentiated according to fixture positioning on the respective system routing, locating "Tee" elements that distribute water flow to fixture water inlets. These were obtained by assigning each Tee node a letter or a number, ensuring clarity about each system sector and the ability to tabulate all necessary parametric information (Figure 4, script 2). The purpose of script 2 is to identify energy losses due to friction, as well as singular losses, changes in elevation, flow rates and velocities as the water is distributed, and calculate the pressure at each fixture. For better pipe section tabulation, the most critical artefact in the system is highlighted according to its available pressure. This starts with pipe sections that take the route to the artefact of the building that has the highest elevation—this is referred to as the critical route since it generally has the lowest pressure in the system. After that, the additional tranches are added (Figure 4, script 3). Once the general information for the project has been defined along with the discretisation and marking of all the pipe sections, the next step is to extract the data to an MS Excel® worksheet to facilitate the respective calculations. This follows the same order that was given for the critical route, and then follows the ensuing sections to the remaining artefacts (Figure 4, scripts 4 and 5).

The results for the second case study for the drinking-water system design (Tables 12 and 13) demonstrate the design's good performance in detecting the fixtures and re-calculating installed flow rates. For the loss summary for cold water in the type 1 retail store (Table 16), the speeds and pressures comply with the normative restrictions. The diameters maintain the pattern of decrease, except for the sections that end in a fixture. Table 22 shows the results obtained for diameters and pressures in each section, along with the results obtained with the traditional method.

The comparison of the values shows that the developed workflow produced similar results to the traditional design approach and performed correctly and in compliance with the RIDAA design standard. No trend can be observed with regard to the pressure results, as the variations were both positive and negative. The relative percentage error ranged from 0.09% to 7.12%. Part of this error could be related to the variability generated by modelling the project on the basis of two-dimensional floor plans (produced using CAD). Due to elevation changes, this variability cannot be completely interpreted, generating differences in elevation that can affect the accuracy of the final pressure calculations. The high level of craftsmanship involved in traditional methods could be another cause of this difference. However, the obtained results ensure—with regard to the design standard—operating conditions that have a certain safety factor established by rounding values, such as the maximum probable flow rate or the singular loss calculations. In addition, the difference may have been influenced by the criteria applied by the professional who performed the calculations for the project.

Table 22. Comparison of diameters and final cold water pressure for type 1 retail store.

Segment		Diameter Using Traditional Method (mm)	Diameter with Automation (mm)	Final Pressure with Traditional Method (mca)	Final Pressure with Automation (mca)	Error between Pressure Results (%)
MAP	RAP	25	25	13.72	13.24	3.50%
RAP	A	25	25	12.57	13.02	3.60%
A	B	25	25	11.19	11.27	0.71%
B	C	19	19	10.69	10.68	0.09%
C	D	19	19	10.52	10.39	1.24%
D	BOLL	13	13	8.05	8.24	2.36%
D	E	19	19	10.35	10.13	2.13%
E	CAL	19	19	8.98	8.54	4.90%
E	LO	13	13	9.05	9.11	0.66%
C	WC	13	13	9.80	10.11	3.16%
B	LP	13	13	8.97	9.61	7.12%

As for the cold-water network sections, the margins of losses and velocities are comparable to those of the hot-water network. Table 23 shows a comparison of the results with the known values.

Table 23. Comparison of diameters and final hot water pressure for type 1 retail store.

Segment		Diameter Using Traditional Method (mm)	Diameter with Automation (mm)	Final Pressure with Traditional Method (mca)	Final Pressure with Automation (mca)	Error between Pressure Results (%)
CAL	F	19	19	7.17	6.49	9.48%
F	G	19	19	6.16	6.35	3.08%
G	LO	19	13	5.85	5.37	8.21%
G	H	19	19	7.06	6.14	13.03%
H	BOLL	13	13	4.58	4.02	12.23%
H	LP	13	13	4.27	4.06	4.92%

In this case, the diameters of the actual project did not completely agree with those obtained through automation, since in the sections leading to an appliance, the criterion of maintaining the 13 mm diameter was assumed to simplify the connection with the outlet taps, avoiding the excessive use of fittings at the terminals of the network. However, using a 19 mm diameter does not represent a significant problem, since it is still an acceptable size for connecting a sanitary fixture. The larger diameter may have been associated with the rather conservative criterion applied by the professional responsible for the project. With regard to the final pressures, the error range varied between 3.08% and 13.03%, values higher than those calculated for the cold-water network. This variability could be associated in part with the difference in the determination of diameters, since larger diameters increase pressure in all the systems. Table 24 compares the actual and calculated results for diameters and pressure losses in the type 2 retail store.

There were no differences in the selected diameters, and the minimum diameter was maintained in the sections with artefacts. On the other hand, the relative errors showed extreme values throughout the network, with a minimum error of 0.58% and a maximum error of 20.04%. The latter value generated an underestimation of the network capacity, since using the automatic method resulted in a higher pressure being obtained in the affected section. Although this error was quite high compared to the actual result, the presented algorithm performed the calculations based on exact numerical information, implying more reliable and accurate results. It can thus be assumed that relative errors with high values suggest an inaccurate interpretation of the two-dimensional plans of the structure.

Table 24. Comparison of diameter and final pressure in a cold water system for type 2 retail store.

Segment		Diameter Using Traditional Method (mm)	Diameter with Automation (mm)	Final Pressure with Traditional Method (mca)	Final Pressure with Automation (mca)	Error between Pressure Results (%)
MAP	RAP	25	25	13.68	13.60	0.58%
RAP	A	25	25	13.12	13.49	2.82%
A	B	25	25	11.86	11.95	0.76%
B	C	19	19	10.41	8.63	17.10%
C	D	19	19	8.69	8.39	3.45%
D	E	19	19	8.37	8.29	0.96%
E	F	19	19	8.32	8.24	0.96%
F	BOLL	13	13	5.09	6.11	20.04%
F	LO	13	13	7.00	7.21	3.00%
E	WC	13	13	7.29	7.61	4.39%
B	CAL	19	19	10.60	10.46	1.32%

As shown in Table 18, and in comparison to the previous results, the maximum daily consumption was optimised, obtaining a head loss of 0.78 mca for a sub-meter of 25 mm. Finally, a comparison of the actual values and calculated values for pressure losses and diameters is shown in Table 25.

Table 25. Comparisons of hot-water network final diameter and pressure for type 2 retail store.

Segment		Diameter Using Traditional Method (mm)	Diameter with Automation (mm)	Final Pressure with Traditional Method (mca)	Final Pressure with Automation (mca)	Error between Pressure Results (%)
CAL	G	25	19	8.44	6.51	22.87%
G	H	19	19	6.34	6.42	1.26%
H	BOLL	19	13	4.40	4.30	2.27%
H	LO	13	13	5.02	5.40	7.57%

When looking at the diameters in Table 25, a clear difference in the first and third segments is evident. Again, this involved the use of larger diameters in the sections that lead to a fixture, as was the case for the "H-BOLL" section. As explained above, this difference was a result of the use of specific criteria for automation. Furthermore, in the case of the "CAL-G" segment, it can be inferred that the designer used a larger diameter in order to increase the pressure in the final segments of the network. However, the automated method shows that the accuracy in the calculation allows diameters to be reduced, as the value of the losses is known with greater reliability. Finally, the relative error with the highest value (22.87%) corresponded precisely to the increase in the load resulting from the use of larger diameters.

For sanitary drainage, the algorithm developed was able to realise the piping design correctly with regard to calculations, modifications and verifications; the performance of the software resulted in a reduction in the time taken to process information and display outcomes. Using programming to create verification tools requires more manual iterations, thus simplifying the project review process and reducing errors.

The scenarios described here assume a properly structured algorithm workflow; although Dynamo provides predefined nodes and simplified actions, building computational programming requires prior knowledge—not only about Dynamo itself but also about code –model interactions. Some of the difficulties found here are largely related to this factor. In some situations, element modifications carried out by automated tasks break connection rules, producing mismatches and separating the geometric and parametric correlation of elements.

Other errors are attributable to the use of MS Excel® in the data insertion stage; when working with lists separated by the system, data distribution must coincide with predefined sheet organisation. Otherwise, data packages can overwrite wrong cells, breaking the calculation logic and displaying incorrect results.

As indicated by the case studies, the proposed approach demonstrates that parametric modelling can host plumbing system attributes correctly, generating subsystems with accuracy and without missing or incorrect information during the modelling stage and parametorization. However, some issues were found in the plumbing fixture selection: predefined elements contain parameters and calculations relating to international codes, and these calculations may need to be excluded, while local regulations may need to be integrated into Dynamo. Without a unified modelling and mechanical equipment standard based on BIM, some necessary data may be omitted, reducing the model accuracy and detail level.

Parametric design with visual programming helps designers in the early design stages, generating a process database that can be used to automate several actions and tasks. The application of these algorithms can be generalised, allowing users to understand their usage in diverse project topologies and obtain exact results if they provide their requirements consistently.

7. Conclusions

In this research, a BIM-based designed workflow for a domestic drinking-water and sewerage system was developed using a semi-automatised design process. This was achieved through the application of a DRSM. This development was integrated with applications that are widely used in the AEC industry: Autodesk® Revit® and MS Excel®. The main advantages of these tools are their interoperability, their potential for customisation and the ease of visual programming.

This paper delivers the following contributions. First, it was found to be possible to adapt BIM tools for DDWSS design projects while taking local standards into account (in this case, Chilean standards). Moreover, this research demonstrates a possible extension of BIM applications over and above typical uses, such as system coordination, clash detection and documentation generation.

The obtained workflow covers a wide part of the design process from an end-to-end perspective: starting with an initial 3D model and a DDWSS draft and proceeding to the designed and validated components displayed in the standardised templates established by the local authority, ready for the request for permission to begin construction. The tool developed, in addition to facilitating the design process by automating many of the tasks involved in it, allows similar results to be obtained to those obtained by the traditional method. Furthermore, the developed workflow reduces subjectivity and errors in the design process by concentrating all design rules and required verification procedures in the code. Therefore, the exploitation of this tool would also help to improve productivity in engineering offices.

Both the DSRM and the workflow obtained allow replication of this automation process in other contexts, so it is hoped that this work will encourage other researchers to expand the uses of BIM (and BIM tools). Moreover, the workflow obtained is suitable for moderately complex developments, as is the case for one- or two-story houses and systems without mechanical equipment, such as pumps. However, the software tools used have the advantage of allowing code customisation, opening the possibility of addressing more complex design scenarios in future studies.

Future studies could also expand knowledge about automatic programming to deal with class detection problems (also known as high and soft collision [42]). These problems increase in complex projects with several disciplines involved. Nowadays, various commercial solutions (such as Navisworks [42], Synchro [43] or Solibri [44]) are among the most efficient clash detection tools. An alternative or complement to these software packages is the use of parametric programming tools. For example, specific Dynamo nodes (such as Bymorph Nodes [45]) can improve the automation of the clash detection process. The integration of these protocols with the DDWSS problems presented in this study will be studied by the authors in future work.

Finally, this approach will hopefully motivate researchers and practitioners to make the design process easier by using BIM automation tools, such as Dynamo scripts, to bridge the digital divide in the AEC industry and maximise the advantages that BIM brings to engineering.

Author Contributions: Conceptualisation, P.A., F.O. and E.A.; methodology, P.A., F.O., E.A., F.M.-L.R. and R.F.H.; software, P.A. and F.O.; formal analysis, P.A., F.O., E.A., F.M.-L.R. and R.F.H.; writing—original draft preparation P.A. and F.O.; writing—review and editing, E.A., F.M.-L.R., R.F.H. and F.L.-G.; visualisation, P.A., F.O. and F.M.-L.R.; supervision, E.A., F.M.-L.R. and R.F.H.; funding acquisition, E.A. and F.L.-G. All authors have read and agreed to the published version of the manuscript.

Funding: This research was funded by Proyecto VRIEA-PUCV, grant number 039.427/2021, and the Grants for the Promotion of Research in the Department of Civil and Building Engineering, UCLM. This research was also funded by Spanish Ministry of Economy and Competitiveness provided through the research project BIA2013-47290-R, BIA2017-86811-C2-1-R and BIA2017-86811-C2-2-R. All these projects were funded with FEDER funds.

Institutional Review Board Statement: Not applicable.

Informed Consent Statement: Not applicable.

Data Availability Statement: Not applicable.

Acknowledgments: The authors wish to thank all organisations that participated in this study and the experts for the insight provided. The authors wish to thank the Technology, Innovation, Management, and Innovation (TIMS) space of the School of Civil Engineering of the Pontificia Universidad Católica de Valparaíso (Chile), where part of the research was carried out.

Conflicts of Interest: The authors declare no conflict of interest.

Appendix A

The files developed in this study are available at the following link for replication. The Revit models of the two case studies and the 10 scripts referenced in Figures 4 and 14 are included. The spreadsheets with results are also included. Available online: https://drive.google.com/drive/folders/104TLNyzGZGcCBm7af3n23wJRLYO8l_XF?usp=sharing (Accessed on 8 September 2022).

References

1. Abdelhameed, W.; Saputra, W. Integration of building service systems in architectural design. *J. Inf. Technol. Constr.* **2020**, *25*, 109–122. [CrossRef]
2. Filho, J.B.P.D.; Angelim, B.M.; Guedes, J.P.; De Castro, M.A.F.; Neto, J.D.P.B. Virtual design and construction of plumbing systems. *Open Eng.* **2016**, *6*, 730–736. [CrossRef]
3. Palomera-Arias, R.; Liu, R. BIM laboratory exercises for a MEP systems course in a construction science and management program. *J. Inf. Technol. Constr.* **2016**, *21*, 188–203.
4. Zhang, J.; Seet, B.C.; Lie, T.T. Building information modelling for smart built environments. *Buildings* **2015**, *5*, 100–115. [CrossRef]
5. Diao, P.H.; Shih, N.J. BIM-based AR maintenance system (BARMS) as an intelligent instruction platform for complex plumbing facilities. *Appl. Sci.* **2019**, *9*, 1592. [CrossRef]
6. Loyola, M. *Encuesta Nacional BIM 2019*; Universidad de Chile-Plan BIM: Santiago, Chile, 2019.
7. Peffers, K.; Tuunanen, T.; Gengler, C.E.; Rossi, M.; Hui, W.; Virtanen, V.; Bragge, J. The Design Science Research Process: A Model for Producing and Presenting Information Systems Research. In Proceedings of the 1st International Conference, DESRIST 2006 Proceedings, Claremont, CA, USA, 24–25 February 2006; Claremont Graduate University: Claremont, CA, USA; pp. 83–106. Available online: http://urn.fi/URN:NBN:fi:jyu-201904092111 (accessed on 8 September 2022).
8. SISS Domestic Drinking Water and Sewerage Systems Chilean Standard Catalog. Available online: https://www.siss.gob.cl/586/w3-article-4152.html (accessed on 8 September 2022).

9. Xie, H.; Tramel, J.M.; Shi, W. Building information modeling and simulation for the mechanical, electrical, and plumbing systems. In Proceedings of the 2011 IEEE International Conference on Computer Science and Automation Engineering, Shanghai, China, 10–12 June 2011; Volume 3, pp. 77–80. [CrossRef]
10. Xiao, Y.Q.; Li, S.W.; Hu, Z.Z. Automatically generating a MEP logic chain from building information models with identification rules. *Appl. Sci.* **2019**, *9*, 2204. [CrossRef]
11. Wang, B.; Yin, C.; Luo, H.; Cheng, J.C.P.; Wang, Q. Fully automated generation of parametric BIM for MEP scenes based on terrestrial laser scanning data. *Autom. Constr.* **2021**, *125*, 103615. [CrossRef]
12. Wang, J.; Wang, X.; Shou, W.; Chong, H.Y.; Guo, J. Building information modeling-based integration of MEP layout designs and constructability. *Autom. Constr.* **2016**, *61*, 134–146. [CrossRef]
13. Chen, Q.; García de Soto, B.; Adey, B.T. Construction automation: Research areas, industry concerns and suggestions for advancement. *Autom. Constr.* **2018**, *94*, 22–38. [CrossRef]
14. Muñoz-La Rivera, F.; Vielma, J.C.; Herrera, R.F.; Carvallo, J. Methodology for Building Information Modeling (BIM) implementation in structural engineering companies (SECs). *Adv. Civ. Eng.* **2019**, *2019*, 0452461. [CrossRef]
15. Teng, Y.; Xu, J.; Pan, W.; Zhang, Y. A systematic review of the integration of building information modeling into life cycle assessment. *Build. Environ.* **2022**, *221*, 109260. [CrossRef]
16. Herrera, R.F.; Morgues, C.; Alarcón, L.F.; Pellicer, E. Analyzing the association between lean design management practices and BIM uses in the design of construction projects. *J. Constr. Eng. Manag.* **2021**, *147*, 1–11. [CrossRef]
17. Leygonie, R.; Motamedi, A.; Iordanova, I. Developments in the built environment development of quality improvement procedures and tools for facility management BIM. *Dev. Built Environ.* **2022**, *11*, 100075. [CrossRef]
18. Collao, J.; Lozano-galant, F.; Lozano-galant, J.A. BIM Visual programming tools applications in infrastructure projects: A state-of-the-art review. *Appl. Sci.* **2021**, *11*, 8343. [CrossRef]
19. Žurić, J.; Zichi, A.; Azenha, M. Integrating HBIM and sustainability certification: A pilot study using GBC historic building certification. *Int. J. Archit. Herit.* **2022**, 1–20. [CrossRef]
20. Potrč Obrecht, T.; Röck, M.; Hoxha, E.; Passer, A. BIM and LCA integration: A systematic literature review. *Sustainability* **2020**, *12*, 5534. [CrossRef]
21. Ren, R.; Zhang, J. A new framework to address BIM interoperability in the AEC domain from technical and process dimensions. *Adv. Civ. Eng.* **2021**, *2021*, 8824613. [CrossRef]
22. Bastos Porsani, G.; de Lersundi, K.; Sánchez-Ostiz Gutiérrez, A.; Fernández Bandera, C. Interoperability between building information modelling (BIM) and building energy model (BEM). *Appl. Sci.* **2021**, *11*, 2167. [CrossRef]
23. Bellido-montesinos, P.; Lozano-galant, F.; Javier, F.; Lozano-galant, J.A. Experiences learned from an international BIM contest: Software use and information work flow analysis to be published in: Journal of Building Engineering. *J. Build. Eng.* **2019**, *21*, 149–157. [CrossRef]
24. Xie, X.; Zhou, J.; Fu, X.; Zhang, R.; Zhu, H.; Bao, Q. Automated rule checking for MEP systems based on BIM and KBMS. *Buildings* **2022**, *12*, 934. [CrossRef]
25. García, D. *Aplicacion de BIM a Instalaciones Hidráulicas en Edificacion*; Universidad de Valladolid: Valladolid, Spain, 2019.
26. Ministerio de Obras Públicas de Chile. *Reglamento de Instalaciones Domiciliarias de Agua Potable y Alcantarillado (RIDAA)*; Biblioteca del Congreso Nacional de Chile: Santiago, Chile, 2009.
27. Council, I.C. *International Plumbing Code*; ICC: Washington, DC, USA, 2012; ISBN 9781580017428.
28. CNP. *Productividad en el Sector de la Construcción*; Santiago de Chile: Santiago, Chile, 2020.
29. Lu, Q.; Wong, Y.H. A BIM-based approach to automate the design and coordination process of mechanical, electrical, and plumbing systems. *HKIE Trans.* **2018**, *25*, 273–280. [CrossRef]
30. Han, J.; Zhou, X.; Zhang, W.; Guo, Q.; Wang, J.; Lu, Y. Directed representative graph modeling of MEP systems using BIM data. *Buildings* **2022**, *12*, 834. [CrossRef]
31. Wei, T.; Chen, G.; Wang, J. Application of BIM technology in water supply and drainage design. In Proceedings of the IOP Conference Series: Earth and Environmental Science; IOP Publishing: Bristol, VA, USA, 2017.
32. Kalasapudi, V.S.; Turkan, Y.; Tang, P. Toward automated spatial change analysis of MEP components using 3D point clouds and as-designed BIM models. In Proceedings of the 2014 2nd International Conference on 3D Vision, Tokyo, Japan, 8–11 December 2014; pp. 145–152. [CrossRef]
33. Hu, Z.Z.; Yuan, S.; Benghi, C.; Zhang, J.P.; Zhang, X.Y.; Li, D.; Kassem, M. Geometric optimization of building information models in MEP projects: Algorithms and techniques for improving storage, transmission and display. *Autom. Constr.* **2019**, *107*, 102941. [CrossRef]
34. Pärn, E.A.; Edwards, D.J.; Sing, M.C.P. Origins and probabilities of MEP and structural design clashes within a federated BIM model. *Autom. Constr.* **2018**, *85*, 209–219. [CrossRef]
35. Vasilev, L. Parametric Modeling in Structural Design. Thesis, LAB University of Applied Sciences, Lahti, Finland, 2020. Available online: https://www.theseus.fi/bitstream/handle/10024/349329/Parametricmodelinginstructuraldesign.pdf?sequence=2 (accessed on 8 September 2022).
36. Stine, D. *Autodesk Revit for Architecture Certified User Exam Preparation*; SDC: Nashville, TN, USA, 2021.
37. Jezyk, M. Dynamo Primer guide. Available online: https://primer.dynamobim.org/ (accessed on 8 September 2022).

38. Wei, L.; Liu, S.; Wei, Q.; Wang, Y. Concept, method and application of computational BIM. *Adv. Intell. Syst. Comput.* **2020**, *1084*, 392–398. [CrossRef]
39. Hofmann, P.; Samp, C.; Urbach, N. Robotic process automation. *Electron. Mark.* **2020**, *30*, 99–106. [CrossRef]
40. Parti, R.; Hauer, S.; Monsberger, M. Process model for BIM-based MEP design. *IOP Conf. Ser. Earth Environ. Sci.* **2019**, *323*, 012045. [CrossRef]
41. Nezamaldin, D. Parametric Design with Visual Programming in Dynamo with Revit: The Conversion from CAD Models to BIM and the Design of Analytical Applications. Master's Thesis, Royal Institute of Technology, Stockholm, Sweden, 2019; 84p.
42. Hu, Y.; Castro-Lacouture, D.; Eastman, C.M. Holistic clash detection improvement using a component dependent network in BIM projects. *Autom. Constr.* **2019**, *105*, 102832. [CrossRef]
43. Ciribini, A.L.C.; Ventura, S.M.; Paneroni, M. Implementation of an interoperable process to optimise design and construction phases of a residential building: A BIM Pilot Project. *Autom. Constr.* **2016**, *71*, 62–73. [CrossRef]
44. Merschbrock, C.; Erik, B. Effective digital collaboration in the construction industry—A case study of BIM deployment in a hospital construction project. *Comput. Ind.* **2015**, *73*, 1–7. [CrossRef]
45. Autodesk Bymorph Nodes for Dynamo BIM. Available online: https://bimorph.com/bimorph-nodes/ (accessed on 2 September 2022).

Article

Methodological and On-Site Applied Construction Layout Plan with Batter Boards Stake-Out Methods Comparison: A Case Study of Romania

Paul Sestras

Faculty of Civil Engineering, Technical University of Cluj-Napoca, 400020 Cluj-Napoca, Romania; psestras@mail.utcluj.ro

Abstract: The layout or stake-out is one of the most important assignments of the surveying engineer, and it is of vital importance in the building process, as the designed geometries of the structure ensure the verticality and the correct positioning inside the terrain. The mission of the surveying engineer involves both legal and technical aspects, and the correct planning of the layout process must take into consideration aspects regarding the site conditions, instrumentation used, the required and achievable accuracies, network design and survey methods used. Given the vast applications of geodesy and topography in different domains and industries, the study incorporates general notions and technical aspects regarding the workflow in cadastre and construction surveying, guidelines for an efficient design of site layout plan with on-site applicability, as well as a novel comparison between four methods of construction lines geometry layout on batter boards. The results of this study aim to further consolidate the importance of accurate and efficient construction layout projects, with comprehensive design plans, methods and instrumentation selection, as well as recommendations. The presented discussions and conclusions are of interest to the geodetic community as well as the construction industry, and due to the pragmatic and experimental nature of the research, incorporates technical notes and original results of professional and academic importance.

Keywords: cadastre; land surveyor; construction surveying; building layout; polar coordinates; stake-out methods; total station

Citation: Sestras, P. Methodological and On-Site Applied Construction Layout Plan with Batter Boards Stake-Out Methods Comparison: A Case Study of Romania. *Appl. Sci.* **2021**, *11*, 4331. https://doi.org/10.3390/app11104331

Academic Editors: Mariusz Szóstak, Marek Sawicki and Jarosław Konior

Received: 21 April 2021
Accepted: 10 May 2021
Published: 11 May 2021

Publisher's Note: MDPI stays neutral with regard to jurisdictional claims in published maps and institutional affiliations.

Copyright: © 2021 by the author. Licensee MDPI, Basel, Switzerland. This article is an open access article distributed under the terms and conditions of the Creative Commons Attribution (CC BY) license (https://creativecommons.org/licenses/by/4.0/).

1. Introduction

The efficient layout planning of a construction site is a fundamental task for any project undertaking, and the survey engineer's responsibility to guide the builders and conduct accurate, safe, time- and cost-efficient layout of the designed structure [1–3]. In theory, any surveyor can attempt a construction survey or precision surveying project. However, due to the difficulties of the work and the unpredictable site conditions and scenarios, the practice is much more difficult than would be expected. While the principals involved in surveying are generally established, the instrumentation used is equally important and the survey engineer must adapt and be prepared for any situation and task [4,5]. Thus, this research attempts to give answers to the following notions: the surveyor engineer's workflow and each legal and technical stage involved in a construction project, the design of a comprehensive site layout plan with on-site applicability and a comparison between four methods of construction lines layout on batter boards.

It is critical for a surveyor to take the time and care, both in the field and when processing data, to avoid or minimize errors and conduct precise and efficient surveys, layout or monitoring projects [5,6]. These applications and tasks integrate the highly sophisticated graphical capabilities of computer-aided design (CAD) platforms, with the geodetic instrumentations used in the field for observations and data collection. Precision surveyors working in the field have a large array of instrumentation at their disposition, with a constant flux of innovation [7–9]. There are numerous instrumentations and technologies used in the industry, such as: total stations, optical and digital levels, GNSS (Global Navigation Satellite System), terrestrial laser scanners, UAV (unmanned aerial vehicle) photogrammetry, airborne or UAV light detection and ranging (LiDAR), and many more geomatics applications. Each one of the instrumentations and techniques have certain advantages or disadvantages, that range from cost, to survey coverage, time efficiency, precision, and learning curve. Although for survey and monitoring projects all of the aforementioned instrumentation can provide viable solutions, in the case of layout projects and precise construction surveying, total stations are the best choice. Total stations have pinpoint accuracy and can yield precise surveying, positioning and observations [10–12]. In spite of the increasing use of new technologies, total stations remain a fundamental instrument for different survey projects, including land, cadastre and especially constructions [13].

Due to the many variables that encroach on the physical world and on the construction sites, surveyors must be extremely cautious with the work management, instrumentation used, data processing and error mitigations [14–16]. In field applications, precision surveying, layout and monitoring projects are used to provide cost savings, offer guidance and expertise to builders, verify design assumptions, reduce risks and even to protect lives [17–19]. Although in theory these are simple and easy benefits to obtain, due to the practical and somehow unpredictable nature of the profession, in practice these benefits may be difficult to obtain due to different site conditions, access problems or budgetary constraints. Thus, the present research serves as a combination of technical notes in the field of geodesy and topography applied in construction, guidelines for efficient layout plan design and on-site implementation, and a pragmatic and experimental comparison for the layout of a construction geometry on batter boards. The results of this study are of interest to the geodetic community as well as the construction industry, for many surveying engineering and land surveying applications. In addition, they can be instructive from an educational point of view, as well as beneficial to the private sector, public sector and academia.

2. Materials and Methods

2.1. Land Surveyor Profession in Romania

Surveying engineers are people who work both in the field taking measurements, as well as in the office, analyzing the measured data and planning maps, technical plans and registering legal documents regarding boundaries of existing land parcels. It is often said that surveyor engineers are the first persons to enter the construction site, and the last to leave it. This is because the inception of any investment construction project starts from the legal documents regarding the location, as well as feasibility studies that involve topographic surveys in order to provide geo-spatial data to the architects, civil engineers and other design engineers.

In Romania, the profession of survey engineering is called topographer or geodetic engineer. Compared to other countries, in Romania this profession combines the knowledge, attributions and responsibility of both land surveyors and building surveyors. In essence, a Romanian topographer or geodetic engineer is the person who consults and works with the legal aspects of property law, such as boundary surveys and cadastre, as well as the technical aspects that involve the construction industry, such as guiding the construction of new structures such as infrastructures or buildings.

In order to practice this profession, a bachelor's degree is required, as well as a number of years of experience in order to become an authorized or licensed surveyor, with one of the four different levels of qualification: C, B, A and D (highest). These levels of qualification require different years of experience, portfolio of survey works and different types of examination and interviews. As a license to practice, a category C surveyor can only work in cadastre and land surveys, categories B and A, beside the previous category C competencies, can design technical topographical plans, building surveys, layout or stake-out of reference points and markers for constructions and, lastly category D, which incorporates the previous competencies, can also create or verify geodetic networks. The authority that grants these licenses to practice is the National Agency for Cadastre and Land Registration, a government institution that oversees the property law and cadastre in the country. A typical workflow for a survey engineer is as follows (Figure 1):

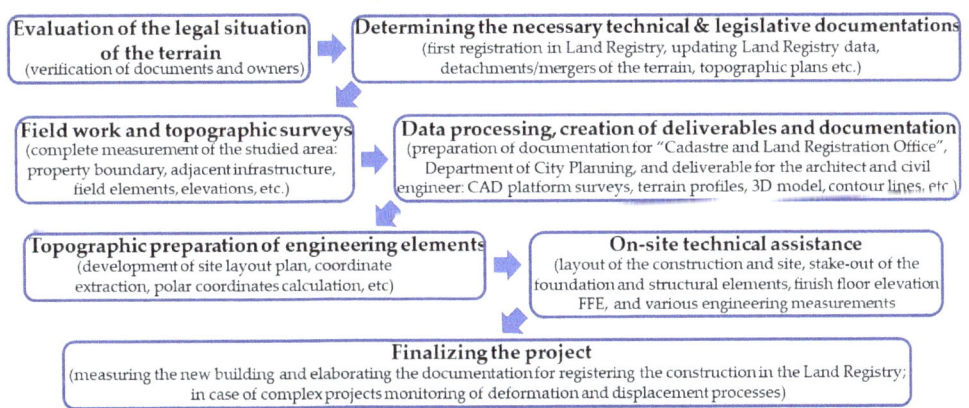

Figure 1. A survey engineer's typical workflow.

The preliminary work consists of consultancy and evaluation of the legal situation of the terrain, as for any building permit it is necessary to have a clear property law, checking property boundaries and assigning a cadastral number. In order to prepare the terrain for investment, besides the land registry data integrity, the shape of the terrain has to be considered, thus there is often documentation for land parcel detachments or mergers. The topographic plan is necessary for both the technical documentation for obtaining the building permit, as well as for the design of the architect and civil engineer, by providing a series of complex works and deliverables such as: computer-aided design (CAD) platform surveys, transverse and longitudinal terrain profiles, 3D modeling of the terrain, contour lines and other elements that are the basis for the architecture, structure and design projects. Works during the construction process include regular site topographic assistance, where the surveyor engineer ensures the connection between the design made by the architect/civil engineer and the builder/contractor. In order to ensure the accuracy of the site position of the designed structural elements, it is necessary to layout or stake-out building reference points as well as construction axes and geometry, markers of finish floor elevations, verifications of various elements and elevations, verticality studies, quantity calculation, monitoring of adjacent buildings and other engineering measurements. Works after completion of constructions include: "As-Built" plan necessary to detect non-conformities between what was built and the project; topographic plans and cadastral works in order to connect the new constructions to utilities and registration in property law; in the case of apartment buildings additional detachment documentations and surveys of interior spaces; and, finally, in the case of complex projects, monitoring of the new building in order to determine its behavior during the period of construction.

2.2. Site Location and Conditions

In the constant expanding and highly populated Cluj-Napoca metropolitan area, Romania (Figure 2), the need for qualitative, safe, time and cost-efficient survey engineering works is imperative. With a general move towards urbanization, the construction industry is booming, land is becoming an increasingly difficult resource to obtain and the construction market is a desideratum [20,21]. The unprecedented urban sprawl phenomenon imposed the expansion of the city limits and the transformation of adjacent villages, agricultural terrains or old industrial parks from Cluj-Napoca into suburbs and residential complexes [22]. Given the current situation, the surveying engineer is a sought-after specialist who can provide multifunctional services for investors such as legal and technical consultancy regarding property law, initial field measurements for the construction design, and on-site technical assistance, up to the final stages of the investment project. Although most of the big construction companies have their own survey engineers, small and medium firms do not and they rely on contracts of service providers with survey engineering companies or licensed individuals. These issues call for a retrospective look at the main technical assistance that survey engineers provide on the construction site (to stake out reference points and markers that will guide the construction of new structures), as well as guidelines for efficient site layout plan creation and the use of on-site batter boards marked with layout lines for future positioning of construction reference points by the builder/contractor.

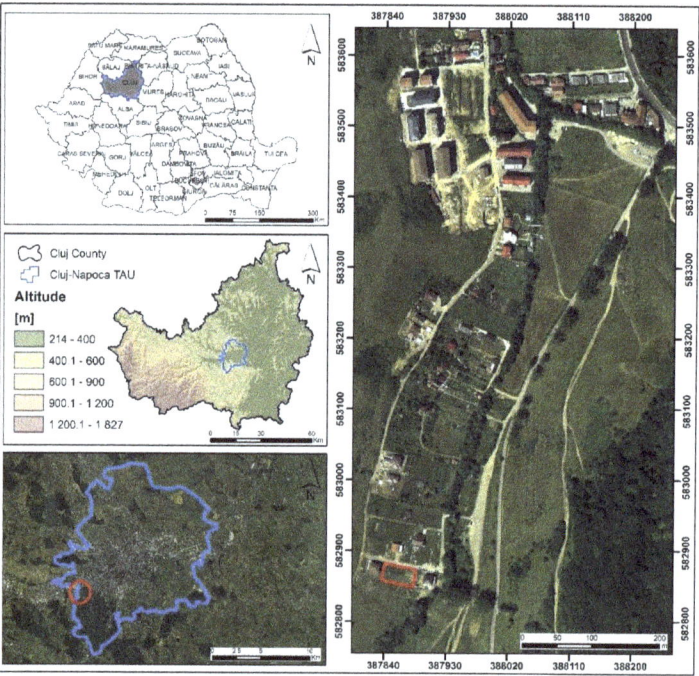

Figure 2. Site location.

The present technical project consists of the construction of 2 duplex houses having four living units, in a newly developed neighborhood consisting mainly of houses on the outskirts of Cluj-Napoca (Figure 2). The terrain has an area of 1147 square meters, the designed duplex houses have a low height regime specific to the region, with ground floor and first floor, and each duplex has a double partition wall between the units, thus having the possibility to be registered in property law as four distinct houses. Each designed unit has a usable area of approximately 120 square meters, 2 parking spaces and a garden of approximately 215 square meters, thus making it a perfect solution for young families.

2.3. Methodological Approach and Instrumentation

The methodological structure pursued to develop the presented study is in accordance with the general line of technical notes and practices in the field. Thus, in the present study, the research direction was divided in two main stages of dissemination: that of evaluating the methodological process of obtaining an efficient site layout plan, with a retrospective look at the design and field-work practiced notions; the second main stage, represents the on-site applied designs, together with a novelty comparison between four methods of batter boards stake-out of the construction layout lines, in order to determine the most qualitative, safe, time and cost-efficient one, as well as highlighting advantages and disadvantages of each method and the possible instrumentation used.

To better understand the workflow of such a project, a graphical abstract of the site layout plan process was created (Figure 3). As previously mentioned, the surveyor engineer's work starts in the office with evaluating the land registry of the terrain, in order to determine the required technical and legislative documentation. Once this stage is complete, an elaborate field measurement is scheduled, with the necessary presence of the land owners or investors. In this topographic survey, the geodetic engineer can either divide his workflow in measurements for cadastral purposes, technical purposes, or combine both of them. Measurements for cadastral purposes are more straightforward, with an emphasis on the boundaries of the property, checking for inadvertences between the measured area and the one in the documents, and accessibility to the terrain. The measurement for technical purposes is much more complex, and requires the data acquisition necessary to create topographic plans for the building permit, as well as deliverables to the architect, civil engineer and other design engineers. It is the survey engineer's duty to accurately and in great detail represent the terrain surface, in order to develop an investment project. The required deliverables differ depending on the type of project, but the most common ones are the following: a CAD platform complete survey of the study area, with the cadastre contours layer; a 3D model of the terrain and contour lines, achieved also in CAD software; transversal or longitudinal profiles, which are necessary especially for infrastructure. Based on these deliverables, together with a clear property law of the terrain and the topographic plan signed and sealed from the Office of Cadastre and Land Registration, the architect and civil engineer can apply for the Building Permit at the local Department of City Planning. This operation can take up to 6 months, depending on the complexity of the project and the urbanistic regulation in the proposed area. Once the Building Permit is obtained, and the owners or investors have also contracted a team of builders, the need for the survey engineer is again required. The survey engineer must obtain from the design team the site/location plan, which details the geometry and location of the designed building inside the terrain, as well as dimensions between the layout lines or building axes, and distances from the boundary limits to the construction. Even though the initial survey may have been undertaken within a geodetic datum and dimensions, the design team, especially the architects, work in millimeter units and different design softwares, thus the site/location plan received must be converted into the correct coordinate system (in the current case, Stereographic 1970). This process is undertaken in the preferred software of the survey engineer (customarily AutoCAD), and involves the correct cadastral contour and the functions of scale, move and rotate. The desideratum is the creation of the Site Layout Plan, a comprehensive design that highlights the layout of the proposed construction inside the terrain, together with the coordinates and the stake-out elements of the building reference points (usually axes intersections). The stake-out elements consist of horizontal angles and distances from known geodetic points or control points, together with the instrumentation and layout method used in order to mark the position of the designed building. These horizontal angles and distances can be calculated from the Cartesian coordinates of the control/geodetical points and the designed points of the building, and can be done manually or by spreadsheet software. Total stations have the processing capabilities of instantly calculating these values, but it is recommended to also have them calculated and displayed on the final site layout plan.

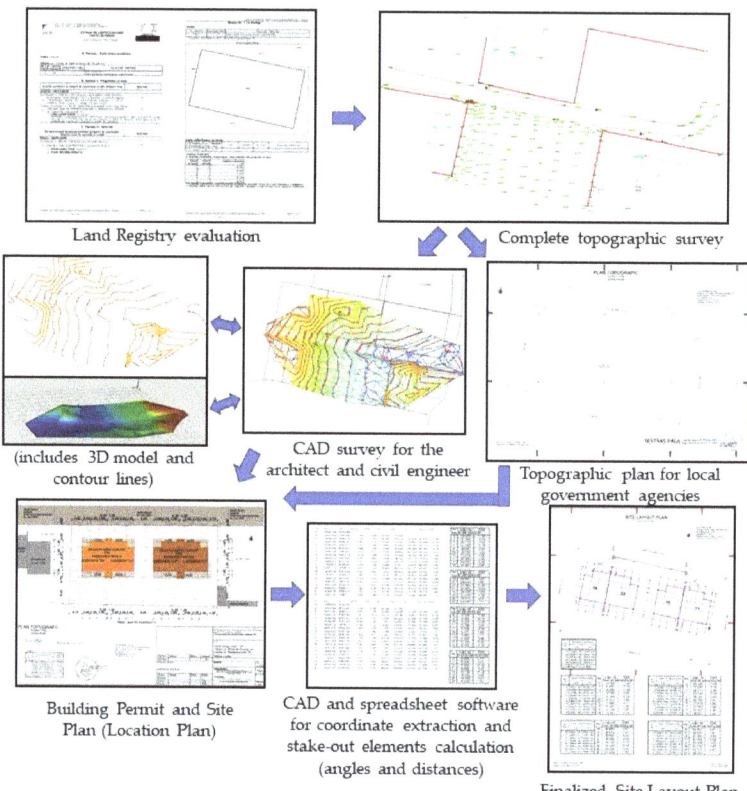

Figure 3. Graphical abstract of the site layout plan process.

In terms of used instrumentation, the total station and GNSS systems are the most used by the survey engineer. GNSS systems are the perfect choice for topographical surveys where the field conditions are optimal (sufficient satellite availability, network RTK services, open field etc.). In recent decades, GNSS systems have been used to obtain geodetic networks or control points, for precise measurements, accurate monitoring processes or construction stake-outs. In the case of classical methods of determining new geodetic points, by means of resection or traverse networks, a current deficiency is the lack of reliable control points or geodetic points, almost completely destroyed in the past decades. The total station is often considered the right hand of the survey engineer, and offers pinpoint accuracy and can yield precise observations, as well as efficient and reliable layout of reference points. The instruments used in this case study were: Leica Viva GS08 Global Navigation Satellite System used in real-time kinematics (RTK) mode for obtaining the control points (St1 and St2) with a horizontal precision between 0.014 m and 0.020 m ensured by online RTK corrections provided by the Romanian Position Determination System (ROMPOS) and the connection to a national permanent GNSS reference station; the total station used was a Leica TS02plus, which has a very good measurement accuracy of angles of 3″ and distances of ±2 mm + 2 ppm.

2.4. Layout of a Project Point and Accuracy Evaluation

Construction layout or stake-out is one of the most important missions of the surveying engineer, with the purpose of ensuring the designed geometries of the engineering structure, by satisfying the required accuracies from the project [2,22–25]. Although modern GNSS systems have the capability of point stake-out with an accuracy of ≈2 cm in the right conditions, construction surveys and layouts are made using total stations. The principle behind a total station stake-out is the polar coordinates method, which consist of calculating and determining the position of a horizontal angle and a distance from a set of two control points (or geodetic points). The two control points are used for instrument stationing and orientation (or bearing), and all calculations regarding the layout of the construction are made in accordance to the established layout network. It was opted for the GNSS technology for the creation of the layout network, because of the efficiency, the lower cost price, and the short time to perform the measurements. In the case of classical methods of determining new geodetic points, by means of resection or traverse networks, the current great deficiency is the lack of reliable control points or geodetic points, almost completely destroyed in the past decades [13].

The layout elements (horizontal angle and a distance) are calculated using the known coordinates of the layout network and the designed coordinates of the construction, as follows (example for layout point H7; Figure 4):

$$tg\theta_{St1-St2} = \frac{\Delta Y_{St1-St2}}{\Delta X_{St1-St2}} \Rightarrow \theta_{St1-St2} = \operatorname{atan}\frac{Y_{St2} - Y_{St1}}{X_{St2} - X_{St1}}, \quad (1)$$

$$tg\theta_{St1-H7} = \frac{\Delta Y_{St1-H7}}{\Delta X_{St1-H7}} \Rightarrow \theta_{St1-H7} = \operatorname{atan}\frac{Y_{H7} - Y_{St1}}{X_{H7} - X_{St1}}, \quad (2)$$

$$\omega_{H7} = (400^g - \theta_{St1-St2}) + \theta_{St1-H7}, \quad (3)$$

$$D_{St1-H7} = \sqrt{\Delta X^2 + \Delta Y^2}, \quad (4)$$

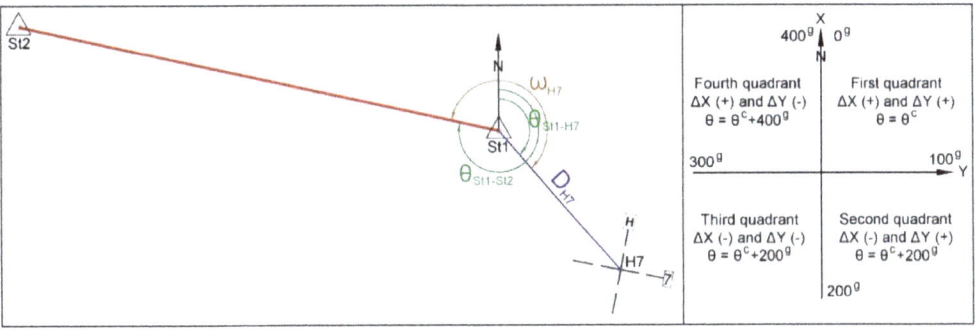

Figure 4. Polar coordinates for layout point H7 and the topographic quadrants for the calculus.

In order to ensure an accurate and efficient layout of the construction points, a retrospective of the measurement errors, required accuracy and the construction tolerances must be taken into account [2,21,22]. The type of building and the building technology are also important aspects, as the components and structural elements used to construct buildings are often fabricated, assembled or built on site, often by hand, in conditions that may be less than ideal. Errorless measurements are impossible, thus it is mandatory to satisfy the recommended accuracies through the "achieved point standard deviation" which is defined by the product of the errors derived from known point coordinates and the layout measurements. In order to assure the required accuracy, the achieved accuracy must be smaller than the required one. Based on the type of construction and the characteristics of the engineering structure, the required point standard deviation ($\pm\mu_P$) should be: $\mu_P = \pm(1-2)$ cm for the majority of layout construction projects that use a monobloc structure with build on site structural elements (reinforced concrete); $\mu_P = \pm(2-5)$ mm for prefabricated and assembled structural elements; $\mu_P = \pm(1-2)$ mm for precise machine guidance and complex structures [2,26,27]. It is the surveying engineer's mission to identify the required accuracy and plan the layout process in accordance. The achieved accuracy if the resultant of error affecting the measurements of the layout elements, expressed as achieved point standard deviation ($\pm \sigma_P$), and must satisfy the required accuracy:

$$|\sigma_P| \leq |\mu_P|, \tag{5}$$

The errors concerning the measurements are classified into three major groups: instrumental errors, personal errors and atmospheric errors [28–30]. Given the present case study, the length of the observations are short and the measurements are made in optimal conditions, as well as through the expertise of the surveying engineers. Thus, personal and atmospheric errors can be considered insignificant and not taken into consideration. The major instrumental errors can be eliminated or significantly reduced by additional checks and calibrations. Given the fact that the measurements are carried out with a new model of the total station with good specifications, it is possible to satisfy the required accuracies of the layout project by using angle and distance reading with only one face of the instrument. By combining the expertise of the survey engineers, the correct checks and calibrations of the instrument, and by using angle readings that not exceed $\pm(5–10)^{cc}$ with layout distances of $\pm(2–3)$ mm, the achieved point standard deviation can easily be between $\sigma_P = \pm(5–10)$ mm, more than enough considering the required point standard deviation ($\pm \mu_P$) for the type of construction and building technology of the present case study.

2.5. Designing the Site Layout Plan and Calculating the Stake-Out Elements

The planning of the layout process is one of the most important tasks for the survey engineer. This layout ensures the horizontal and vertical geometries of the engineering structure, and takes into account the instrumentation and the required accuracies previously mentioned. The established site layout plan is compiled using the cumulative cadastral information, land survey, building permit and design plans of the construction. A comprehensive design must highlight the layout of the proposed construction inside the terrain, as well as the inclusion of the coordinates and the calculated layout elements of the building reference points. The stake-out elements consist of horizontal angles and distances from known geodetic points or control points and, although total stations are capable of coordinate geometry calculation (COGO functions), the site layout plan should incorporate these values in order to carry out the on-site project regardless of the instrumentation. This is because many construction companies or construction survey engineers use electronic theodolites such as the Leica Builder series, which are reliable instrumentations that measure angles and distances, but do not have the processing power and the functions of coordinate geometry calculation. Thus, a comprehensive site layout plan (Figure 5) should include all the information necessary to layout the designed structure (the extracted coordinates of each construction layout point and the stake-out horizontal angle and distance from the station points).

In order to obtain an automation of the calculations of horizontal angles and distances, a spreadsheet type software can be used. The measured (geodetic network/control points) and the designed coordinates (construction layout points/axes intersections) are extracted from the CAD platform and exported in the spreadsheet. The calculus is similar to that presented in Section 2.4, with the following formulas used:

$$\theta = \text{ATAN}(\Delta Y / \Delta X) * 200/\text{PI}() + \text{IF}(\Delta X < 0{,}200, \text{IF}(\Delta Y < 0{,}400)), \tag{6}$$

$$\omega = (\theta_A - \theta_B) + \text{IF}((\theta_A - \theta_B) < 0{,}400), \tag{7}$$

$$D = \text{SQRT}(\Delta X^2 + \Delta Y^2), \tag{8}$$

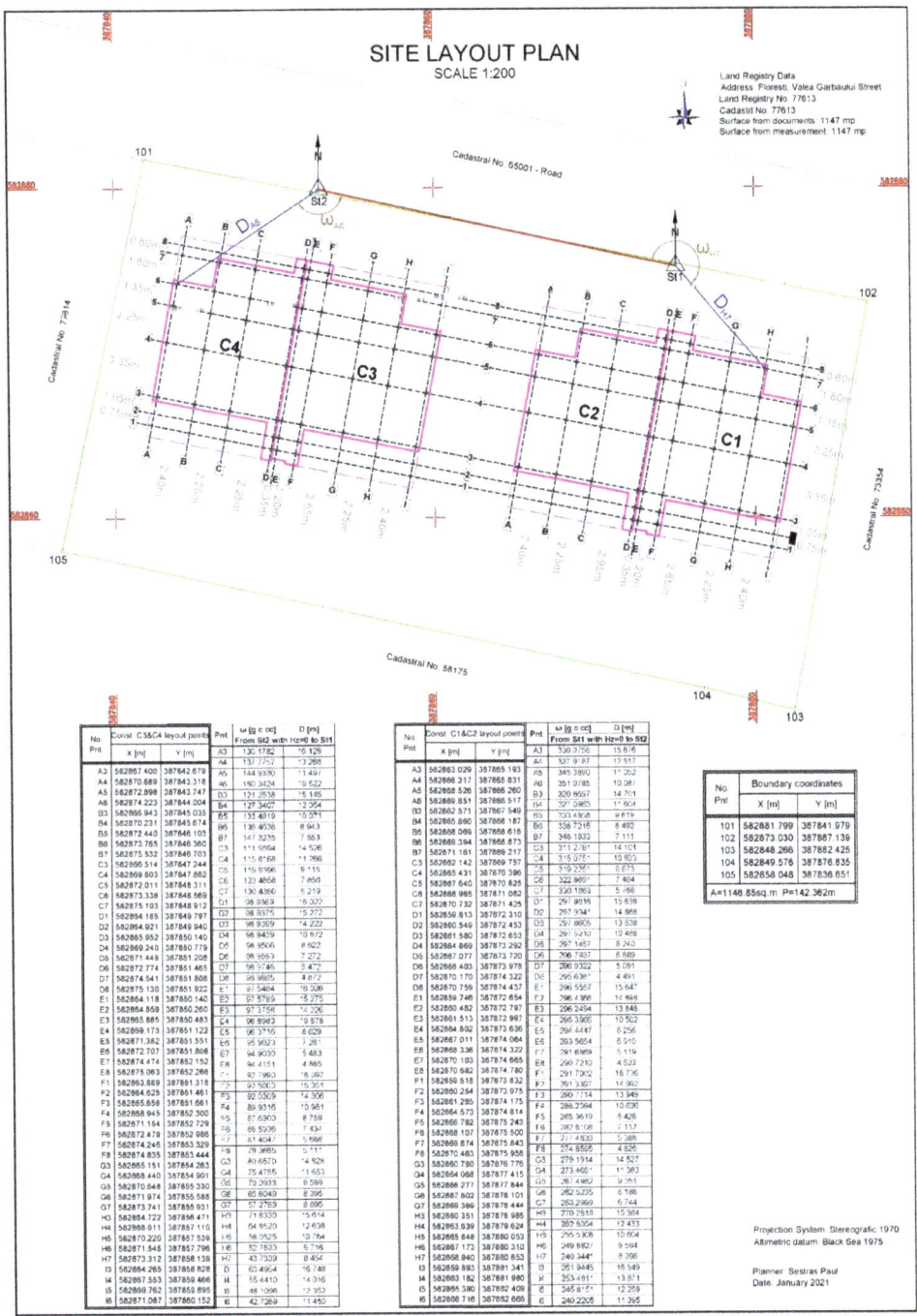

Figure 5. Site layout plan.

2.6. Batter Boards Importance and Layout Lines Marking Methods

Batter boards are temporary wooden frameworks constructed and displayed around the in-site layout, used to suspend the layout strings for a foundation or a structural element from the ground floor. Their placement is crucial for building with the correct designed geometries. Batter boards consist of two vertical wood poles and a horizontal crosspiece screwed to the verticals (Figure 6). The height of the boards must be over the height of the finish floor elevation, and if possible, all horizontal crosspieces must have the same elevation. Batter boards are commonly set beyond the corners of a planned foundation, but they can also be continuous and cover the whole perimeter of the construction [1,4]. These batter boards are then used in order to layout or stake-out on them the axes of the construction, and by intersecting construction twine between two axes to indicate the precise location of a construction layout point.

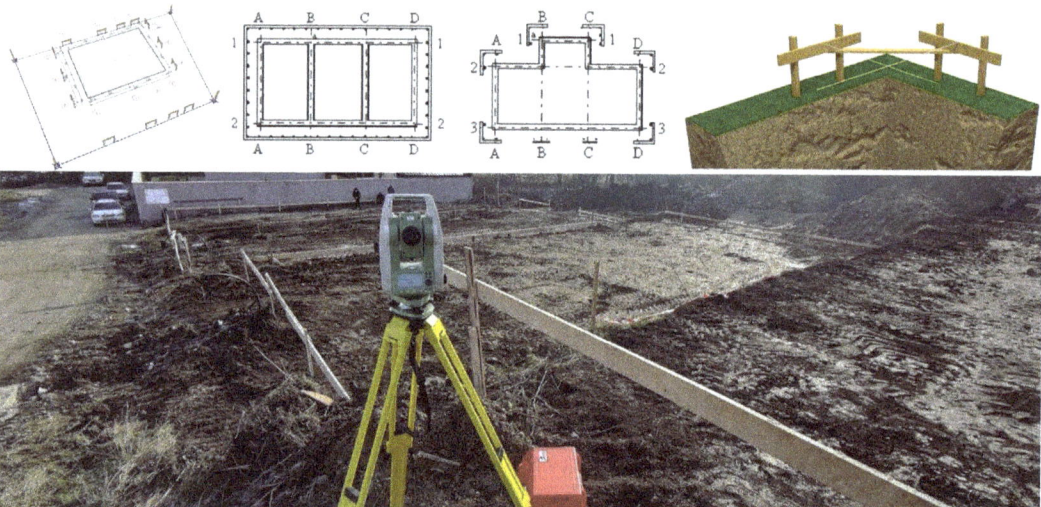

Figure 6. Examples of batter boards (**top**); the present in-site conditions (**bottom**).

Batter boards are very popular for small to medium scale construction sites. Survey engineers commonly use them in order to layout the construction axes, in order for the builder to further determine construction layout point without the help of instrumentation or survey expertise. This practice can be applied for smaller scale construction sites, because the intersection of construction twine in order to determine a layout point can be achieved for the foundation and the structural elements on the ground floor (columns, beams, load-bearing walls). This is because the elevation of the batter boards is usually at one meter above ground level, and it would be impossible to use them on an upper floor. Due to the high demand of construction survey and layout, the batter boards holding layout lines is a very popular solution [1,24].

There are several methods of axes layout on these wood structures, depending on the instrumentation used and the calculus capability of the user and apparatus. Thus, the present case study contains a novelty comparison between four methods of batter boards stake-out of the construction layout lines, in order to determine the most efficient way possible, as well as a comprehensive evaluation of the advantages and disadvantages of each method and the possible instrumentation used. The four methods consist of: the classical optical method using a theodolite; the survey of the batter boards with manual calculation of coordinate intersection between the wood plank and the construction axes; the survey of the batter boards with CAD implementation and extraction of coordinate intersection between the wood plank and the construction axes; reference line or layout line function of the total station. These four methods will be further presented and evaluated in the next chapter, with a dissemination of the results, discussion and concluding remarks.

3. Results

3.1. On-Site Design Points Layout

The on-site layout of the construction design points was carried out using the geodetic network established in the vicinity; respectively, control points St1 and St2. Each were used as a station point for the instrument, and the other one for orientation (bearing). The stake-out process can be made using the special COGO functions of the total station, or in the case of electrical theodolites, basic measurements of horizontal angles and distances are required. Total stations have built-in programs, usually named "Stake-out" or "Layout", where you can select from the internal memory or manually enter the coordinate [1,4]. With the internal processing power, they instantly display the values of the horizontal angles at which the user has to rotate the instrument on the horizontal axis in order to be on the right direction, and the distance of the point on the established direction. Based on repeated measurements of angles and distances, and a good coordination between the survey engineer and the technical assistant wielding the reflector, the correct positioning of the design points are marked on the ground. In the case of electrical theodolites such as the Leica Builder series, the process is similar, with the exception of manually inserting the value of Hz = 0^g (horizontal angle) in the direction of the second control point (used for orientation/bearing), with further manual positioning based on the calculated layout element present in the tables of the site layout plan.

In the next figure (Figure 7) it can be observed the general workflow of a construction layout point (e.g., A6) marked on-site, by following the steps previously mentioned. By careful rotation of the instrument on the horizontal axis, in order to not exceed $\pm(5-10)^{cc}$, and with layout distances of $\pm(2-3)$ mm, the achieved point standard deviation is relatively low, and more than enough for this type of project. The same steps are taken for all the important layout points, which the builder requests to be marked on site (usually the construction corners). For a better visualization of the site and the layout of the constructions inside the terrain, a georeferenced orthophoto was made using a UAV system (Figure 8). Because further earthworks for the foundation will be established, wooden stakes and topographic nails used to signal the layout points will be destroyed or removed from site. Thus, the batter boards are used in order for the survey engineer to display the construction layout lines that will help the builder redetermine the position of the design points and the construction geometry.

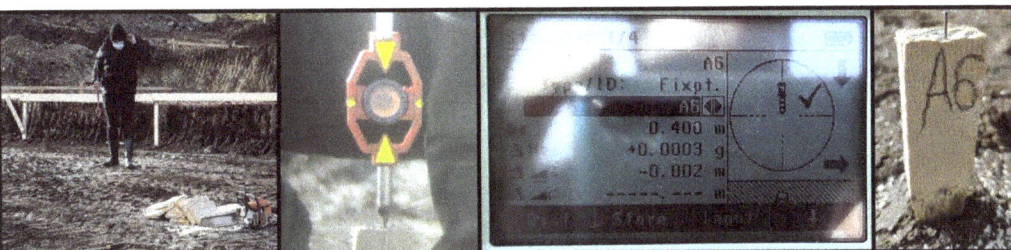

Figure 7. General workflow of a construction layout point marked on-site.

Figure 8. Layout site with an unmanned aerial vehicle (UAV) orthophoto layer.

3.2. Results and Discussions Regarding the Batter Boards Layout Methods

3.2.1. Classical Method with Theodolite Instrument

The classical method of batter boards layout is based on the initial construction layout points on the ground. These points must be marked accordingly, with wood stakes and topographic nails, using stake-out methods and calculus as previously mentioned. After the construction layout points on the ground is complete, the survey engineer must leave the station point and move the theodolite on one of the points marked on the ground. After positioning and centering the instrument on the newly marked construction point, the user must target with the moveable telescope another of the construction layout points marked on the ground, but it is mandatory for the point to be on the same construction axes (geometry line). It is recommended to insert on the instrument menu the horizontal angle value at "0" (Hz = 0^g), in order to notice any displacements or instrumental movements at the next stages. By lifting the theodolite telescope, the survey engineer must target the batter board in front of him, and guide the technical assistant with the reflector left and right on the respective wood plank until they determine a position collinear with their direction, and mark it accordingly. Then, the telescope is tilted 200^g until it faces the opposite batter board; it must be checked if the angular reading remains approx. Hz = 0^g; and the guidance process between the engineer and assistant is repeated in order to determine and mark the collinear point on the respective batter board. After that construction line is laid out, the team can rotate the instrument in order to create a right angle (at 100^g or 300^g, depending on the position) and target another construction layout point on the ground. The process is repeated, thus each next stationing on a construction layout point can assure the marking on the batter boards of two construction lines (axes). The presented figure (Figure 9) illustrates the established workflow, with stationing on point A6, targeting point A3 in order to lay out on the batter boards construction line A–A, and then rotating the instrument and targeting point I6 in order to lay out construction line 6–6.

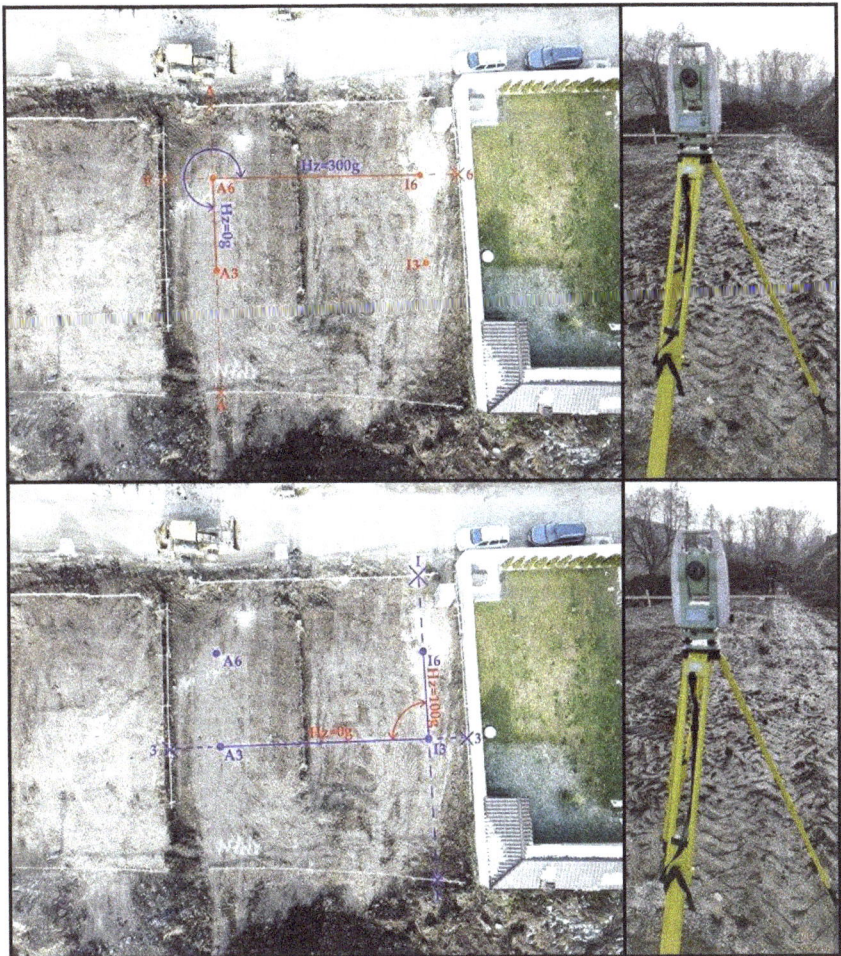

Figure 9. Typical workflow for theodolite method construction line layout.

The advantages of this method are that it can be performed with any optical instrument related to the theodolite, that includes any type of classical theodolite, electrical theodolite and total stations. Also, because it is an optical and mechanically applied method, the chances of error are low (considering that the ground layout points were correctly determined). Another advantage is the fact that the position of the construction layout points on the ground are verified when targeting points at a right angle (obtaining readings of 100^g and 300^g). The disadvantages of this method are numerous, because it has a low efficiency in terms of time. Also, due to the fact that it is necessary to station on points inside the construction site, which are usually on rough terrain or in the excavation for the foundations, there are dangers and inconveniences.

3.2.2. Survey of the Batter Boards and Manual Calculation

This method is based on the survey of the batter boards, in order to determine the coordinates of each side. This survey can be undertaken using a total station or an electronic theodolite with electronic distance measurement (EDM), and the survey can be carried out from one of the control points on site. The principle is to measure and obtain the coordinates of the wood plank at the middle of each side, in order to obtain a line or direction for each batter board (Figure 10). The calculus consists of determining the coordinate of the intersection between the construction line (axes) and the line representing the batter board. This can be obtained in a number of different ways, the easiest being to use an intersection of orientations (bearings). The orientations (bearings) will be calculated based on the designed coordinates (construction layout points) and the measured coordinates (each side of the wood plank). For example, in the case of construction line A–A and the measured batter board of points 10–11, the two bearings will be calculated as θ_{A3-A6} and θ_{10-11} as previously mentioned in Section 2.4. We consider the new point 201 as the intersection between the two bearings, and in order to determine its coordinates, the bearing between point 10 and point 201 is needed, which is the same as the bearing of the batter board θ_{10-11} (point 201 is collinear on line 10–11). Also, the bearing between point A3 and point 201 is needed, which is the same as the bearing of the construction line A3–A6, calculated from the designed coordinates (Figure 11). Thus, the coordinates of point 201 can be obtained using the formulas:

$$X_{201} = \frac{Y_{10} - Y_{A3} - X_{10}\text{tg}\theta_{10-201} + X_{A3}\text{tg}\theta_{A3-201}}{\text{tg}\theta_{10-201} - \text{tg}\theta_{A3-201}}, \tag{9}$$

$$Y_{201} = Y_{A3} + (X_{201} - X_{A3})\text{tg}\theta_{A3-201}, \tag{10}$$

Figure 10. Batter boards survey.

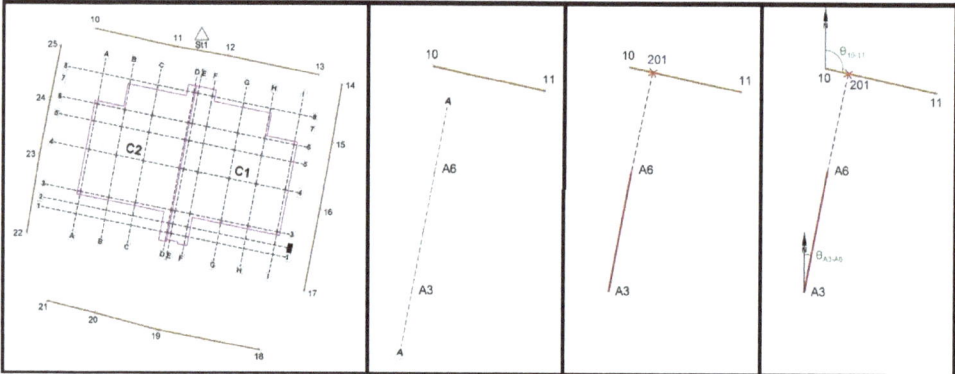

Figure 11. Construction lines and batter boards intersection.

The advantages of this method are that it can be performed with older models of total stations that do not have special functions (e.g., reference line/layout line) or with total stations where these functions are blocked and it is necessary to purchase additional software packages. It is an engineering alternative to the classical method, the actual layout on the batter board is easy and accessible to anyone, as it is a regular stake-out based on the polar coordinates method (horizontal angle and distance). The disadvantages of this method are again numerous, because it has a low efficiency in terms of time. Also, numerous errors can occur when measuring the wood planks and manually calculating these intersecting coordinates. Another disadvantage is the need for qualitative batter boards, because if the wood planks are bent and the survey is at each side, the resulting intersection of coordinates will be outside the wood section.

3.2.3. Survey of the Batter Boards and Computer-Aided Design (CAD) Implementation

This method is based on the same methodology and principle as the previous one, with the survey of the batter boards, in order to determine the coordinates of each side. The difference is that instead of a manual calculus of the intersecting coordinate, the survey data is exported inside CAD software, the same one used for establishing the site layout plan, and the batter boards are represented as polylines. Using the geometry functions, the construction line (axes) is extended until they intersect the new polyline (batter board), and the intersecting coordinate is extracted and inserted in the total station (Figure 12). The advantages of this method are that it can be performed with older models of total stations that do not have special functions (e.g., reference line/layout line) or with total stations where these functions are blocked and it is necessary to purchase additional software packages. Also, the actual layout on the batter board is easy and accessible to anyone, as it is a regular stake-out based on the polar coordinates method (horizontal angle and distance). The disadvantages of this method are numerous, because it has a low efficiency in terms of time. Also, numerous errors can occur when measuring the wood planks and the insert-export in CAD software, as well as the need to bring a laptop on-site. Another disadvantage is the need for qualitative batter boards, because if the wood planks are bent and the survey is at each side, the resulting intersection of coordinates will be outside the wood section.

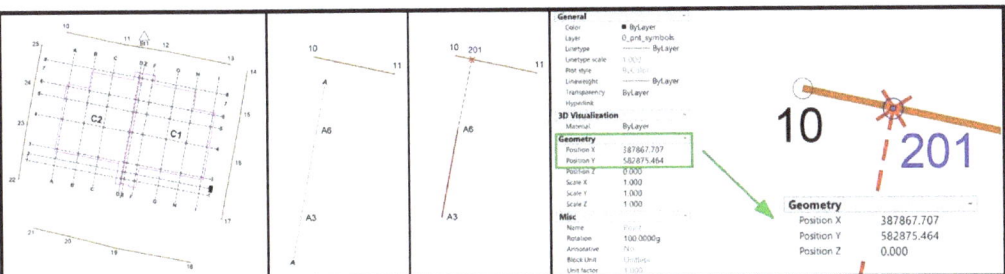

Figure 12. Intersection coordinate extraction.

3.2.4. Reference Line Function

This method is the most popular current choice for on-site layout in a construction survey. It is a special function or program inside the total station, and it is commonly called reference line or layout line (depending on the total station brand). This function creates a reference line between two of the characteristic points of the construction (measured or designed), and all future measurements are displayed with respect to this line. The displayed values are ΔL (line) and ΔO (offset), in regards to a newly observed (measured) point. These values represent the length/distance (ΔL) from the first reference point to the observed (measured) one, and the offset (ΔO) of the observed (measured) point from the reference line (displacement from collinearity). These values serve vital applications in the field that simplify most of the layout applications, and provide invaluable assistance to the survey engineer. By knowing these values, it is very accessible to lay out points at different lengths from each other or from a certain position, or lay out points collinear to a certain direction or at certain designed offsets. In the case of batter boards layout, the methodology is to create a reference line represented by a construction line (axes), and with repeated measurements to locate a collinear point on the batter board with the value of ΔO = 0.000 m. The values ΔL and ΔO are calculated from the designed and measured coordinates, and are based on trigonometry, as presented in Figure 13 and the following formulas:

$$\omega_{P'} = \theta_{A3-P'} - \theta_{A3-A6}, \tag{11}$$

$$\Delta O = D_{A3-P'} \cdot \sin \omega_{P'}, \tag{12}$$

$$\Delta L = D_{A3-P'} \cdot \cos \omega_{P'}, \tag{13}$$

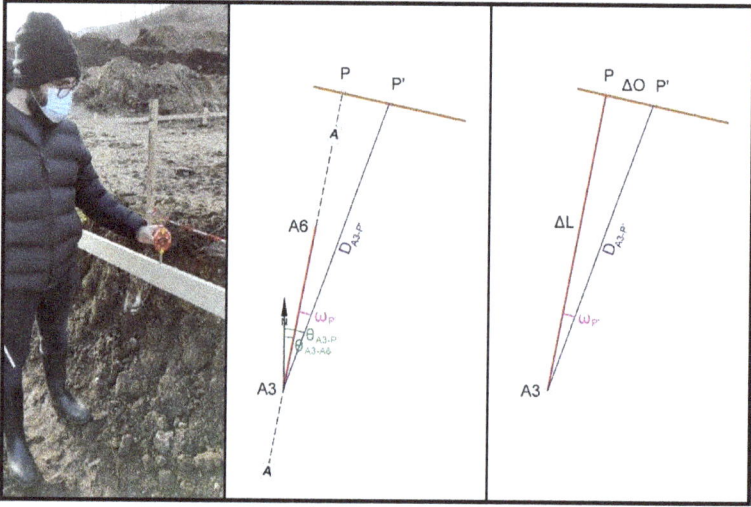

Figure 13. Reference line method applied on field (**left**) and general concept (**right**).

The main advantage of this method is the efficiency in terms of execution time. It is the most popular layout method and can be used on any construction site and under any conditions. In terms of disadvantages, this method requires a newer model total station or to buy certain software packages. Also, it requires a better experience and teamwork between the survey engineer and assistant, as well as a better orientation in space than a normal polar coordinate stake-out.

4. Discussion

For the comparison between the layout lines methods, a trial-and-error method was implemented in the field in order to analyze the different geometries obtained. Because some of the methods can be considered old-fashioned with low efficiency, the precision and time were compared regarding the stake-out of four construction lines (axes), respectively eight markings on batter boards. Each method yielded feasible results, resulting in collinear markings on the batter boards, with very few and small deviations (as shown in the last picture of Figure 14). The second and third methods both produce the same coordinate for stake-out and the topographic nail representing this method is the middle one; the interior one was marked using the classical method and the exterior one was determined using reference line function. In terms of time efficiency, the methods were timed, achieving routinely predictable results: the batter boards survey with manual calculation of coordinates took longest, with roughly 2 h of work; the batter boards survey with CAD implementation as well as the classical theodolite method took approximately 1 h; the reference line method was the most efficient and easy to put into practice, with 15 min of implementation.

Figure 14. Construction twine and batter boards markings used for layout points positioning; obtained markings of the 4 methods of batter board stake-out (**right**).

The desideratum was to mark reliable points on each batter board, in order for the builders to further position construction layout points without instrumentation and expertise. As shown in Figure 14, this is obtained using construction twine by stretching it between the points (nails) on the wood planks, and a mechanical projection with a plummet on the ground from the twine intersection. Although not as precise as a layout with a total station, this established method is viable for smaller to medium-scale projects, and can ensure a required point standard deviation of $\mu_P = \pm(1-2)$ cm. These layout points are used by the builders for guidance regarding the excavation area, foundation and structural elements positioning, and assures the verticality of the erected building.

In future, building information modelling (BIM) will also have a significant impact on the work of surveyor engineers, as it is currently regarded as a major paradigm shift in the construction industry, especially for civil engineers and architects [31]. Traditional building design models depend to a large extent on two-dimensional technical drawings, and a BIM platform expands the three primary spatial dimensions (width, height and depth), integrating multiple advantages due to their economic benefits in design and construction phases [32,33]. Future perspectives include the establishment and integration of the site layout plan with the BIM design in order to achieve sustainable and productive practice regarding the construction industry.

5. Conclusions

The efficient layout planning of buildings in a construction site is a fundamental task for undertaking any project. In an attempt to enhance the general practice of layout planning of construction sites, the paper presents a retrospective for the surveyor engineer's role in the construction industry, a design for a comprehensive site layout plan with on-site applicability and introduces a novel comparison between four methods of construction lines layout on batter boards. The presented general knowledge further cements the importance of geodetic specialists in the field, and the designed layout plan serves as guidelines for existing and future engineers, as well as researchers in the field. In this context, safety and freedom from hazard concerns are key factors for a productive construction survey and layout project. The comparison highlights the viable implementation of each of the four methods of batter board construction lines layout, especially in regard to the instrumentation used and the on-site conditions, with the undisputed recommendation of the reference line method due to its overall efficiency and accuracy. Given the current development and expanding implementations of BIM that also impact the work of surveyors, future investigations will be made in order to integrate the site layout plan inside the platform.

Funding: This research received no external funding.

Institutional Review Board Statement: Not applicable.

Informed Consent Statement: Not applicable.

Data Availability Statement: Not applicable.

Acknowledgments: The author would like to thank the Academic Editor and anonymous reviewers for their helpful and valuable comments and suggestions.

Conflicts of Interest: The author declares no conflict of interest.

References

1. Bondrea, M.V.; Naș, S.; Fărcaș, R.; Dîrja, M.; Sestraș, P. Construction Survey and Precision Analysis Using RTK Technology and a Total Station at Axis Stake-Out on a Construction Site. In Proceedings of the 16th International Multidisciplinary Scientific GeoConference SGEM2016, Albena, Bulgaria, 28 June–7 July 2016; Volume 2, pp. 155–161.
2. Baykal, O.; Tari, E.; Coşkun, M.Z.; Erden, T. Accuracy of point layout with polar coordinates. *J. Surv. Eng.* **2005**, *131*, 87–93. [CrossRef]
3. Huang, C.; Wong, C.K. Optimisation of site layout planning for multiple construction stages with safety considerations and requirements. *Automat. Constr.* **2015**, *53*, 58–68. [CrossRef]
4. Sestraș, P.; Bondrea, M.V.; Sălăgean, T.; Dîrja, M.; Cîmpeanu, S.M. Engineering Survey for Excavated Volume Calculation in a Construction Site Using a Total Station. In Proceedings of the 16th International Multidisciplinary Scientific GeoConference SGEM2016, Albena, Bulgaria, 28 June–7 July 2016; Volume 2, pp. 247–254.
5. Cosarca, C. Considerations on the tolerances and precisions in engineering measurements. In *Recent Advances in Geodesy and Geomatics Engineering-Proceedings of the 1st European Conference of Geodesy Geomatics Engineering*; Technical University for Civil Engineering Bucharest: Bucharest, Romania, 2013; Volume 1, pp. 8–18.
6. Herban, S.I.; Vîlceanu, C.B.; Grecea, C. Road-Structure Monitoring with Modern Geodetic Technologies. *J. Surv. Eng.* **2017**, *4*, 143. [CrossRef]
7. Sestraș, P.; Roșca, S.; Bilașco, Ș.; Naș, S.; Buru, S.M.; Kovacs, L.; Spalević, V.; Sestras, A.F. Feasibility Assessments Using Unmanned Aerial Vehicle Technology in Heritage Buildings: Rehabilitation-Restoration, Spatial Analysis and Tourism Potential Analysis. *Sensors* **2020**, *20*, 2054. [CrossRef]
8. Drewes, H.; Kuglitsch, F.G.; Adám, J.; Rózsa, S. The geodesist's handbook 2016. *J. Geod.* **2016**, *90*, 907–1205. [CrossRef]
9. Hope, C.J.; Chuaqui, M. Precision surveying monitoring of shoring and structures. In Proceedings of the 7th FMGM 2007: Field Measurements in Geomechanics, Boston, MA, USA, 24–27 September 2007; Volume 1, pp. 1–12.
10. Artese, S.; Perrelli, M. Monitoring a Landslide with High Accuracy by Total Station: A DTM-Based Model to Correct for the Atmospheric Effects. *Geosciences* **2018**, *8*, 46. [CrossRef]
11. Afeni, T.B.; Cawood, F.T. Slope Monitoring using Total Station: What are the Challenges and How Should These be Mitigated? *S. Afr. J. Geomat.* **2013**, *2*, 41–53.
12. Horemuž, M.; Andersson, J.V. Analysis of the precision in free station establishment by RTK GPS. *Surv. Rev.* **2011**, *43*, 679–686. [CrossRef]
13. Sestraș, P.; Bilașco, Ș.; Roșca, S.; Dudic, B.; Hysa, A.; Spalević, V. Geodetic and UAV Monitoring in the Sustainable Management of Shallow Landslides and Erosion of a Susceptible Urban Environment. *Remote Sens.* **2021**, *13*, 385. [CrossRef]
14. Elbeltagi, E.; Hegazy, T.; Eldosouky, A. Dynamic layout of construction temporary facilities considering safety. *J. Constr. Eng.* **2004**, *130*, 534–541. [CrossRef]
15. Ning, X.; Lam, K.C.; Lam, M.C.K. A decision-making system for construction site layout planning. *Automat. Constr.* **2011**, *20*, 459–473. [CrossRef]
16. Cheng, M.Y.; O'Connor, J.T. ArcSite: Enhanced GIS for construction site layout. *J. Constr. Eng.* **1996**, *122*, 329–336. [CrossRef]
17. Biljecki, F.; Heuvelink, G.B.M.; Ledoux, H.; Stoter, J. The effect of acquisition error and level of detail on the accuracy of spatial analyses. *Cartogr. Geogr. Inf. Sci.* **2018**, *45*, 156–176. [CrossRef]
18. Berk, S.; Ferlan, M. Accurate area determination in the cadaster: Case study of Slovenia. *Cartogr. Geogr. Inf. Sci.* **2018**, *45*, 1–17. [CrossRef]
19. Hanus, P.; Pęska-Siwik, A.; Szewczyk, R. Spatial analysis of the accuracy of the cadastral parcel boundaries. *Comput. Electron. Agric.* **2018**, *144*, 9–15. [CrossRef]
20. Dolean, B.-E.; Bilașco, Ș.; Petrea, D.; Moldovan, C.; Vescan, I.; Roșca, S.; Fodorean, I. Evaluation of the Built-Up Area Dynamics in the First Ring of Cluj-Napoca Metropolitan Area, Romania by Semi-Automatic GIS Analysis of Landsat Satellite Images. *Appl. Sci.* **2020**, *10*, 7722. [CrossRef]
21. Sestraș, P.; Sălăgean, T.; Bilașco, Ș.; Bondrea, M.V.; Naș, S.; Fountas, S.; Cîmpeanu, S.M. Prospect of a GIS based digitization and 3D model for a better management and land use in a specific micro-areal for crop trees. *Environ. Eng. Manag. J.* **2019**, *18*, 1269–1277. [CrossRef]

22. Matei, I.; Pacurar, I.; Rosca, S.; Bilasco, S.; Sestras, P.; Rusu, T.; Jude, E.T.; Tăut, F.D. Land Use Favourability Assessment Based on Soil Characteristics and Anthropic Pollution. Case Study Somesul Mic Valley Corridor, Romania. *Agronomy* **2020**, *10*, 1245. [CrossRef]
23. Kavanagh, B.F.; Slattery, D.K. *Surveying: With Construction Applications*; Pearson: Upper Saddle River, NJ, USA, 2010.
24. Coșarcă, C. *Măsurători Inginerești: Aplicații în Domeniul Construcțiilor*, 1st ed.; Matrix: Bucharest, Romania, 2011; pp. 22–93.
25. Schofield, W. *Engineering Surveying: Theory and Examination Problems for Students*; Elsevier: London, UK, 2001.
26. Anderson, J.M.; Mikhail, E.M. *Surveying—Theory and Practice*, 7th ed.; McGraw-Hill: New York, NY, USA, 2012.
27. Uren, J.; Price, W.F. *Surveying for Engineers*; Macmillan International Higher Education: New York, NY, USA, 2010.
28. Horemuž, M.; Jansson, P. Optimum establishment of total station. *J. Surv. Eng.* **2017**, *2*, 143. [CrossRef]
29. Sun, H. Precision analysis of free-station positioning in total station. *Adv. Mater. Res.* **2013**, *694*, 1281–1285. [CrossRef]
30. Jasińka, E. Determining the influence of position accuracy and measurements on the accuracy of the point determined by tacheometric measurement. In Proceedings of the 18th International Multidisciplinary Scientific GeoConference SGEM2018, Albena, Bulgaria, 2–8 July 2018; Volume 18, pp. 717–722.
31. Kim, K.P.; Freda, R.; Nguyen, T.H.D. Building Information Modelling Feasibility Study for Building Surveying. *Sustainability* **2020**, *12*, 4791. [CrossRef]
32. Bryde, D.J.; Broquetas, M.; Volm, J.M. The project benefits of Building Information Modelling (BIM). *Int. J. Proj. Manag.* **2013**, *31*, 971–980. [CrossRef]
33. Durdyev, S.; Mbachu, J.; Thurnell, D.; Zhao, L.; Hosseini, M.R. BIM Adoption in the Cambodian Construction Industry: Key Drivers and Barriers. *ISPRS Int. J. Geo-Inf.* **2021**, *10*, 215. [CrossRef]

Article

Improving Bridge Expansion and Contraction Installation Replacement Decision System Using Hybrid Chaotic Whale Optimization Algorithm

Zian Xu and Minshui Huang *

School of Civil Engineering and Architecture, Wuhan Institute of Technology, Wuhan 430073, China; xuzian@wit.edu.cn
* Correspondence: huangminshui@tsinghua.org.cn

Abstract: Bridge expansion and contraction installation (BECI) has proved to be an essential component of the bridge structure due to its stability, comfort, and durability benefits. At present, traditional replacement technologies for modular type, comb plate type, and seamless type BECIs are widely applied worldwide. Nevertheless, it is unfortunate that the research conducted on decision-making (DM) approaches for the technical condition assessment and the optimal replacement plan selection of existing BECIs remain scarce, which results in the waste of resources and the increase in cost. Therefore, a BECI technical condition assessment approach, which contains specific on-site inspection regulations with both qualitative and quantitative descriptions, is proposed in this research, and a corresponding calculation program has been developed based on the MATLAB platform, which provides the basis for the necessity of replacement. Simultaneously, the hybrid chaotic whale optimization algorithm is designed and performed to improve and automate the process of optimal replacement plan selection under the assistance of the analytic hierarchy process (AHP), where both the achievement in consistency modification and the reservation of initial information are perused, and its superiority and effectiveness are verified via the comparative experimental analysis. The improved BECI replacement decision system is established, and the corresponding case study demonstrates that the proposed system in this research proves reasonable and feasible. The improved system can effectively assist bridge managers in making more informed operation and maintenance (O and M) decisions in actual engineering projects.

Keywords: bridge expansion and contraction installation (BECI); decision making (DM); technical condition assessment; analytic hierarchy process (AHP); whale optimization algorithm; Tent chaotic mapping; Lévy flight

Citation: Xu, Z.; Huang, M. Improving Bridge Expansion and Contraction Installation Replacement Decision System Using Hybrid Chaotic Whale Optimization Algorithm. *Appl. Sci.* **2021**, *11*, 6222. https://doi.org/10.3390/app11136222

Academic Editor: Mariusz Szóstak

Received: 18 June 2021
Accepted: 1 July 2021
Published: 5 July 2021

Publisher's Note: MDPI stays neutral with regard to jurisdictional claims in published maps and institutional affiliations.

Copyright: © 2021 by the authors. Licensee MDPI, Basel, Switzerland. This article is an open access article distributed under the terms and conditions of the Creative Commons Attribution (CC BY) license (https://creativecommons.org/licenses/by/4.0/).

1. Introduction

With the rapid increase in technical obstacles caused by the damage of bridge expansion and contraction installations (BECIs), related resources waste, economic burdens, traffic hazards, and social arguments have attracted considerable attention, which emphasizes the significance of systematic operation and maintenance (O and M) for BECIs.

Accordingly, the replacement of BECI has played a crucial role in the field of bridge O and M, where the decision making (DM) and optimization of the O and M plan are highly prized. In accordance with installation types and construction characteristics, the replacement technologies for BECIs can be generally divided into modular type replacement, comb plate type replacement, and seamless type replacement [1]. Referring to the research of Huang et al. [2], a BECI replacement decision system is established based on the analytic hierarchy process (AHP) [3], where the design requirements, construction requirements, management requirements, and scopes of application are integrated as decision criteria to provide a persuasive mathematical model for the selection of an optimal replacement plan. As a potential multi-criteria decision-making (MCDM) method, AHP has higher applicability and effectiveness compared with other common ones, such as the analytic network process (ANP), the preference ranking organization method for enrichment evaluations (PROMETHEE), and the simple additive weighting (SAW) [4]. Additionally, under the assistance of the Delphi method, which is capable of providing a strong basis for the construction of hierarchy model, the AHP is widely applied to evaluate the complexity of projects [5].

Nevertheless, the AHP is always confronted with the challenge of consistency test due to massive decision criteria. In many studies, the framework of AHP consistency modification is usually divided into two stages. Initially, a mathematical model for modification is established in accordance with the characteristics of comparison matrices, where the least square method is effectively employed [6,7], and the iterative algorithm is designed and performed to handle it subsequently [8]. Similarly, much research has been conducted successfully in such a framework based on the meta-heuristic algorithms, such as particle swarm optimization (PSO) [9], genetic algorithm (GA) [10], and ant colony optimization (ACO) [11], where the modification method proves reasonable and feasible and could also be extended to multiple engineering optimization problems [12–14].

Derived from the bubble-net hunting behavior of humpback whales, the whale optimization algorithm (WOA) [15] was proposed as a meta-heuristic algorithm to handle optimization problems in 2016. As an emerging and prevailing meta-heuristic algorithm, the superiority and effectiveness of WOA are verified by being tested with 29 mathematical optimization problems and six structural design problems [16], which leads to the wide application of WOA in engineering optimization problems, such as electrical engineering, civil engineering, classification, clustering, image processing, mechanical engineering, control engineering, robot path, networks, industrial engineering, task scheduling, and other engineering applications [17–28]. However, similarly to other meta-heuristic algorithms, the WOA is confronted with the problem of slow convergence speed, which motivates chaos theory to be introduced into the WOA optimization process, and the results prove that the chaotic mappings are capable of enhancing the performance of WOA, especially Tent chaotic mapping [29]. Originally, the Lévy flight strategy was introduced to develop the cuckoo search (CS) algorithm, which is a specific class of random walk in which the step lengths are distributed based on a heavy power law tails, and it proves efficient to assist the algorithm to perform a global search [30,31]. In order to improve the global search performance of WOA, a set of studies has been conducted on the hybridization of WOA and Lévy flight [32–34], where the results demonstrate that this strategy can effectively enhance the convergence accuracy, speed, and stability of WOA, even to handle a large-scale global optimization problem.

Much research has been conducted on decision making and plan optimization, whose achievements play a significant guiding role in later study and production. However, to date, the existing developed system utilized in decision making for BECI replacement remains insufficient to handle the large-scale MCDM optimization problem. Meanwhile, the approach of technical condition assessment of BECIs is constrained by the lack of on-site inspection regulations, which brings technical obstacles to the actual operation, and thus, motivates this article. Furthermore, the significance of automation in the DM system should never be ignored.

Based on the results of existing research, a BECI technical condition assessment approach, which contains specific on-site inspection regulations with both qualitative and quantitative descriptions, is proposed in this research, and a corresponding calculation program is developed based on the MATLAB platform, which provides the basis for the necessity of replacement. Simultaneously, the hybrid chaotic whale optimization algorithm is designed and performed to improve and automate the process of optimal replacement plan selection under the assistance of the analytic hierarchy process (AHP), where both the achievement in consistency modification and the reservation of initial information are perused, and its superiority and effectiveness are verified via the comparative experimental analysis. Consequently, a performance-based optimal replacement plan is selected by the improved decision system to guide later bridge managers, which also promotes the application of computer science technologies in the field of bridge O and M.

2. Objective

The improved BECI replacement decision system established in this article is aimed at achieving the objectives of two aspects, namely, realizing the standardization of BECI on-site inspection and the automations of both the technical condition assessment and the replacement plan selection. Accordingly, the proposed decision system can be divided into two stages, as illustrated in Figure 1.

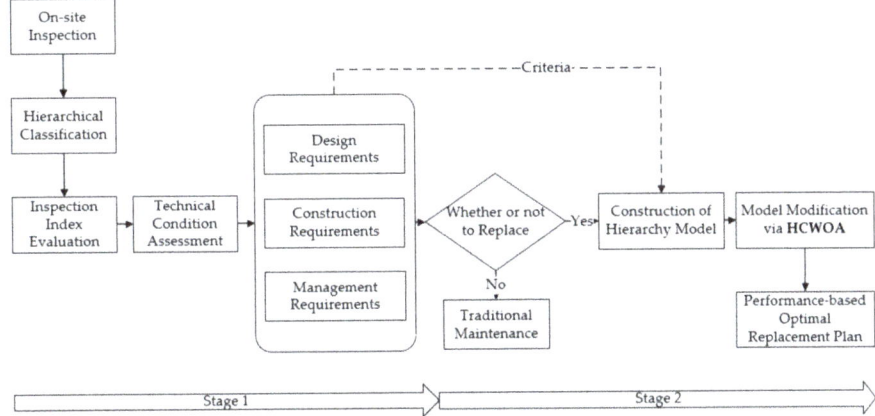

Figure 1. Framework of improved BECI replacement decision system.

Stage 1: Technical condition assessment. The on-site inspection is conducted on the target BECI, which will be hierarchically classified according to various installation types. The inspection index evaluation is performed to provide the basis of progressive assessment calculation, and "whether or not to replace" is determined.

Stage 2: Selection of replacement plan based on HCWOA. Twelve kinds of design, construction, and management requirements are introduced as the decision criteria, where the hierarchy model for selecting the performance-based optimal replacement plan is constructed via AHP. The HCWOA is designed and utilized to modify and handle the proposed mathematical model, where the optimal plan is selected precisely.

3. Methodology

3.1. Technical Condition Assessment

First and foremost, in order to provide a basis for the necessity of BECI replacement, the technical condition assessment should be performed. In the light of the transportation industry standard of the People's Republic of China [35,36], the framework of the BECI technical condition assessment approach is established and demonstrated in Figure 2.

3.1.1. Hierarchical Classification

In terms of the design mentality of this method, hierarchical classification is conducted after the on-site inspection to divide the inspected BECI into three layers from top to bottom: main structure and accessory structure; components; members. Referring to related standards of the transportation industry of the People's Republic of China [1], 6 kinds of modular type, comb plate type, and seamless type BECIs are introduced to this method as specific assessment objects, which are widely applied worldwide. The hierarchical classification of BECIs is shown in Table 1.

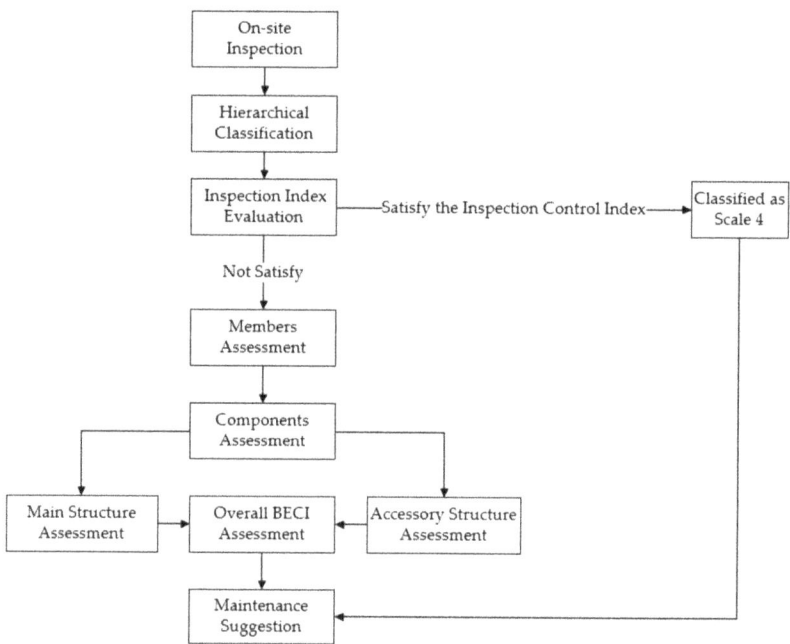

Figure 2. Framework of BECI technical condition assessment.

Table 1. Hierarchical classification of BECIs.

Installation	Part	No.	Component	Quantity of Members
MA Modular type	Main structure	1	Side beam	2
		2	Anchorage concrete	2
	Accessory structure	3	Rubber sealing tape	1
MB Modular type	Main structure	1	Side beam	2
		2	Intermediate beam	Quantity of intermediate beams
		3	Bearing system	Quantity of support beams
		4	Displacement control system	1
	Accessory structure	5	Elastic support	1
		6	Anchorage concrete	2
		7	Rubber sealing tape	Quantity of intermediate beams +1
SC Comb plate type	Main structure	1	Comb plate	Quantity of comb plate units
		2	Anchor bolt	Quantity of comb plate units
		3	Anchorage concrete	2
	Accessory structure	4	Drainage device	1
SSA, SSB Comb plate type	Main structure	1	Fixed comb plate	Quantity of fixed comb plate units
		2	Movable comb plate	Quantity of movable comb plate units
		3	Stainless steel sliding plate	Quantity of comb plate units
		4	Anchor bolt	Quantity of comb plate units
		5	Multidirectional displacement device	Quantity of movable comb plate units
		6	Anchorage concrete	2
	Accessory structure	7	Drainage device	1
Seamless type	Main structure	1	Elastic expansion body	1
	Accessory structure	2	Steel cover plate	1
		3	Nail	1
		4	Foam plate	1

3.1.2. Inspection Index Evaluation

Subsequently, in the process of on-site inspection, inspectors are also required to evaluate the inspection indices of each member. Therefore, a standardized inspection index evaluation approach of BECIs with both specific qualitative and quantitative regulations are proposed for the determination of inspection index scale. Corresponding to various types of BECIs, the regulations of inspection index evaluation are shown in Tables 2–6, respectively. In particular, the scale will be defined as 1 when the technical condition of inspection index remains excellent.

Table 2. Inspection index evaluation of the BECI of MA modular type.

Inspection Object	Inspection Index	Scale	Qualitative Description	Quantitative Description
Side beam	Evenness	2	Slightly uneven	Elevation difference ≤ 10 mm
		3	Obviously uneven, local fractures occur	10 mm < elevation difference ≤ 30 mm
		4	Excessively uneven, vehicle bumping, and side sinking occur	Elevation difference > 30 mm
	Spacing	2	Slightly narrow	50 mm < spacing ≤ design value
		3	Obviously narrow	10 mm < spacing ≤ 50 mm
		4	Excessively narrow or wide	Spacing > design value, or spacing ≤ 10 mm
Anchorage concrete	Fracture	2	Local fractures occur	Fractures ≤ 3, fracture width ≤ 2 mm, damaged area ≤ 10% anchorage concrete area
		3	Multiple fractures occur	3 < fractures ≤ 5, 2 mm < fracture width ≤ 10 mm, 10% < damaged area ≤ 20% anchorage concrete area
		4	Serious fractures occur, anchorage function fails	Fractures > 5, fracture width > 10 mm, damaged area > 20% anchorage concrete area
Rubber sealing tape	Aging	2	Slightly aging, local fractures occur	Fractures ≤ 3, damaged area ≤ 20%
		3	Obviously aging, multiple fractures occur	3 < fractures ≤ 5, 20% < damaged area ≤ 50%
		4	Excessively aging, serious fractures occur	Fractures > 5, damaged area > 50%

Table 3. Inspection index evaluation the BECI of MB modular type.

Inspection Object	Inspection Index	Scale	Qualitative Description	Quantitative Description
Side beam, intermediate beam	Evenness	2	Slightly uneven	Elevation difference ≤ 15 mm
		3	Obviously uneven, local fractures occur	15 mm < spacing ≤ 30 mm
		4	Excessively uneven, vehicle bumping, and side sinking occur	30 mm < spacing ≤ 100 mm
	Spacing	2	Slightly narrow	50 mm < spacing ≤ design value
		3	Obviously narrow	10 mm < spacing ≤ 50 mm
		4	Excessively narrow or wide	Spacing > design value, or spacing ≤ 10 mm
Bearing system	Bolt looseness, hanger damage	2	Local anchor bolts are unfixed, local hanger damages occur	Proportion of unfixed bolts ≤ 10%, or 1 hanger damage
		3	Multiple anchor bolts are unfixed, multiple hanger damages occur	10% < proportion of unfixed bolts ≤ 30%, or 2 hanger damages
		4	Massive anchor bolts are unfixed, massive hanger damages occur	Proportion of unfixed bolts > 30%, or hanger damages > 2
Displacement control system	Spring deformation, bolt looseness, hinge damage	2	Serious spring deformations occur, local bolts are unfixed, local hinges are damaged	Spring deformation > 20% design value, or proportion of unfixed bolts ≤ 10%, or 1 hinge damage
		3	Local springs are damaged, multiple bolts are unfixed, multiple hinges are damaged	Proportion of damaged springs ≤ 10%, or 10% < proportion of unfixed bolts ≤ 30%, or 2 hinge damages
		4	Multiple springs are damaged, massive bolts are unfixed, massive hinges are damaged	Proportion of damaged springs 10%, or proportion of unfixed bolts > 30%, or hinge damages > 2

Table 3. Cont.

Inspection Object	Inspection Index	Scale	Qualitative Description	Quantitative Description
Elastic support	Support deformation	2	Slight support deformations occur	Support deformations ≤ allowable value
		3	Obvious support deformations occur	Support deformations > allowable value
		4	Serious support deformations occur	Proportion of damaged supports > 30%
Anchorage concrete	Fracture	2	Local fractures occur	Fractures ≤ 3, fracture width ≤ 2 mm, damaged area ≤ 10%
		3	Multiple fractures occur	3 < fractures ≤ 5, 2 mm < fracture width ≤ 10 mm, 10% < damaged area ≤ 20%
		4	Serious fractures occur, anchorage function fails	Fractures > 5, fracture width > 10 mm, damaged area > 20%
Rubber sealing tape	Aging	2	Slightly aging, local fractures occur	Fractures ≤ 3, damaged area ≤ 20%
		3	Obviously aging, multiple fractures occur	3 < fractures ≤ 5, 20% < damaged area ≤ 50%
		4	Excessively aging, serious fractures occur	Fractures > 5, damage area > 50%

Table 4. Inspection index evaluation of the BECI of SC comb plate type BECI.

Inspection Object	Inspection Index	Scale	Qualitative Description	Quantitative Description
Comb plate	Evenness	2	Slightly uneven, local fractures occur	Damaged area ≤ 10%, 1 mm < elevation differences ≤ 2 mm
		3	Obviously uneven, multiple fractures occur	10% < damaged area ≤ 30%, 2 mm < elevation differences ≤ 4 mm
		4	Seriously uneven, serious fractures differences occur	Damaged area > 30%, elevation differences > 4 mm
Anchor bolt	Looseness	2	Local anchor bolts are unfixed	Proportion of damaged bolts ≤ 10%
		3	Multiple anchor bolts are unfixed	10% < proportion of damaged bolts ≤ 30%
		4	Massive anchor bolts are unfixed	Proportion of damaged bolts > 30%
Anchorage concrete	Fracture	2	Local fractures occur	Fractures ≤ 3, fracture width ≤ 2 mm, damaged area ≤ 10% anchorage concrete area
		3	Multiple fractures occur	3 < fractures ≤ 5, 2 mm < fracture width ≤ 10 mm, 10% anchorage concrete area < damaged area ≤ 20% anchorage concrete area
		4	Serious fractures occur, anchorage function fails	Fractures > 5, fracture width > 10 mm, damaged area > 20% anchorage concrete area
Drainage device	Fracture	2	Local fractures occur	Fractures ≤ 3, damaged area ≤ 10% total area
		3	Multiple fractures occur	3 < fractures ≤ 5, 10% < damaged area ≤ 30% total area
		4	Serious fractures occur	Fractures > 5, damaged area > 30% total area

Table 5. Inspection index evaluation of the BECI of SSA, SSB comb plate type.

Inspection Object	Inspection Index	Scale	Qualitative Description	Quantitative Description
Fixed comb plate, movable comb plate	Evenness	2	Slightly uneven, local fractures occur	Damaged area $\leq$ 10%, 1 mm < elevation differences $\leq$ 2 mm
		3	Obviously uneven, multiple fractures occur	10% < damaged area $\leq$ 30%, 2 mm < elevation differences $\leq$ 4 mm
		4	Seriously uneven, serious fractures differences occur	Damaged area > 30%, elevation differences > 4 mm
Stainless steel sliding plate	Abrasion	2	Local abrasions occur	Abrasion area $\leq$ 10% total area
		3	Multiple abrasions occur	10% < abrasion area $\leq$ 30% total area
		4	Massive abrasions occur	Abrasion area > 30% total area
Anchor bolt	Looseness	2	Local anchor bolts are unfixed	Proportion of damaged bolts $\leq$ 10%
		3	Multiple anchor bolts are unfixed	10% < proportion of damaged bolts $\leq$ 30%
		4	Massive anchor bolts are unfixed	Proportion of damaged bolts > 30%
Multidirectional displacement device	Flexibility	2	Individual unit is inflexible	1 inflexible unit
		3	Partial units are inflexible	2 inflexible units
		4	Multiple units are inflexible, device function fails	Inflexible units > 2
Anchorage concrete	Fracture	2	Local fractures occur	Fractures $\leq$ 3, fracture width $\leq$ 2 mm, damaged area $\leq$ 10% anchorage concrete area
		3	Multiple fractures occur	3 < fractures $\leq$ 5, 2 mm < fracture width $\leq$ 10 mm, 10% anchorage concrete area < damaged area $\leq$ 20% anchorage concrete area
		4	Serious fractures occur, anchorage function fails	Fractures > 5, fracture width > 10 mm, damaged area > 20% anchorage concrete area
Drainage device	Fracture	2	Local fractures occur	Fractures $\leq$ 3, damaged area $\leq$ 10% total area
		3	Multiple fractures occur	3 < fractures $\leq$ 5, 10% < damaged area $\leq$ 30% total area
		4	Serious fractures occur	Fractures > 5, damaged area > 30% total area

Table 6. Inspection index evaluation of the BECI of seamless type.

Inspection Object	Inspection Index	Scale	Qualitative Description	Quantitative Description
Elastic expansion body	Fracture	2	Local fractures occur	Damaged area $\leq$ 10% total area
		3	Multiple fractures occur	10% < damaged area $\leq$ 30% total area
		4	Serious fractures occur, anchorage function fails	Damaged area > 30% total area
Steel cover plate	Abrasion	2	Local abrasions occur	Abrasion area $\leq$ 10% total area
		3	Multiple abrasions occur	10% < abrasion area $\leq$ 30% total area
		4	Massive abrasions occur	Abrasion area > 30% total area
Nail	Corrosion	2	Local corrosions occur	Corrosions $\leq$ 3
		3	Multiple corrosions occur	3 < corrosions $\leq$ 5
		4	Massive corrosions occur	Corrosions > 5
Foam plate	Fracture	2	Local fractures occur	Damaged area $\leq$ 10% total area
		3	Multiple fractures occur	10% < damaged area $\leq$ 30% total area
		4	Serious fractures occur	Damaged area > 30% total area

3.1.3. Calculation of Progressive Assessment

In accordance with the obtained technical condition scales of each inspection index of each member in on-site inspection process, numerical progressive assessment is carried out from bottom to top. Specifically, the numerical calculations of the technical condition indices of members, components, main structure and accessory structure, and the overall BECI will be conducted successively, where the necessity of replacement will be settled in consequence.

1. Technical condition assessment of BECI members

   The technical condition assessment of BECI members is calculated as follows:

   $$\begin{array}{l} \mathrm{MMCI}_{i-l}(\mathrm{AMCI}_{i-l}) = 100 - \sum\limits_{x=1}^{k} N_x \\ N_e = \mathrm{DP}_{i-l-j}, \ e = 1 \\ N_e = \frac{\mathrm{DP}_{i-l-j}}{100 \times \sqrt{e}}(100 - \sum\limits_{r=1}^{e-1} N_r)(\text{where } j = e), \ e \geq 2 \end{array} \quad (1)$$

   where MMCI and AMCI are the condition indices of members, which belong to the main structure and accessory structure, respectively; i, l, and j denote component i, member l, and inspection index j; k is the quantity of inspection indices; x, e, and r are introduced variables; DP is the deduct point, which is determined by the technical condition scales of each inspection index according to Table 7.

2. Technical condition assessment of BECI components

   The technical condition assessment of BECI components is calculated as follows:

   $$\begin{array}{l} \mathrm{MCCI}_i = \overline{\mathrm{MMCI}}_i - (100 - \mathrm{MMCI}_{i,\ \min})/t \\ \mathrm{ACCI}_i = \overline{\mathrm{AMCI}}_i - (100 - \mathrm{AMCI}_{i,\ \min})/t \\ \mathrm{MCCI}_i(\mathrm{ACCI}_i) = 0, \ \text{when } \mathrm{MCCI}_i(\mathrm{ACCI}_i) < 0 \end{array} \quad (2)$$

   where MCCI and ACCI are the condition indices of components, which belong to the main structure and accessory structure, respectively; i denotes component i; t, as is illustrated in Table 8, is the correction coefficient, which is a variable with the quantity of members and introduced to neutralize the adverse effect of individual members on overall component technical condition assessment.

3. Technical condition assessment of BECI main structure and accessory structure

   The technical condition assessment of BECI main structure and accessory structure is calculated as follows:

   $$\mathrm{MSCI}(\mathrm{ASCI}) = \sum_{i=1}^{c} \mathrm{MCCI}_i(\mathrm{ACCI}_i) \times W_i \quad (3)$$

   where here MSCI and ASCI are the condition indices of main structure and accessory structure; i denotes component i; c represents the quantity of components; as is shown in Table 9, W_i is the calculation weight of each component, which is established via an analytic hierarchy process method under the guidance of the transportation industry standard of the People's Republic of China [35,36].

4. Technical condition assessment of overall BECI

   The technical condition assessment of overall BECI is calculated as follows:

   $$S_I = \mathrm{MSCI} \times W_{MS} + \mathrm{ASCI} \times W_{AS} \quad (4)$$

   where S_I denotes the summative condition index of overall BECI; W_{MS} and W_{AS} are the calculation weights of the main structure and accessory structure, respectively, which are valued according to Table 10.

5. Technical condition scale classification of BECI

   In accordance with the computation results of S_I, the technical condition scales S_S of BECI are classified and corresponding maintenance suggestions are proposed, which is exhibited in Table 11.

6. Inspection control index of BECI

   In particular, the technical condition scale of BECI will be classified as 4 without following up subsequent assessment process if the results of on-site inspection satisfy

the inspection control indices, which implies that a BECI replacement is necessary to be developed immediately. The inspection control indices of BECI are proposed as follows:
- More than 2 fractures occur in side beam or intermediate beam;
- The damaged area of comb plate exceeds 30% and more than 3 defective units occur;
- The damaged area of elastic expansion body exceeds 30%.

Compared with traditional inspection, this approach allows an exact number to be output when on-site inspectors find that the BECI is between two scales and it is difficult to completely make judgment, which proves the most significant improvement.

Table 7. Deduct point of inspection index.

The Highest Scale That Can Be Achieved by Inspection Index	Scale of Inspection Index			
	1	2	3	4
4	0	25	50	100

Table 8. Value of correction coefficient t.

Quantity of Members	t	Quantity of Members	t
1	∞	11	7.9
2	10	12	7.7
3	9.7	13	7.5
4	9.5	14	7.3
5	9.2	15	7.2
6	8.9	16	7.08
7	8.7	17	6.96
8	8.5	18	6.84
9	8.3	19	6.72
10	8.1	20	6.6

Table 9. Calculation weight of the component.

Installation	Part	No.	Component	W_i
MA Modular type	Main structure	1	Side beam	0.75
		2	Anchorage concrete	0.25
	Accessory structure	3	Rubber sealing tape	1
MB Modular type	Main structure	1	Side beam	0.27
		2	Intermediate beam	0.27
		3	Bearing system	0.10
		4	Displacement control system	0.10
		5	Elastic support	0.10
		6	Anchorage concrete	0.16
	Accessory structure	7	Rubber sealing tape	1
SC Comb plate type	Main structure	1	Comb plate	0.52
		2	Anchor bolt	0.24
		3	Anchorage concrete	0.24
	Accessory structure	4	Drainage device	1
SSA, SSB Comb plate type	Main structure	1	Fixed comb plate	0.25
		2	Movable comb plate	0.35
		3	Stainless steel sliding plate	0.05
		4	Anchor bolt	0.10
		5	Multidirectional displacement device	0.12
		6	Anchorage concrete	0.13
	Accessory structure	7	Drainage device	1
Seamless type	Main structure	1	Elastic expansion body	1
	Accessory structure	2	Steel cover plate	0.40
		3	Nail	0.40
		4	Foam plate	0.20

Table 10. Weights of main structure and accessory structure.

Part	Weight
Main structure	0.8
Accessory structure	0.2

Table 11. Technical condition scale classification.

Technical Condition Scale S_S	Technical Condition Index S_I	Qualitative Description	Maintenance Suggestion
1	[90, 100]	Healthy	Normal maintenance
2	[75, 90)	Slightly damaged	Individual members repair or replacement
3	[60, 75)	Obviously damaged	Essential components replacement
4	[0, 60)	Seriously damaged	BECI replacement

3.1.4. Software Development

Derived from the composition design, the "Bridge Expansion and Contraction Installation Assessment (BECIA)" software was developed based on the MATLAB platform to assist decision makers in making replacement decision efficiently. The software provides decision makers with a friendly user interface (Figure 3) where the users are only required to input the inspection index scales succinctly, and the technical condition index and scale of overall BECI will be output through automatic computation. Compared with traditional assessment, the development of software implies that both project managers and on-site inspector are entitled to participate in decision making directly, which proves the most significant improvement.

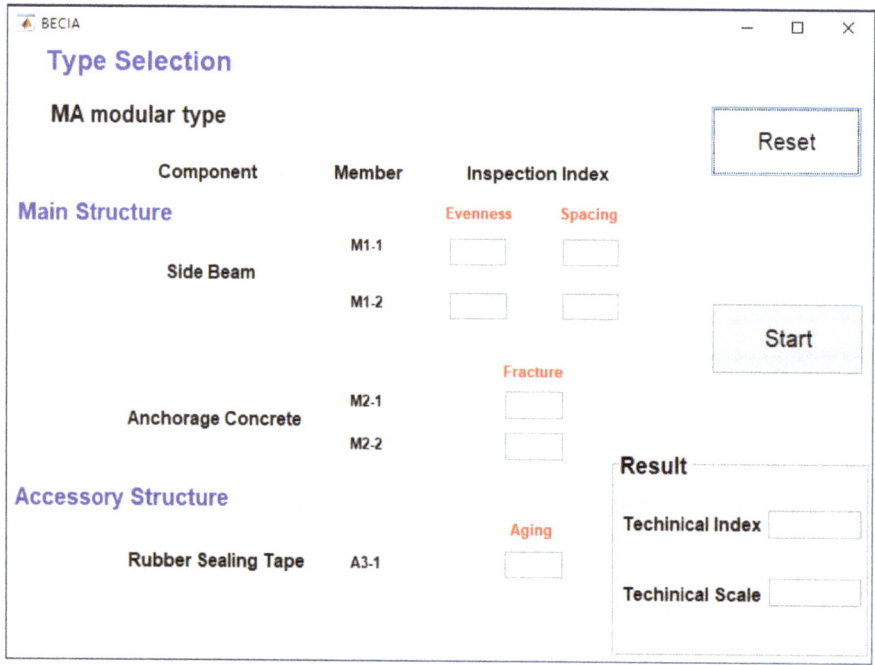

Figure 3. User interface of BECIA software.

3.2. Selection of Replacement Plan Using Hybrid Chaotic Whale Optimization Algorithm

The above-mentioned BECI technical condition assessment approach proves capable of being utilized to determine the necessity of replacement. Specifically, if the technical condition scale of overall BECI is defined as 4 or the inspection control index is satisfied, BECI replacement will be carried out and the analytic hierarchy process method will be employed to select the performance-based optimal replacement plan. In particular, the hybrid chaotic whale optimization algorithm is introduced originally to modify and automate the standard AHP.

3.2.1. Analytic Hierarchy Process

Under the suggestions of experts selected from relevant research field via the Delphi method [37], the hierarchy structure model of performance-based optimal replacement plan selection is constructed and illustrated in Figure 4, which takes 12 kinds of design, construction, and management requirements into consideration.

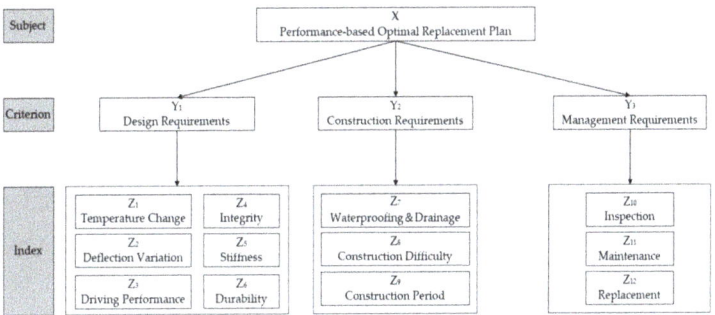

Figure 4. Hierarchy structure model of the performance-based replacement method selection.

In accordance with the 1–9 evaluation scale method [38], pairwise comparison matrices will be determined via the Delphi method. To put it in practical terms, the comparison matrix X is exhibited as an example:

$$X = (x_{ij})_{3\times 3} = \begin{pmatrix} 1 & Y_1/Y_2 & Y_1/Y_3 \\ Y_2/Y_1 & 1 & Y_2/Y_3 \\ Y_3/Y_1 & Y_3/Y_2 & 1 \end{pmatrix} \quad (5)$$

where Y_1/Y_2 is the relative priority of criterion Y_1 with respect to criterion Y_2, whose specific value is determined by the results of the above-mentioned 1–9 evaluation scale method, and the specific evaluation regulations are demonstrated in Table 12.

Table 12. Evaluation scale classification.

Evaluation Scale	Comparison of the Priority
1	Y_1 and Y_2 are equally important
3	Y_1 is slightly important compared with Y_2
5	Y_1 is obviously important compared with Y_2
7	Y_1 is highly important compared with Y_2
9	Y_1 is extremely important compared with Y_2
2, 4, 6, 8	Intermediate values used to represent compromise

After the construction of comparison matrices, the asymptotic normalization coefficient method is utilized to realize the process of level simple sequence (LLS), where the eigenvector of comparison matrix w_i is defined as the hierarchical weight vector and calculated as follows [3]:

$$M_i = \prod_{j=1}^{n} x_{ij}, \quad \overline{w_i} = \sqrt[n]{M_i}, \quad w_i = \frac{\overline{w_i}}{\sum_{j=1}^{n} \overline{w_j}} \quad (i = 1, 2 \ldots n) \qquad (6)$$

where M_i is the chain-multiplication result of x_{ij}, and n is the order of comparison matrix.

Simultaneously, the related maximum eigenvalue is $\lambda_{max} = \sum_{i=1}^{n} \frac{(Xw)_i}{nw_i}$, where $(Xw)_i$ is the part i of Xw.

In accordance with obtained λ_{max}, the consistency ratio (CR) and consistency index (CI) are introduced to perform a consistency test, which demonstrates the rationality of the constructed comparison matrix and can be calculated as follows:

$$CR = \frac{CI}{RI}$$
$$CI = \frac{\lambda_{max} - n}{n - 1} \qquad (7)$$

where RI is the random consistency index that corresponds to the order of comparison matrix and can be defined as Table 13 illustrates.

Table 13. Random consistency index.

N	1	2	3	4	5	6	7	8
RI	0.00	0.00	0.58	0.90	1.12	1.24	1.32	1.41

If $CR < 0.10$, the matrix consistency is acceptable. Otherwise, the matrix is inconsistent and is required to be modified. In particular, the matrix is defined as the complete consistency matrix when CR is equal to 0.

3.2.2. Modification Strategy of Inconsistent Comparison Matrix

In terms of high-order matrices that were initiated under the circumstance of human factors, it is difficult to fulfil the requirements of consistency test, thus motivating the modification of inconsistent comparison matrix. In consideration of both the characteristics of the matrix with complete consistency and the relationship between the initiated matrix and the induced matrix, the mathematical model is established through the least square method, which leads to a multi-objective optimization problem. $M = (m_{ij})_{n \times n}$ and $X = (x_{ij})_{n \times n}$ are introduced as the initiated matrix and the induced matrix, respectively. If X is supposed to be a complete consistency matrix, Equation (8) will be established as follows [9]:

$$x_{ij} = w_i / w_j \qquad (8)$$

Hence, the first least square mathematical model is proposed as follows, which aims to improve the consistency:

$$\min \sum_{i=1}^{n} \sum_{j=1}^{n} (x_{ij} - w_i / w_j)^2$$
$$\text{s.t.} \quad \sum_{i=1}^{n} w_i = 1, w_i > 0,$$
$$x_{ij} = 1 / x_{ij},$$
$$i, j = 1, 2, 3 \ldots n. \qquad (9)$$

However, not only does the consistency of comparison matrix need to be enhanced, but also the suggestions of experts should be respected. Therefore, the second least square mathematical model is constructed as follows, which aims to reserve the original information of initiated matrix:

$$\min \sum_{i=1}^{n} \sum_{j=1}^{n} (x_{ij} - m_{ij})^2$$
$$s.t. \quad x_{ij} \in [(1-\mu)m_{ij}, (1+\mu)m_{ij}], \quad (10)$$
$$0 < \mu < 1,$$
$$i, j = 1, 2, 3...n.$$

Eventually, as Equation (11) illustrates, the above-mentioned mathematical models are integrated through weighted method, which turns the modification of inconsistent comparison matrix into a single-objective optimization problem with $n(n+1)/2$ variables due to the symmetry of comparison matrix.

$$\min L = \sum_{i=1}^{n} \sum_{j=1}^{n} \left[k_1 (x_{ij} - m_{ij})^2 + k_2 (x_{ij} - w_i/w_j)^2 \right]$$
$$s.t. \quad \sum_{i=1}^{n} w_i = 1, w_i > 0,$$
$$k_1 + k_2 = 1, k_1, k_2 > 0, \quad (11)$$
$$x_{ij} = 1/x_{ij},$$
$$x_{ij} \in [(1-\mu)m_{ij}, (1+\mu)m_{ij}], 0 < \mu < 1,$$
$$i, j = 1, 2, 3...n.$$

3.2.3. Hybrid Chaotic Whale Optimization Algorithm

In accordance with the established mathematical model, the hybrid chaotic whale optimization algorithm is introduced to solve the objective function minL, where both the modified comparison matrix and hierarchical weights will be output automatically.

1. Standard whale optimization algorithm.

Derived from the bubble-net hunting behavior of humpback whales (Figure 5), the standard whale optimization algorithm was proposed as a meta-heuristic algorithm to handle optimization problems in 2016, which contains 3 principal operations as follows:

- Encircling prey.

The humpback whales are capable of recognizing the location of prey, which indicates the solution space in WOA, and encircle it. Additionally, the current best candidate solution is assumed to be the target prey or close to the optimum, which motivates other search agents to update their positions towards the best one. This behavior is simulated by Equation (12) as follows:

$$\vec{D} = \left| \vec{C} \cdot \vec{X^*}(t) - \vec{X}(t) \right|$$
$$\vec{X}(t+1) = \vec{X^*}(t) - \vec{A} \cdot \vec{D} \quad (12)$$

where t denotes the current iteration, and $\vec{D}$ and $\vec{X}$ are position vectors. In particular, $\vec{X^*}$ is the position vector of the best solution acquired so far which should be updated in each iteration if a better solution occurs. $\vec{A}$ and $\vec{C}$ are the coefficient vectors, which can be calculated as follows:

$$\vec{A} = 2\vec{a} \cdot \vec{r} - \vec{a}$$
$$\vec{C} = 2 \cdot \vec{r} \quad (13)$$

where $\vec{a}$ is decreased from 2 to 0 linearly over the course of iterations in both exploration and exploitation phases, and $\vec{r}$ is a random vector in $[0, 1]$.

- Bubble-net attacking (exploitation phase).

In order to model the bubble-net attacking behavior of humpback whales mathematically, the shrinking encircling mechanism and spiral updating position strategy are introduced, where the mechanism is realized by setting random values for $\vec{A}$ in $[-1,1]$, and the spiral updating position strategy is described, as Equation (15) demonstrates:

$$\begin{aligned}\vec{X}(t+1) &= \vec{D'} \cdot e^{bl} \cdot \cos(2\pi l) + \vec{X^*}(t) \\ \vec{D'} &= \left|\vec{X^*}(t) - \vec{X}(t)\right|\end{aligned} \quad (14)$$

where $\vec{D'}$ indicates the distance of the whale to the prey, and b is a constant for defining the logarithmic spiral shapes, random number $l \in [-1,1]$.

It is assumed that there is a probability of 50% to switch between either the encircling prey mode or the spiral model to update the position of whales, which can be mathematically expressed as follows:

$$\begin{cases}\vec{X}(t+1) = \vec{X^*}(t) - \vec{A} \cdot \vec{D} & if\ p < 0.5 \\ \vec{X}(t+1) = \vec{D'} \cdot e^{bl} \cdot \cos(2\pi l) + \vec{X^*}(t) & if\ p \geq 0.5\end{cases} \quad (15)$$

where random number $p \in [0,1]$.

- Search for prey (exploration phase).

The humpback whales search randomly according to the position of each other, which is simulated through the similar approach based on the variation of $\vec{A}$. The exploration phase and the exploitation phase differ in that the position of search agent is updated by a randomly chosen search agent rather than by the best search agent at present. To put it in practical terms, the whales will swim randomly outside the shrinking encircling area when $\left|\vec{A}\right| > 1$, which emphasizes the global search of WOA and can be described as follows:

$$\begin{aligned}\vec{D} &= \left|\vec{C} \cdot \vec{X}_{rand}(t) - \vec{X}(t)\right| \\ \vec{X}(t+1) &= \vec{X}_{rand}(t) - \vec{A} \cdot \vec{D}\end{aligned} \quad (16)$$

where $\vec{X}_{rand}(t)$ is a random position vector of a search agent chosen from the current population.

2. Tent chaotic mapping strategy.

The random-based optimization algorithm using chaotic variables instead of random variables are defined as the chaotic optimization algorithm, which is capable of carrying out global searches at both higher speed and accuracy than stochastic searches that depend on probabilities due to the non-repetition and ergodicity of chaos. Additionally, compared with other chaotic maps, Tent chaotic map proves efficient to enhance the performance of standard WOA when the process of population initialization is conducted. Hence, the Tent chaotic mapping strategy is introduced and expressed as follows [39]:

$$x_{k+1,d} = \begin{cases}\frac{x_{kd}}{0.7} & x_{kd} < 0.7 \\ \frac{10}{3}(1-x_{kd}) & x_{kd} \geq 0.7\end{cases} \quad (17)$$

where k denotes the search agent k, d indicates the dimension d and $x_{kd} \in [0,1]$.

Specifically, a random vector is generated as the position vector of the first search agent in the solution space of optimization problem. Then, other position vectors of the rest search agents are calculated by Equation (17), where the chaotic set is formed. Ultimately, the obtained position vectors are mapped to the solution space, which can be realized by Equation (18).

$$y_{kd} = lb_d + x_{kd} \cdot (ub_d - lb_d) \quad (18)$$

where lb_d and ub_d are the lower boundary and upper boundary of the optimization problem, and y indicates the mapped position of the search agent.

3. Lévy flight strategy.

Confronted with high-dimensional and multi-modal optimization problems, the performance of standard WOA remains insufficient due to the dependence on randomness, which results in the risk of local optimum. Originally, the Lévy flight strategy was introduced to develop the cuckoo search (CS) algorithm, which proves efficient with both the high speed and accuracy of global convergence. Therefore, in order to prevent the optimization algorithm from trapping in the local optimum and enhance the global search performance, the Levy flight strategy is conducted to be hybridized with the chaotic WOA, which is utilized to update the position of whales again at the later stage of each iteration and can be demonstrated as follows [40]:

$$\vec{X}(t+1) = \vec{X}_{rand}(t) - \alpha \cdot sign[rand - 1/2] \oplus Lévy(s) \quad (19)$$

where $sign[rand - 1/2]$ takes only 3 values: $-1, 0$, or 1, the product $\oplus$ denotes the entry wise multiplication, and α is the step size related to the scales of optimization problems, which can be described as follows:

$$\alpha = 0.01 \left[\vec{X}_{rand}(t) - \vec{X}(t) \right] \quad (20)$$

The *Lévy* flight strategy proves a non-Gaussian random process, where the step lengths follow the *Lévy* distribution and can be expressed as follows:

$$Lévy(s) \sim |s|^{-1-\beta}, 0 < \beta \leq 2 \quad (21)$$

where β is an index, s indicates the random step length of *Lévy* flight, which can be obtained by Mantegna's algorithm [41]:

$$s = \frac{\mu}{|v|^{1/\beta}} \quad (22)$$

where β is set to be 1.5, μ and v obey the normal distribution and can be acquired as follows:

$$\mu \sim N(0, \sigma_\mu^2), v \sim N(0, \sigma_v^2)$$
$$\sigma_\mu = \left\{ \frac{\Gamma(1+\beta)\cdot \sin(\pi\beta/2)}{\Gamma[(1+\beta)/2]\beta \cdot 2^{(\beta-1)/2}} \right\}^{1/\beta}, \sigma_v = 1 \quad (23)$$

where Γ is the standard Gamma function.

In summary, the optimization flowchart of the hybrid chaotic whale optimization algorithm is established in the light of above theories, which is exhibited in Figure 6. Additionally, both the modified comparison matrix and hierarchical weights will be output automatically when the optimal solution of objective function minL is satisfied or the max iteration is achieved.

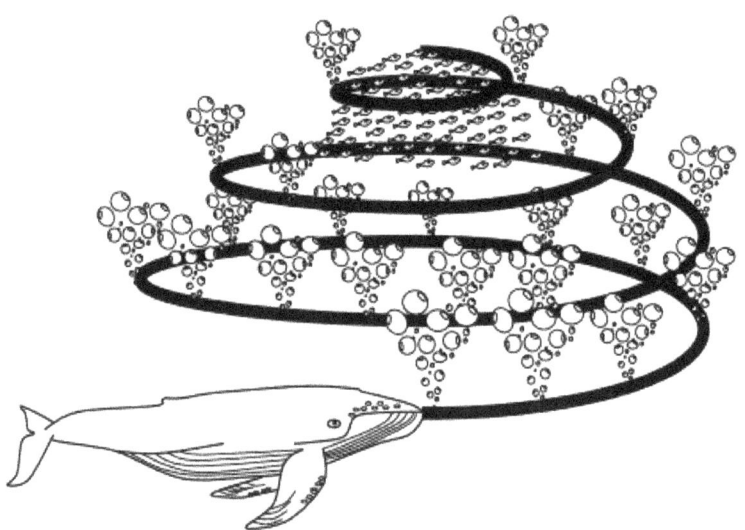

Figure 5. Bubble-net hunting behavior of humpback whales.

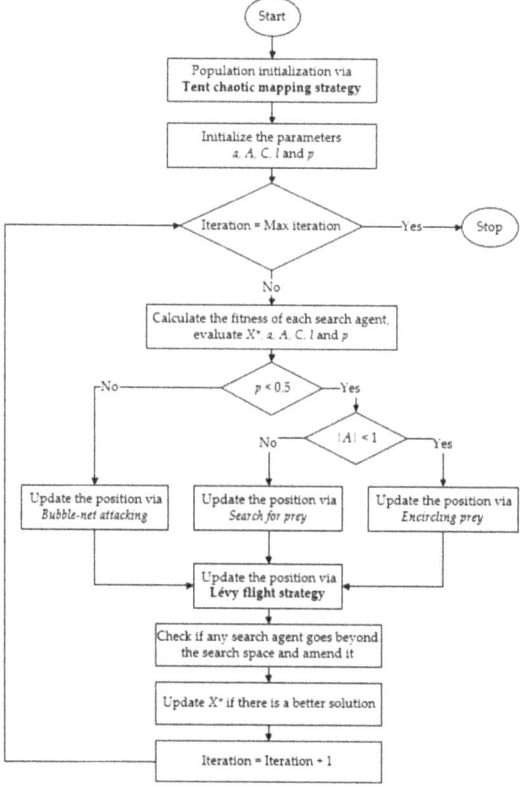

Figure 6. Optimization flowchart of hybrid chaotic whale optimization algorithm.

3.2.4. Performance Coefficient Determination

In accordance with the obtained hierarchical weights, a mathematical model is established to determine the performance coefficient of each replacement plan, where the performance-based optimal replacement plan with the highest performance coefficient will be selected as follows:

$$P = (Y_1 \cdot \sum_{i=1}^{6} Z_i \cdot I_i + Y_2 \cdot \sum_{i=7}^{9} Z_i \cdot I_i + Y_3 \cdot \sum_{i=10}^{12} Z_i \cdot I_i)/100 \qquad (24)$$

where P denotes the performance coefficient of the replacement plan, Y and Z are the hierarchical weights of criteria and indices in AHP, i represents index i, and I indicates the performance coefficients of the indices, which are obtained under the suggestion of experts via the Delphi method and range from 1 to 100.

4. Case Study
4.1. Research Background

In this article, the BECI O and M project of Guo Bridge is introduced as a case study, which is located in Taihe County, Fuyang City, Anhui Province, China and belongs to the provincial highway S254.

Guo bridge is in service with a total length of 20.6 m and a width of 8.6 m, where the MA modular type BECI is applied with an expansion amount of 60 mm. Under the circumstances of heavy-load vehicles and long-term environmental impact, the performance of BECI in Guo Bridge has become unfavorable, thus motivating the technical condition assessment process. In October 2018, an on-site inspection was conducted on Guo bridge, where the image data were acquired and demonstrated in Figure 7.

(a) Overview of BECI

(b) Elevation difference in side beam

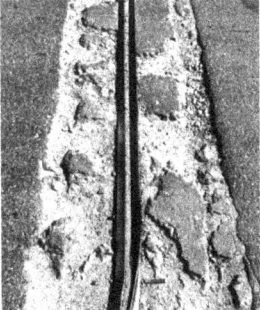

(c) Fracture in anchorage concrete

(d) Aging of rubber sealing tape

Figure 7. BECI of Guo Bridge.

4.2. Technical Condition Assessment

Above all, the BECI of Guo Bridge is hierarchically classified into three layers and numbered in accordance with Table 1. In addition, combined with the results of on-site inspection and the regulations proposed in Table 2, the inspection index evaluation can be carried out.

4.2.1. Inspection Index Evaluation

The elevation differences of 25 mm occur in both the left and right side beams, where the index evaluation scales can be defined as 3; the spacing of 51 mm between side beams proves slightly narrow, which is more than 50 mm but less than the design value (60 mm) and indicates the inspection index evaluation scale is 2. Referring to Table 7, the deduct points can be obtained:

$$DP_{1-1-1} = DP_{1-2-1} = 50, \quad DP_{1-1-2} = DP_{1-2-2} = 25$$

More than five serious fractures occurred in both the left and right anchorage concrete areas, and the damaged areas both exceed 20% of total area, where the index evaluation scales should be determined as 4. Referring to Table 7, the deduct points are exhibited as follows:

$$DP_{2-1-1} = DP_{2-2-1} = 100$$

There are three fractures that occurred in the rubber sealing tape, where the damaged area exceeds 20% of total area but less than 50%. The inspection index evaluation scale of aging should be defined as 3. Similarly, the deduct point is acquired:

$$DP_{3-1-1} = 50$$

4.2.2. Calculation of Progressive Assessment

The calculation of progressive assessment can be performed according to the deduct points gained above.

1. Technical condition assessment of BECI members.

- Members of side beam.

According to Equation (1), the technical condition indices (keep two decimal place) of the left and right side beam are calculated as follows:

$$N_1 = DP_{1-1-1} = 50$$
$$N_2 = \frac{DP_{1-1-2}}{100 \times \sqrt{e}}(100 - \sum_{r=1}^{e-1} N_r) = \frac{25}{100 \times \sqrt{2}}(100 - \sum_{r=1}^{1} N_1) = 8.84$$
$$MMCI_{1-1} = 100 - \sum_{e=1}^{2} N_e = 100 - (50 + 8.84) = 41.16$$
$$MMCI_{1-2} = MMCI_{1-1} = 41.16$$

- Members of anchorage concrete.

In the same way, the technical condition indices of the left and right anchorage concrete are obtained:

$$N_1 = DP_{2-1-1} = DP_{2-2-1} = 100$$
$$MMCI_{2-1} = MMCI_{2-2} = 100 - \sum_{e=1}^{1} N_e = 100 - (100) = 0$$

- Members of rubber sealing tape.

There is only one member of rubber sealing tape in MA modular type BECI. Therefore, the technical condition index is calculated as follows:

$$N_1 = DP_{3-1-1} = 50$$
$$AMCI_{3-1} = 100 - \sum_{e=1}^{1} N_e = 100 - (50) = 50$$

2. Technical condition assessment of BECI components.

- Side beam.

According to Table 8, the quantity of members of the side beam is two, which indicates the correction coefficient t is defined as 10. Under the guidance of Equation (2), the technical condition index of side beam can be calculated:

$$\overline{MMCI}_1 = (MMCI_{1-1} + MMCI_{1-2})/2 = 41.16$$
$$MCCI_1 = \overline{MMCI}_1 - (100 - MMCI_{1,min})/t = 41.16 - (100 - 41.16)/10 = 35.28$$

- Anchorage concrete.

In the same way, the technical condition index of anchorage concrete is acquired:

$$\overline{MMCI}_2 = (MMCI_{2-1} + MMCI_{2-2})/2 = 0$$
$$MCCI_2 = \overline{MMCI}_2 - (100 - MMCI_{2,min})/t = 0 - (100 - 0)/10 = -10 < 0$$
$$MCCI_2 = 0$$

- Rubber sealing tape.

Next comes the technical condition index of anchorage concrete:

$$\overline{AMCI}_3 = AMCI_3/1 = 50$$
$$ACCI_3 = \overline{AMCI}_3 - (100 - AMCI_{3,min})/t = 50 - (100 - 50)/\infty = 50$$

3. Technical condition assessment of BECI main structure and accessory structure.

In accordance with Table 9, the calculation weights of side beam, anchorage concrete, and rubber sealing tape are gained, which are 0.75, 0.25, and 1, respectively. The technical condition indices of the main structure and accessory structure are calculated by Equation (3) as follows:

$$MSCI = \sum_{i=1}^{c} MCCI_i \times W_i = 35.28 \times 0.75 + 0 \times 0.25 = 26.46$$
$$ASCI = \sum_{i=1}^{c} ACCI_i \times W_i = 50 \times 1 = 50$$

4. Technical condition assessment of overall BECI.

Referring to Table 10 and the technical indices gained above, the technical condition index of overall BECI is acquired by Equation (4):

$$S_I = SMCI \times W_{MS} + SACI \times W_{AS} = 26.46 \times 0.8 + 50 \times 0.2 = 31.17$$

4.2.3. Operation of BECIA Software

Simultaneously, the BECIA software is conducted on technical condition assessment of the BECI of Guo Bridge, where only the inspection index scales are required to be input, and the technical condition index and scale of overall BECI will be acquired automatically, as Figure 8 illustrates.

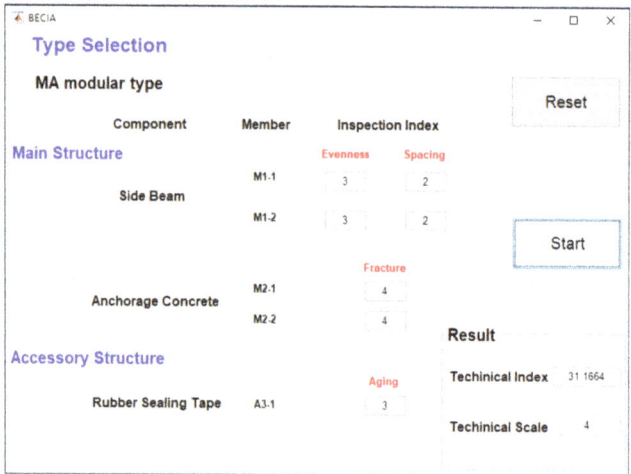

Figure 8. Operation interface of BECIA software.

Consequently, the results gained through BECIA are completely consistent with that of traditional manual calculation, which indicates that on-site inspectors are capable of participating in decision making independently via BECIA.

In conclusion, the technical condition scale of overall BECI of Guo Bridge is defined as 4 according to Table 11, which implies that the BECI are seriously damaged, and the replacement should be carried out immediately, thus leading to the selection of a replacement plan.

4.3. Selection of Replacement Plan

To begin with, in order to provide scientific guidance on the replacement plan selection, 16 experts selected from the expert database of Anhui Provincial Highway Institute were invited anonymously to solicit their opinions on research issues and phenomena via the Delphi method, where the candidate replacement plans are proposed in consideration of the allowable expansion amount of original BECI (60 mm) and the transportation industry standard of the People's Republic of China [1].

- Plan A: MA modular type with the allowable expansion amount from 20 mm to 80 mm.
- Plan B: SC comb plate type with the allowable expansion amount from 60 mm to 240 mm.

4.3.1. Construction of Comparison Matrix

Under the suggestions of the expert group, the priority of each index, criterion, and subject of AHP hierarchical model was distinguished through the 1–9 evaluation scale method, and the comparison matrices are constructed as follows:

$$X = \begin{pmatrix} 1 & 1 & 2 \\ 1 & 1 & 2 \\ 1/2 & 1/2 & 1 \end{pmatrix}, Y_1 = \begin{pmatrix} 1 & 6 & 6 & 4 & 4 & 3 \\ 1/6 & 1 & 1 & 1/5 & 1/5 & 1/3 \\ 1/6 & 1 & 1 & 1/5 & 1/5 & 1/3 \\ 1/4 & 5 & 5 & 1 & 1 & 1/4 \\ 1/4 & 5 & 5 & 1 & 1 & 1/4 \\ 1/3 & 3 & 3 & 4 & 4 & 1 \end{pmatrix}, Y_2 = \begin{pmatrix} 1 & 1/3 & 2 \\ 3 & 1 & 4 \\ 1/2 & 1/4 & 1 \end{pmatrix}, Y_3 = \begin{pmatrix} 1 & 1/2 & 1 \\ 2 & 1 & 2 \\ 1 & 1/2 & 1 \end{pmatrix}$$

In accordance with Equation (6), the eigenvectors and maximum eigenvalues of comparison matrices are obtained:

$$w_X = (0.40, 0.40, 0.20)^T \quad \lambda_{maxX} = 3.15$$
$$w_{Y_1} = (0.42, 0.04, 0.04, 0.13, 0.13, 0.24)^T \quad \lambda_{maxY_1} = 6.66$$
$$w_{Y_2} = (0.24, 0.63, 0.14)^T \quad \lambda_{maxY_2} = 3.06$$
$$w_{Y_3} = (0.25, 0.50, 0.25)^T \quad \lambda_{maxY_3} = 3.00$$

Next comes the result of the consistency test:

$$CI_X = 0 R \quad I_X = 0.58 \quad CR_X = 0 < 0.10$$
$$CI_{Y_1} = 0.13 \quad RI_{Y_1} = 1.24 \quad CR_{Y_1} = 0.11 > 0.10$$
$$CI_{Y_2} = 0.03 \quad RI_{Y_2} = 0.58 \quad CR_{Y_2} = 0.05 < 0.10$$
$$CI_{Y_3} = 0 \quad RI_{Y_3} = 0.58 \quad CR_{Y_3} = 0 < 0.10$$

Unfortunately, as a high-order matrix, Y_1 proves incapable of meeting the requirements of consistency test, which indicates that the acquired hierarchical weight w_{Y_1} is not acceptable and modification should be performed via the hybrid chaotic whale optimization algorithm.

4.3.2. Modification of Inconsistent Comparison Matrix

In the light of above-mentioned strategies, the modification of inconsistent comparison matrix Y_1 is initiated. In order to achieve better consistency on the premise of reserving initial information, the parameters in Equation (11) are defined as follows, where the 15 elements in the upper right half above the trace of matrix and six hierarchical weights are integrated to turn the modification into a 21-dimentional optimization problem.

$$\min L = \sum_{i=1}^{n} \sum_{j=1}^{n} \left[0.3(x_{ij} - y_{1,ij})^2 + 0.7(x_{ij} - w_i/w_j)^2 \right]$$
$$s.t. \sum_{i=1}^{n} w_i = 1, w_i > 0,$$
$$x_{ij} = 1/x_{ij},$$
$$x_{ij} \in [0.7y_{1,ij}, 1.3y_{1,ij}],$$
$$i, j = 1, 2, 3...n.$$

Simultaneously, the proposed hybrid chaotic whale optimization algorithm is conducted to handle this optimization problem with MATLAB R2016a, where the modified comparison matrix and corresponding hierarchical weights are output automatically. It is worth mentioning that the function value gained after 1000 iterations will be output as the optimal fitness value due to its uncertainty in advance. Additionally, the modified comparison matrix is exhibited as follows:

$$Y_1 = \begin{pmatrix} 1 & 6.1514 & 6.1457 & 3.8134 & 3.9181 & 2.7247 \\ 0.1626 & 1 & 0.9832 & 0.2600 & 0.2600 & 0.3507 \\ 0.1627 & 1.0170 & 1 & 0.2600 & 0.2600 & 0.4333 \\ 0.2622 & 3.8462 & 3.8462 & 1 & 1.0591 & 0.3250 \\ 0.2552 & 3.8462 & 3.8462 & 0.9442 & 1 & 0.3250 \\ 0.3670 & 2.8514 & 2.3079 & 3.0769 & 3.0769 & 1 \end{pmatrix}$$

$$CI_{Y_1} = 0.082 \, RI_{Y_1} = 1.24 \, CR_{Y_1} = 0.066 < 0.10$$

Consequently, all the hierarchical weights prove acceptable and are collected as follows:

$$w_X = (0.40, 0.40, 0.20)^T$$

$$w_{Y_1} = (0.45, 0.07, 0.07, 0.12, 0.12, 0.17)^T$$

$$w_{Y_2} = (0.24, 0.62, 0.14)^T$$

$$w_{Y_3} = (0.25, 0.50, 0.25)^T$$

4.3.3. Performance Coefficient Determination

Referring to Equation (24) and gained hierarchical weights, the performance coefficient of each replacement plan can be determined and illustrated, as Table 14 shows.

Table 14. Performance coefficients determination of replacement plans.

Layer	Item					Hierarchical Weight							
Criterion	Y	Y_1					Y_2			Y_3			
	w	0.40					0.40			0.20			
Index	Z	Z_1	Z_2	Z_3	Z_4	Z_5	Z_6	Z_7	Z_8	Z_9	Z_{10}	Z_{11}	Z_{12}
	w	0.45	0.07	0.07	0.12	0.12	0.17	0.24	0.62	0.14	0.25	0.50	0.25
Plan A	I	85	80	80	90	90	85	80	85	80	90	90	90
	P	0.854											
Plan B	I	90	90	85	85	80	85	90	80	75	80	80	80
	P	0.835											

Ultimately, plan A with the higher performance coefficient of 0.854 will be adopted to guide the BECI replacement, which implies that the MA modular type BECI will be selected instead of the SC comb plate type.

5. Comparative Experimental Analysis

A set of comparative experiments were designed and performed to verify the accuracy and effectiveness of the proposed hybrid chaotic whale optimization algorithm for solving the minL in the case study. The basic particle swarm optimization (PSO) algorithm and standard whale optimization algorithm (WOA) are introduced as the competitive algorithms, where the former is a classical representative meta-heuristic algorithm and the latter is an emerging prevailing one. The evolution curves of the fitness value for minL are illustrated in Figure 9, where the superiority of HCWOA with both considerable convergence speed and excellent convergence accuracy for solving minL is demonstrated.

In the phase of population initialization, the Tent chaotic mapping strategy is introduced to HCWOA, which achieved the least fitness value at the first iteration compared with other rivals and led to a fast convergence. Significantly, in the period of iteration 200 to 300, the HCWOA hybridized with the Lévy flight strategy never failed to keep the momentum in perusing further convergence, while the standard WOA was convergent, which demonstrates that the Lévy flight strategy is capable of preventing WOA from trapping in local optimum and improving the convergence accuracy substantially.

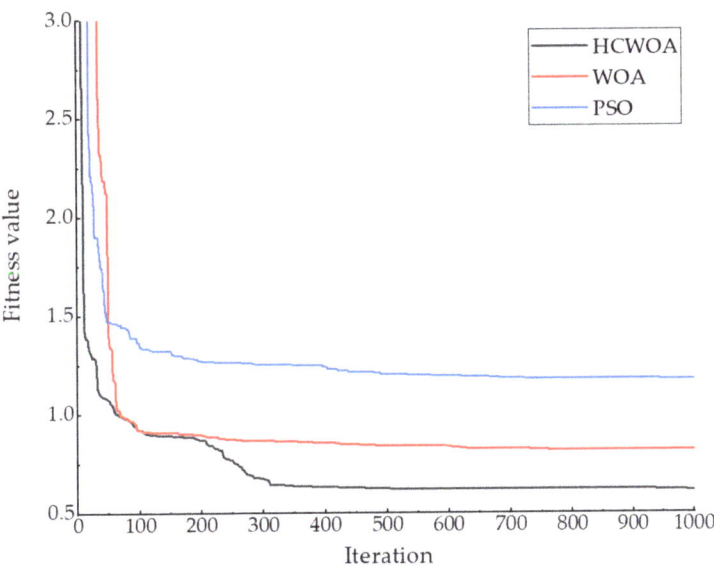

Figure 9. Evolution curves of the fitness value for minL.

Corresponding to the optimal fitness value of each optimization algorithm, the modified comparison matrices are generated automatically, and the results of consistency tests are gained as follows:

$$Y_{1,HCWOA} = \begin{pmatrix} 1 & 6.1514 & 6.1457 & 3.8134 & 3.9181 & 2.7247 \\ 0.1626 & 1 & 0.9832 & 0.2600 & 0.2600 & 0.3507 \\ 0.1627 & 1.0170 & 1 & 0.2600 & 0.2600 & 0.4333 \\ 0.2622 & 3.8462 & 3.8462 & 1 & 1.0591 & 0.3250 \\ 0.2552 & 3.8462 & 3.8462 & 0.9442 & 1 & 0.3250 \\ 0.3670 & 2.8514 & 2.3079 & 3.0769 & 3.0769 & 1 \end{pmatrix}$$

$CI_{Y_1,HCWOA} = 0.082 \quad RI_{Y_1,HCWOA} = 1.24 \quad CR_{Y_1,HCWOA} = 0.066 < 0.10$

$$Y_{1,WOA} = \begin{pmatrix} 1 & 6.1156 & 6.1009 & 4.1316 & 3.7958 & 2.8451 \\ 0.1635 & 1 & 1.1734 & 0.2056 & 0.2030 & 0.3003 \\ 0.1639 & 0.8522 & 1 & 0.2047 & 0.2355 & 0.3941 \\ 0.2420 & 4.8638 & 4.8852 & 1 & 0.9961 & 0.3250 \\ 0.2634 & 4.9261 & 4.2463 & 1.0039 & 1 & 0.2854 \\ 0.3515 & 3.3300 & 2.5374 & 3.0769 & 3.5039 & 1 \end{pmatrix}$$

$CI_{Y_1,WOA} = 0.105 \quad RI_{Y_1,WOA} = 1.24 \quad CR_{Y_1,WOA} = 0.085 < 0.10$

$$Y_{1,PSO} = \begin{pmatrix} 1 & 6.1623 & 6.1083 & 3.7820 & 2.8000 & 2.4107 \\ 0.1623 & 1 & 0.9884 & 0.2450 & 0.2087 & 0.3278 \\ 0.1637 & 1.0117 & 1 & 0.1982 & 0.2361 & 0.3471 \\ 0.2644 & 4.0816 & 5.0454 & 1 & 0.9020 & 0.2431 \\ 0.3571 & 4.7916 & 4.2355 & 1.1086 & 1 & 0.2459 \\ 0.4148 & 3.0506 & 2.8810 & 4.1135 & 4.0667 & 1 \end{pmatrix}$$

$CI_{Y_1,PSO} = 0.112 \quad RI_{Y_1,PSO} = 1.24 \quad CR_{Y_1,PSO} = 0.090 < 0.10$

In conclusion, the results of comparative experimental analysis are summarized and expressed in Table 15, where the superiority and effectiveness of the proposed HCWOA for solving minL and modifying inconsistent comparison matrix are verified by comparing with the standard WOA and the basic PSO. The HCWOA optimization algorithm with the best optimal fitness value proves efficient to reduce the consistency ratio of initial matrix to 60%, which is capable of assisting the initial matrix to pass the consistency test automatically and efficiently. Compared with the standard WOA, the HCWOA proposed in this research under the assistance of the Tent chaotic mapping strategy and the Lévy flight strategy proves capable of being equipped with excellent convergence accuracy, speed, and stability, even to handle a large-scale global optimization problem. Furthermore, the parameters μ and k in Equation (11) can be adjusted appropriately when the better consistency of the comparison matrix is required or more initial information of the matrix is necessary to be reserved.

Table 15. Result of comparative experimental analysis.

Item	Initial Matrix	Optimization Algorithm		
		HCWOA	WOA	PSO
Optimal fitness value	/	0.6131	0.8166	1.1767
Consistency ratio	0.110	0.066	0.085	0.090

6. Summary and Conclusions

In this research, a BECI technical condition assessment approach, which contains specific on-site inspection regulations with both qualitative and quantitative descriptions, is proposed, and an improved BECI replacement plan decision system is constructed under the assistance of the HCWOA. In the first stage of the decision process, on-site inspection was conducted on the target BECI, which were hierarchically classified according to various installation types, and the inspection index evaluation was performed to provide the basis of progressive assessment calculation. A BECI technical condition assessment software was developed based on the MATLAB platform by utilizing the proposed assessment approach and algorithm, where the automation of assessment could be realized and "whether or not to replace" was also determined. In the second stage, 12 kinds of design, construction, and management requirements were introduced as the criteria, where the hierarchy model for selecting the performance-based optimal replacement plan was constructed via AHP. The HCWOA was designed and employed to modify and handle the established mathematical model, where the optimal plan can be selected precisely. The case study demonstrates that the BECI replacement decision system improved by this research proves reasonable and feasible. Simultaneously, the comparative experiments verify the superiority and effectiveness of the developed HCWOA for modifying and solving the proposed mathematical problem, where the consistency ratio of initial matrix is capable of being reduced to 60% and pass the consistency test effectively. The improved decision system proves to be a reliable DM tool in the field of bridge O and M.

Based on the findings and compared with a traditional decision system, the improved BECI replacement decision system established in this article cannot only achieve the standardization of BECI on-site inspection but also attain the automations of both the technical condition assessment and the replacement plan selection. Accordingly, both the project managers and on-site inspectors are entitled to participate in the decision-making process directly, which reduces the O and M cost effectively and proves the most significant improvement. In conclusion, this research not only improves and automates the processes of DM system, but also emphasizes the centrality of people themselves. The efficiency improvement and resource conservation of DM are pursued for those who are the decision makers themselves; hence, every member in the system has the opportunity to be empowered and motivated to make decisions.

Notably, only the performances of BECI are considered in the process of replacement plan selection, while the economic cost during the operation period is excluded. Consequently, in order to enhance the integrity of improved system, further research should be conducted on the selection of economic-based optimal replacement plan of BECI. In addition, the forecasting of traffic and the environmental elements such as weather and temperature, which could be variable in the near future, should also be taken into consideration as crucial indices in the DM model, where the sensors arranged inside the construction could help in data monitoring and collection. Furthermore, the developed HCWOA is only performed to handle the mathematical model in this research, which could also be extended to higher-dimensional or multi-objective optimization problems.

Author Contributions: Conceptualization, Z.X. and M.H.; methodology, Z.X. and M.H.; software, Z.X. and M.H.; validation, M.H.; formal analysis, Z.X.; investigation, Z.X. and M.H.; resources, M.H.; data curation, Z.X.; writing—original draft preparation, Z.X.; writing—review and editing, M.H. and Z.X.; visualization, Z.X. and M.H.; supervision, M.H.; project administration, M.H.; funding acquisition, M.H. and Z.X. Both authors have read and agreed to the published version of the manuscript.

Funding: The paper is funded by the 2017 Transportation Technological Progress Plan Project of Anhui Province, China (Rapid replacement techniques of expansion and contraction installations for highway bridges) and the Graduate Innovative Fund of Wuhan Institute of Technology (CX2020113).

Institutional Review Board Statement: Not applicable.

Informed Consent Statement: Not applicable.

Data Availability Statement: The data used to support the findings of this study are available from the corresponding author upon request.

Conflicts of Interest: The authors declare that there is no conflict of interest regarding the publication.

References

1. Ministry of Transport of the People's Republic of China (MOT). *General Technical Requirements of Expansion and Contraction Installation for Highway Bridge*; China Communications Press: Beijing, China, 2016. (In Chinese)
2. Huang, M.; Xu, Z.; Li, L.; Lei, Y. Construction and Application of Bridge Expansion and Contraction Installation Replacement Decision System Based on the Analytic Hierarchy Process. *Materials* **2020**, *13*, 4177. [CrossRef]
3. Saaty, T.L. How to make a decision: The Analytic Hierarchy Process. *Eur. J. Oper. Res.* **1990**, *48*, 9–26. [CrossRef]
4. Harputlugil, T.; Prins, M.A.; Topcu, I. Conceptual framework for potential implementations of multi criteria decision making (MCDM) methods for design quality assessment. In Proceedings of the Management and Innovation for a Sustainable Built Environment, CIB International Conference, Amsterdam, The Netherlands, 20–23 June 2011.
5. Vidal, L.A.; Marle, F.; Bocquet, J.C. Using a Delphi process and the Analytic Hierarchy Process (AHP) to evaluate the complexity of projects. *Expert Syst. Appl.* **2011**, *38*, 5388–5405. [CrossRef]
6. Liu, X.; Pan, Y.; Xu, Y.; Yu, S. Least square completion and inconsistency repair methods for additively consistent fuzzy preference relations. *Fuzzy Set. Syst.* **2012**, *198*, 1–19. [CrossRef]
7. Ahmed, F.; Kilic, K. Fuzzy Analytic Hierarchy Process: A performance analysis of various algorithms. *Fuzzy Set. Syst.* **2019**, *362*, 110–128. [CrossRef]
8. Wang, H.; Kou, G.; Peng, Y. An Iterative Algorithm to Derive Priority from Large-Scale Sparse Pairwise Comparison Matrix. *IEEE Trans. Syst. Man Cybern. Syst.* **2021**. [CrossRef]
9. Yang, I.; Wang, W.; Yang, T. Automatic repair of inconsistent pairwise weighting matrices in analytic hierarchy process. *Automat. Constr.* **2012**, *22*, 290–297. [CrossRef]
10. Lin, C.; Wang, W.; Yu, W. Improving AHP for construction with an adaptive AHP approach (A^3). *Automat. Constr.* **2008**, *17*, 180–187. [CrossRef]
11. Girsang, A.S.; Tsai, C.; Yang, C. Ant algorithm for modifying an inconsistent pairwise weighting matrix in an analytic hierarchy process. *Neural Comput. Appl.* **2015**, *26*, 313–327. [CrossRef]
12. Huang, M.; Cheng, S.; Zhang, H.; Mustafa, G.; Lu, H. Structural Damage Identification under Temperature Variations Based on PSO–CS Hybrid Algorithm. *Int. J. Struct. Stab. Dyn.* **2019**, *19*, 1950139. [CrossRef]
13. Huang, M.; Li, X.; Lei, Y.; Gu, J. Structural damage identification based on modal frequency strain energy assurance criterion and flexibility using enhanced Moth-Flame optimization. *Structures* **2020**, *28*, 1119–1136. [CrossRef]
14. Huang, M.; Mustafa, G.; Zhu, H. Vibration-Based Structural Damage Identification under Varying Temperature Effects. *J. Aerosp. Eng.* **2018**, *31*, 04018014. [CrossRef]

15. Mirjalili, S.; Lewis, A. The Whale Optimization Algorithm. *Adv. Eng. Softw.* **2016**, *95*, 51–67. [CrossRef]
16. Gharehchopogh, F.S.; Gholizadeh, H. A comprehensive survey: Whale Optimization Algorithm and its applications. *Swarm. Evol. Comput.* **2019**, *48*, 1–24. [CrossRef]
17. Xiong, G.; Zhang, J.; Shi, D.; He, Y. Parameter extraction of solar photovoltaic models using an improved whale optimization algorithm. *Energ. Convers. Manag.* **2018**, *174*, 388–405. [CrossRef]
18. Moodi, Y.; Mousavi, S.R.; Ghavidel, A.; Sohrabi, M.R.; Rashki, M. Using Response Surface Methodology and providing a modified model using whale algorithm for estimating the compressive strength of columns confined with FRP sheets. *Constr. Build. Mater.* **2018**, *183*, 163–170. [CrossRef]
19. Mafarjaa, M.M.; Mirjalili, S. Hybrid Whale Optimization Algorithm with simulated annealing for feature selection. *Neurocomputing* **2017**, *260*, 302–312. [CrossRef]
20. Reddy, M.P.; Babu, M.R. Implementing self adaptiveness in whale optimization for cluster head section in Internet of Things. *Cluster. Comput.* **2019**, *22*, 1361–1372. [CrossRef]
21. Aziz, M.A.; Ewees, A.A.; Hassaniend, A.E. Whale Optimization Algorithm and Moth-Flame Optimization for multilevel thresholding image segmentation. *Expert Syst. Appl.* **2017**, *83*, 242–256. [CrossRef]
22. Srivastava, A.; Das, D.K.; Rai, A.; Raj, R. Parameter Estimation of a Permanent Magnet Synchronous Motor using Whale Optimization Algorithm. In Proceedings of the 2018 Recent Advances on Engineering, Technology and Computational Sciences (RAETCS), Allahabad, India, 6–8 February 2018; pp. 1–6.
23. Zhang, X.; Liu, Z.; Miao, Q.; Wang, L. Bearing fault diagnosis using a whale optimization algorithm-optimized orthogonal matching pursuit with a combined time–frequency atom dictionary. *Mech. Syst. Signal Process.* **2018**, *107*, 29–42. [CrossRef]
24. Dao, T.; Pan, T.; Pan, J. A Multi-Objective Optimal Mobile Robot Path Planning Based on Whale Optimization Algorithm. In Proceedings of the 2016 IEEE 13th International Conference on Signal Processing (ICSP), Chengdu, China, 6–10 November 2016; pp. 337–342.
25. Horng, M.; Dao, T.; Shieh, C.; Nguyen, T. A Multi-Objective Optimal Vehicle Fuel Consumption Based on Whale Optimization Algorithm. In *Advances in Intelligent Information Hiding and Multimedia Signal Processing*; Springer: Cham, Switzerland, 2017; pp. 371–380.
26. Khalilpourazari, S.; Pasandideh, S.H.; Ghodratnama, A. Robust possibilistic programming for multi-item EOQ model with defective supply batches: Whale Optimization and Water Cycle Algorithms. *Neural Comput. Appl.* **2019**, *31*, 6587–6614. [CrossRef]
27. Sreenu, K.; Sreelatha, M. W-Scheduler: Whale optimization for task scheduling in cloud computing. *Cluster Comput.* **2019**, *22*, 1087–1098. [CrossRef]
28. Huang, M.; Cheng, X.; Lei, Y. Structural damage identification based on substructure method and improved whale optimization algorithm. *J. Civ. Struct. Health Monit.* **2021**, *11*, 351–380. [CrossRef]
29. Kaur, G.; Arora, S. Chaotic Whale Optimization Algorithm. *J. Comput. Des. Eng.* **2018**, *5*, 275–284. [CrossRef]
30. Yang, X.S.; Suash, D. Cuckoo search via levy flights. In *World Congress Nature Biologically Inspired Computer (NaBIC)*; IEEE Publication: New York, NY, USA, 2009; pp. 210–214.
31. Kamaruzaman, A.F.; Zain, A.M.; Yusuf, S.M.; Udin, A. Levy flight algorithm for optimization problems—A literature review. *Appl. Mech. Mater.* **2013**, *421*, 496–501. [CrossRef]
32. Abdel-Basset, M.; El-Shahat, D.; Sangaiah, A.K. A modified nature inspired meta-heuristic whale optimization algorithm for solving 0-1 knapsack problem. *Int. J. Mach. Learn. Cyb.* **2019**, *10*, 495–514. [CrossRef]
33. Abdel-Basset, M.; Abdle-Fatah, L.; Sangaiah, A.K. An improved Lévy based whale optimization algorithm for bandwidth-efficient virtual machine placement in cloud computing environment. *Cluster Comput.* **2019**, *22*, 8319–8334. [CrossRef]
34. Sun, Y.; Wang, X.; Chen, Y.; Liu, Z. A modified whale optimization algorithm for large-scale global optimization problems. *Expert Syst. Appl.* **2018**, *114*, 563–577. [CrossRef]
35. Ministry of Transport of the People's Republic of China (MOT). *Highway Performance Assessment Standards*; China Communications Press: Beijing, China, 2018. (In Chinese)
36. Ministry of Transport of the People's Republic of China (MOT). *Standards for Technical Condition Evaluation of Highway Bridges*; China Communications Press: Beijing, China, 2011. (In Chinese)
37. Ocampoa, L.; Ebisab, J.A.; Ombeb, J.; Escotob, M.G. Sustainable ecotourism indicators with fuzzy Delphi method—A Philippine perspective. *Ecol. Indic.* **2018**, *93*, 874–888. [CrossRef]
38. Cao, W.D.; Wang, A.T.; Yu, D.S.; Liu, S.T.; Hou, W. Establishment and implementation of an asphalt pavement recycling decision system based on the analytic hierarchy process. *Resour. Conserv. Recycl.* **2019**, *149*, 738–749. [CrossRef]
39. Gandomi, A.H.; Yang, X.-S.; Talatahari, S.; Alavi, A.H. Firefly algorithm with chaos. *Commun. Nonlinear Sci.* **2013**, *18*, 89–98. [CrossRef]
40. Liu, M.; Yao, X.; Li, Y. Hybrid whale optimization algorithm enhanced with Lévy flight and differential evolution for job shop scheduling problems. *Appl. Soft Comput.* **2019**, *87*, 105954. [CrossRef]
41. Mantegna, R.N. Fast, accurate algorithm for numerical simulation of Lévy stable stochastic processes. *Phys. Rev. E* **1994**, *49*, 4677. [CrossRef] [PubMed]

Article
Modelling of Decision Processes in Construction Activity

Elżbieta Szafranko * and Jolanta Harasymiuk

Institute Geodesy and Civil Engineering, Faculty of Geoengineering, University of Warmia and Mazury in Olsztyn, 10-719 Olsztyn, Poland; jolanta.harasymiuk@uwm.edu.pl
* Correspondence: elasz@uwm.edu.pl; Tel.: +48-89-523-47-18

Citation: Szafranko, E.; Harasymiuk, J. Modelling of Decision Processes in Construction Activity. *Appl. Sci.* **2022**, *12*, 3797. https://doi.org/10.3390/app12083797

Academic Editors: Asterios Bakolas and Muhammad Junaid Munir

Received: 17 February 2022
Accepted: 7 April 2022
Published: 9 April 2022

Publisher's Note: MDPI stays neutral with regard to jurisdictional claims in published maps and institutional affiliations.

Copyright: © 2022 by the authors. Licensee MDPI, Basel, Switzerland. This article is an open access article distributed under the terms and conditions of the Creative Commons Attribution (CC BY) license (https:// creativecommons.org/licenses/by/ 4.0/).

Abstract: Construction activity with a huge variety of structures, forms and conditions underlying the implementation of construction projects, require special management approach. Decisions are most often made at the planning and preparation stage of a construction project. The literature on the subject includes descriptions of decision support methods and models, including single-criteria and multi-criteria models, operations research and fuzzy models. Different models can be used in different situations. The article contains an analysis of model approaches proposed in the literature, confronted with decision-making processes in engineering practice. The study covered 34 construction projects and 15 companies operating in the construction industry. Several decision situations have been considered. The research carried out in accordance with the seven-stage research process has shown that although the various methods proposed in the cited sources can be used in the implementation of engineering projects, they require modification to suit the specificity of engineering practice. The results of the research are the decision support models proposed by the authors, adapted to the conditions in which construction projects are implemented. In the case of small and relatively simple construction projects, simplified models are usually used, where the use of the last steps of verifying the results and improving the applied model is limited. Large and more complex construction projects were often accompanied by a decision support system consisting of more stages, and in these cases, it turned out to be important to obtain feedback and to refine the decision model accordingly. Research has shown that in large projects it is important to obtain feedback. This is due to, inter alia, from the fact that the implementation of these projects involves much greater financial resources than in small and medium-sized projects. Decision-makers take much more care to verify the correctness of the model, because the effects of wrongly made decisions can be much more severe than in the case of small and medium-sized enterprises. If it is necessary to make strategic decisions related to the future of a given company, attention was paid to models in which the starting point was to clearly define the goal and collect a complete set of information about the decision-making environment. Various analytical and research methods were used, but feedback was always needed to improve the final solution. The observations obtained during the research helped the authors to develop decision support models dedicated to engineering practice that may be useful in the implementation of construction projects.

Keywords: decision-making process; decision modelling in construction activities; decisions in civil engineering

1. Introduction

There are many decision support systems used in business activities, but the very process of making a decision consists of a few stages. The literature provides information about various approach methods and models, which differ in the process of data collection and preparation, the stage of analysing the problem, or the final stage associated with the actual making of a decision [1,2]. There are models including three, eight or even as many as 11 steps in a decision-making process. Different approaches are used depending on the type of business, the decision being made, and the circumstances surrounding the decision. A decision-making process made in the well-recognised context, with a large amount of data, will look differently from one occurring in an uncertain situation. Different forms of business activities are characterised by different degrees of complexity of the executed processes, and decisions made under different conditions may affect different situations [3–5].

Construction activities are related to the implementation of various facilities and construction works. When we speak of the construction industry, we mean, on the one hand, the implementation of construction projects and, on the other hand, the functioning of enterprises and institutions involved in their implementation. The construction activity has some specific features. First of all, the vast majority of production is carried out outdoors and many contractors participate in the execution of works. The construction object is large compared to the products of other sectors of the economy, requires large investments and the benefits are deferred in time. Objects are implemented according to individual projects, in unique conditions with the participation of many industries. The large variety of civil structures also deserves attention. Starting from single and multi-family residential buildings, through public utility buildings for various purposes, to infrastructure structures such as roads, bridges, transmission lines and water structures. The construction process consists of several stages. At each of them, there are various participants in the involvement in the creation of the civil structures. Decisions on the works undertaken are mostly made in the planning and design phase under conditions of uncertainty with limited amounts of data. In many situations, we have to assess possible solutions and choose the variant that best meets the expectations related to the future construction (e.g., location, design concepts, material and technological solutions, preparation of tender offers, etc.) [6–8].

Due to the complexity of situations in which decisions in construction activity are made, the decision-making process can be supported with mathematical methods, systems and models, which allow the user to impose some order over the process and facilitate the achievement of the aim, which is to arrive at a decision [9,10].

As it is difficult to find methods and approaches dedicated to construction activities in the literature, the authors of this article attempted to examine the decision-making processes in 34 implemented projects and 15 construction companies. As a result of the research, the models used in construction activities were specified in the original, proprietary approach. Research methods were used based on the analysis of source materials, questionnaire research, interviews and in-depth interviews

The article in Section 2 presents a literature review taking into account the issues of decision-making in engineering activities, various decision models are presented, and the review is summarized in tables. Section 3 presents the methodology, details and results of the research. The next Section 4 contains a discussion of the results, and Section 5 presents a summary and conclusions.

2. Decisions—A Review of the Literature

A decision means selecting one of the available options [11]. If there are no options, there is no decision to make either. There are several methods which can lead to choosing the right solution to a problem. An example is a quantitative decision-making technique. Its underlying principle is to rely exclusively on data, which are analysed in order to generate facts. If the number of facts that justify taking a given action outweigh facts supporting other possible solutions, it becomes the decisive factor. However, this perception of a decision-making process excludes any subjective evaluation, intuition, premonition or bias, while forcing one to present the justification of their decision, which entails the demonstration of all underlying presumptions, restrictions or limitations [9].

A different situation emerges when a problem is being described with qualitative parameters. Qualitative methods apply to cases composed of hardly measurable phenomena, which are compared via descriptive analysis of different parameters. Assessment scales adopted in this approach often raise doubts, and their application needs to adhere to detailed guidelines.

It is advisable to follow previously defined principles while making decisions. The absence of such guidelines may cause chaos and eventually lead to an erroneous perception of the problem and incorrect solution. Decisions are always subject to certain limitations, due to preceding events. The main criteria, which can be evident or hidden, can be grouped as a set of aims, a system of priorities, a course of alternative measures, consequences of each alternative solution, and a system of selection criteria [12].

2.1. Decisions and Their Structure

A decision problem is defined through decision variants and assessment criteria. According to literature sources [13,14] decision problems can be divided into problems with single-layer, hierarchical and fuzzy structures. A decision system is often a single-layer system with a closed decision problem [14,15]. Figure 1 illustrates a model of the first-type structure. D_i are decision areas, d_{ij} are elementary decisions in particular areas, where $i = \{1, 2, \ldots, n\}$ and $j = \{1, 2, \ldots, m\}$. Final decisions eliminate one another within the same decision area.

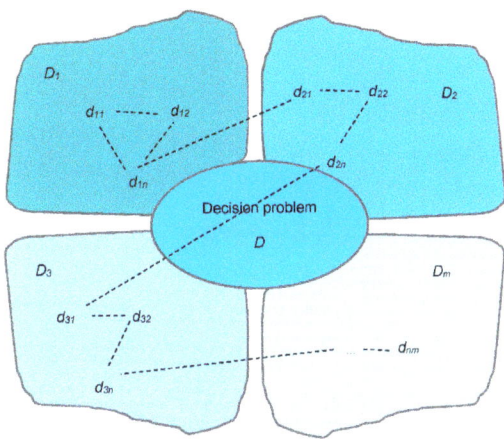

Figure 1. Model of a flat structure decision process.

In most cases, this structure does not reflect a decision situation because decisions are taken in many areas, which means there are mutual relationships between decision problems D_i. In many situations, these relationships may contribute to the quality of the final solution [16–19] Dependences between elementary decisions are depicted by a hierarchical structure of a decision problem, in which no sooner that all decision criteria of the lower-level K_{nm} have been considered and solved can higher-level decisions be made and the final choice of decision variant V_i made (Figure 2).

The third group of decision-making situations consists of fuzzy models, where problems are shown as dependent on links and relations between different-level decision makers [20,21]. The mentioned relations include cooperation, coordination, or competition. Such relations differ in terms of information transmitted between the system's constituents, and with respect to set boundaries of effects produced by particular decision criteria. Cooperation necessitates sharing information, while competition is associated with information non-linearity and disproportions [22,23]. The shared characteristic for both forms of relations is non-linearity and non-continuity of decision processes (Figure 3).

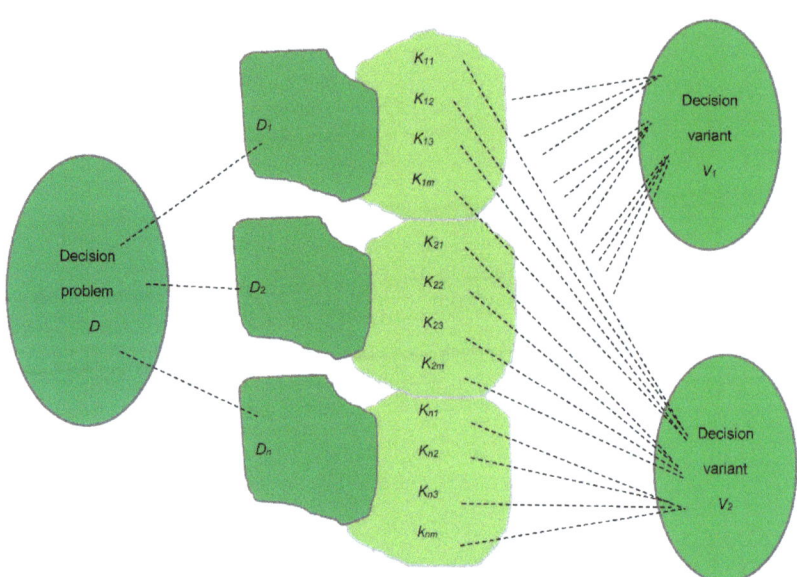

Figure 2. Model of a decision-making process with a hierarchical structure.

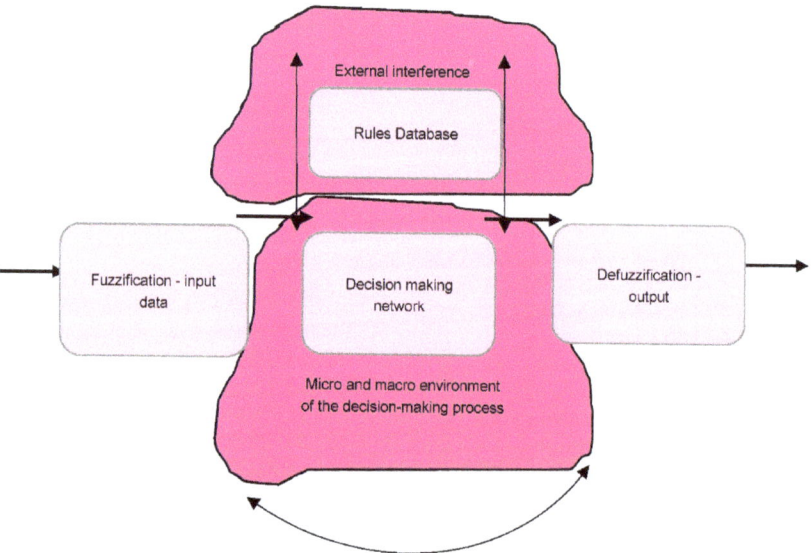

Figure 3. Model of a fuzzy decision process.

2.2. A Decision Process

Business enterprises operate in an increasingly difficult environment, as the competition intensifies and markets are being made devoid of rules or principles. Big corporations more and more often dictate conditions in which businesses are run, both locally and globally. This tendency appears in all sectors, including the civil engineering segment [24,25]. A decision-making process is undergoing significant changes due to progressing globalisation. At all stages illustrated schematically in Figure 4, it is recommended to refer continually to the ever changing nearer and further environment. Increasingly often, the cutting edge information technologies, seen as tools of globalisation, are a source of information and datasets supporting the process of arriving at a decision [26–28]. Making a decision is a process that encompasses such activities as expressing the need for a decision, collecting and processing data to support the decision, measuring the outcomes and finally evaluating the execution of the chosen option and the extent to which it has satisfied the assessment criteria established at the onset of the process (Figure 4).

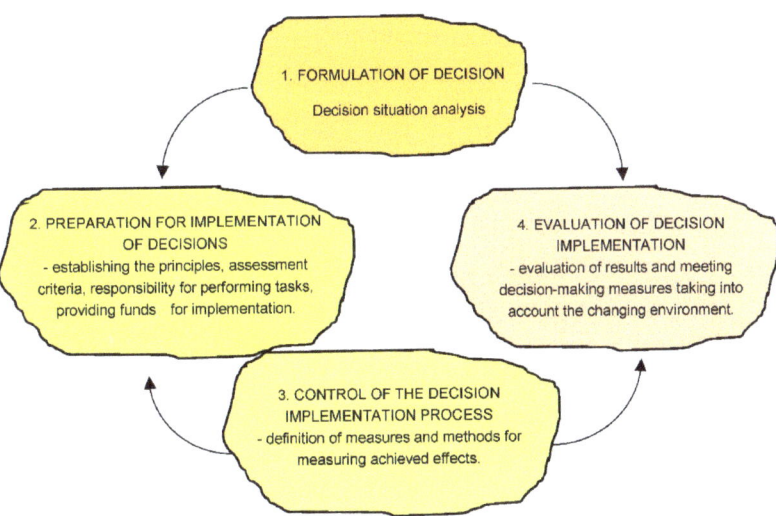

Figure 4. The process of making and implementing a decision.

The factors that polarise conditions under which decisions are made include: decoupling of global demand from the purchasing power of local markets, narrowing of the investment horizon as a consequence of competitive struggle, and enterprises becoming more market-driven [23,29,30]. All these circumstances cause growing uncertainty, evoked by difficulty in predicting market changes and, on the other hand, by greater propensity to face investment risk in order to earn profit. In a situation where global competition is experienced, there is a tendency towards making uncontrollable decisions as well as taking steps that lead to changes in relationships between market participants. It is more and more common to observe that negative consequences of erroneous decisions are shifted over contractors and subcontractors of development projects [23,29,30].

In every process of making a decision there are decision makers, a set of feasible solutions, a set of environmental components [31], the function of the usefulness of a solution [32] and uncertainty with respect to conditions in which actions will be pursued [4]. A decision maker is a person who is responsible for the consequences of the solution eventually accepted. A set of feasible solutions should not violate certain limits and ought to be doable. This set must not be empty and needs to contain at least two elements. The set of environmental components describes the conditions in which the decision-making process occurs, and which either directly or indirectly affect the decisions that are made [22].

The objective function is a way to assign numerical values to the values of a solution being searched for. If usefulness can be described numerically, attaining high accuracy, this is a suitable criterion for making a decision. By assigning certain contractual values to decisions, we reduce the decision-making process to a situation where the option with the highest usefulness is chosen. If a decision maker is unable to evaluate the probability of the occurrence of particular events, the external environment is said to be uncertain.

2.3. Models of a Decision-Making Process

The approaches used in a decision-making process depend on the type of a problem. The most popular methods are: linear programming, dynamic programming, integer programming, the Bayes' decision theorem, game theory, and probability theory. These methods belong to the quantitative research domain. As events are described through sets of norms, principles and algorithms, the above methods are classified as prescriptive models. However, prescriptive models may not always offer satisfying solutions. Due to some subjective circumstances, such as qualifications and skills of decision makers, as well as objective ones, arising from external and internal conditions, prescriptive models do not always work well. In certain situations, descriptive models (describing and explaining the situation) are more useful. An analysis is then conducted starting from single cases up to generalised conclusions.

Depending on an analysed situation, different decision support models are applied. The subsequent part of this article will present issues connected with the application of prescriptive models. The models are described in literature (Table 1).

Table 1. Models of a decision-making process—literature review.

No.	Subject	Authors	Publications, Titles
1.	Monocriterial Models (Monolayer)	Bolesta-Kukułka K. [33]	Managerial decisions in management theory and practice (pl). Scientific Publishers of the Faculty of Management at the University of Warsaw (2000).
		Berredo, R. C., Cruz, E. C., Ekel, P. Y., Junges, M. F. D., Contijo, M. M., Pereira Jr, J. G., and Popov, V. A. [34]	Monocriteria and multicriteria optimization of network configuration in distribution systems. WSEAS Int. Conference on Power Engineering Systems (2005).
		Kasharin, D. V. [35]	Intelligent decision support systems in the design of mobile micro hydropower plants and their engineering protection. In Proceedings of the First International Scientific Conference "Intelligent Information Technologies for Industry" (IITI'16) Springer, Cham, (2016).
2.	Multi-criterial Models	Opricovic, S., and Tzeng, G. H. [36]	Defuzzification within a multicriteria decision model. International Journal of Uncertainty, Fuzziness and Knowledge-Based Systems, (2003).
		Barker, T. J., and Zabinsky, Z. B. [37]	A multicriteria decision making model for reverse logistics using analytical hierarchy process. Omega, (2011).
		Cheng, M. Y., Hsiang, C. C., Tsai, H. C., and Do, H. L. [38]	Bidding decision making for construction company using a multi-criteria prospect model. Journal of Civil Engineering and Management, (2011).
3.	Sutherland's Model	Sutherland, J. W. [39]	Administrative decision-making: Extending the bounds of rationality. New York: Van Nostrand Reinhold (1977).
		Guerry, A. D., Ruckelshaus, M. H., Arkema, K. K., Bernhardt, J. R., Guannel, G., Kim, C. K., and Wood, S. A. [29]	Modeling benefits from nature: using ecosystem services to inform coastal and marine spatial planning. International Journal of Biodiversity Science, Ecosystem Services and Management, (2012).
		Chatterjee, P., Banerjee, A., Mondal, S., Boral, S., and Chakraborty, S. [40]	Development of a hybrid meta-model for material selection using design of experiments and EDAS method. Engineering Transactions, (2018).
4.	Holt's Model	Hutchins, M. J., and Sutherland, J. W. [41]	An exploration of measures of social sustainability and their application to supply chain decisions. Journal of cleaner production, (2008).
		Clithero, J. A. [42]	Improving out-of-sample predictions using response times and a model of the decision process. Journal of Economic Behavior and Organization, (2018).
		Karimi, S., Papamichail, K. N., and Holland, C. P. [43]	The effect of prior knowledge and decision-making style on the online purchase decision-making process: A typology of consumer shopping behavior. Decision Support Systems, (2015).
		Zhang, X., Wu, Y., Shen, L., and Skitmore, M. [44]	A prototype system dynamic model for assessing the sustainability of construction projects. International Journal of Project Management, (2014).
5	Operational Research Model	Tamošaitiene, J., Bartkiene, L., and Vilutiene, T. [30]	The new development trend of operational research in civil engineering and sustainable development as a result of collaboration between German-Lithuanian-Polish scientific triangle. Journal of Business Economics and Management, (2010).
		Turskis, Z., Gajzler, M., and Dziadosz, A. [45]	Reliability, risk management, and contingency of construction processes and projects. Journal of Civil Engineering and Management, (2012).
		Vukomanovic, M., and Radujkovic, M. [46]	The balanced scorecard and EFQM working together in a performance management framework in construction industry. Journal of Civil Engineering and Management, (2013).

Table 1. *Cont.*

No.	Subject	Authors	Publications, Titles
6.	Cybernetic Model	Bozeman, D. P., and Kacmar, K. M. [47]	A cybernetic model of impression management processes in organizations. Organizational behavior and human decision processes, (1997).
		Cheng, M. Y., and Roy, A. F. [28]	Evolutionary fuzzy decision model for construction management using support vector machine. Expert Systems with Applications, (2010).
		Mohammadi, F., Sadi, M. K., Nateghi, F., Abdullah, A., and Skitmore, M. [48]	A hybrid quality function deployment and cybernetic analytic network process model for project manager selection. Journal of Civil Engineering and Management, (2014).
7.	Fuzzy Data Model	Adeli, H. [49]	Neural networks in civil engineering. Civil and Infrastructure Engineering, (2001).
		Kazimieras Zavadskas, E., Antucheviciene, J., Adeli, H., and Turskis, Z. [50]	Hybrid multiple criteria decision making methods: A review of applications in engineering. Scientia Iranica, (2016).
		Antucheviciene, J., Kala, Z., Marzouk, M., and Vaidogas, E. R. [51]	Solving civil engineering problems by means of fuzzy and stochastic MCDM methods: current state and future research. Mathematical Problems in Engineering, (2015).

The authors who present decision models suggest different procedures, referred in the literature as decision models. Table 2 presents models that are used in the broadly understood engineering activity and in scientific research related to this field.

Table 2. Models of decision-making processes.

I. Single-Criterion Models:	II. Multiple-Criteria Models:
1. Discovering the difficulty. 2. Identifying the problem. 3. Determining the criterion applied to evaluate problem solution variants. 4. Setting a list of solutions. 5. Describing effects of the implementation of each solution. 6. Selecting the best solution. 7. Implementing the decision.	1. Identifying the problem. 2. Identifying decision criteria. 3. Assigning weights to criteria. 4. Elaborating alternative solutions. 5. Evaluating alternative solutions. 6. Selecting the best solution. 7. Implementing the chosen solution. 8. Evaluating the efficiency of the decision implemented.
III. Sutherland's model:	**IV. Holt's model:**
1. The need to make a decision (goal). 2. Primary information (opinions, theories). 3. Empirical studies. 4. Building a model. 5. Generating solutions. 6. Selecting criteria for evaluation. 7. Evaluation of variants. 8. Selection of the solution. 9. Making a decision. 10. Implementation. 11. Feedback to correct the model.	1. Identification of the problem. 2. Analysis of the context. 3. Definition of the problem. 4. Elaboration of solutions. 5. Evaluation of variant solutions. 6. Selection of a solution. 7. Implementation. 8. Evaluation of effects.
V. Model based on operational studies:	**VI. Cybernetic decision model:**
1. Building a model (describing the situation with the mathematical language). 2. Solving the problem presented in the form of a mathematical model. 3. Verification of the model—possible corrections. 4. Monitoring—feedback and correction of the decision made.	1. Input—primary, raw information. 2. Transformation—a decision-making process. 3. Output—secondary information in the form of a decision.
VII. Fuzzy data model:	
1. Data collection stage-input signals 2. The fuzzification stage 3. The stage of fuzzy inference	1. The stage of defuzzification 2. Making a decision

The Table 2 shows two major groups of models—one where the process begins by identifying the problem, and the other where the procedure is based on building a decision model. The first group includes methods which take advantage of single-criterion and multiple-criteria approaches whose aim is to make an assessment of several variants or alternative solutions. In these models, the assessment leads to the identification of the best solution, its implementation and possibly the evaluation of the effects attained.

The second group consists of methods based on building a mathematical model, which through the mathematical language presents the decision situations described previously. These methods require the user to prepare primary information, to identify the goal and the measure that will be applied to determine to what degree this aim is reached. The final stage provides feedback that enables the user to correct the model or decision if necessary.

The models presented in the table differ from one another in both the approach to the problem being solved and the subsequent procedures, but in each case the completed procedure leads to the solution of a problem. Different approaches may apply to different decision situations, in different fields of business activity.

An analysis of the literature related to decision making and research carried out in construction companies and during erection of civil structures revealed the lack of decision models adapted to the specificity of construction activities. The article undertakes research on source materials in order to find an answer to the question of how the decision-making processes in construction look similar to in practice, what stages they consist of and whether they fit into the models described in the literature, and as a result of the research, models of decision-making processes taking into account the specificity of this industry were proposed.

The purpose of this article is to develop a decision support model dedicated to the construction activity.

3. Research Methodology, Course and Results

3.1. Research Methods

Developers, investors and construction companies were questioned about decision-making issues in order to identify decision problems which appear in construction activity. From among all companies based in north-eastern Poland, those that have been involved in the construction of facilities representing very diverse types of engineering structures over the last five years have been selected. The preliminary selection was accomplished on the basis of information obtained in offices of district and provincial authorities, including issued building permits.

The first part of the interview concerned determining who and when made decisions. Groups of decision-makers related to the projects were established, and then at which stage these decisions were made (preparation and planning stage, design stage, execution and completion stage). The frequency of making decisions in subsequent stages and in specific decision groups was also examined. The main problem that has been studied is the way of making decisions including preparation stages (e.g., problem identification, context analysis, collecting output data), creating a decision model, defining criteria or sending feedback to correct the assumptions underlying the decision.

The subsequent stage in the research consisted of contacting the selected respondents and having preliminary interviews in order to find out who and how made decisions regarding the ongoing development projects. The character, type and frequency of decisions made were also investigated. These preliminary interviews allowed us to distinguish a group of 34 civil structures under construction. This research stage employed the following methods: an analysis of information sources, probes, surveys, and in-depth interviews. The study covered the years 2015–2019.

The analysed objects were divided into small, medium and large construction projects. The classification of construction objects to these groups takes into account the nature of the object and looks different for road, bridge and environmental protection structures. On the one hand, the size and complexity of the facility is evidenced by the number of subcontractors and, as a rule, there are about 5–6 of them for small facilities, about 10–12 medium ones and more than 14. However, due to the nature of the facilities and the degree of compilation, there are deviations. The value and volume of works was assessed based on the opinions of contractors and investors. In order to obtain their opinion, a list of parameters determining the classification of the construction to a specific group was prepared. These are, among others the size and complexity of the structure and related construction works, the size of the construction site, including the number and size of auxiliary factories, the distances between important points on the construction site, the complexity of the processes carried out during the construction of facilities. Most of the components of the analysis are immeasurable phenomena, therefore the linguistic and descriptive assessment methods were used, which, based on the description of the phenomenon, allow them to be assessed on the adopted 5-point scale. The scores were summed up and assigned to one of the three size groups for the final grade.

The research, the purpose of which was to develop decision models tailored to the specificity of construction activities, was carried out in 7 stages, including previously planned activities. The diagram of the research process presenting the research methodology is presented in Figure 5.

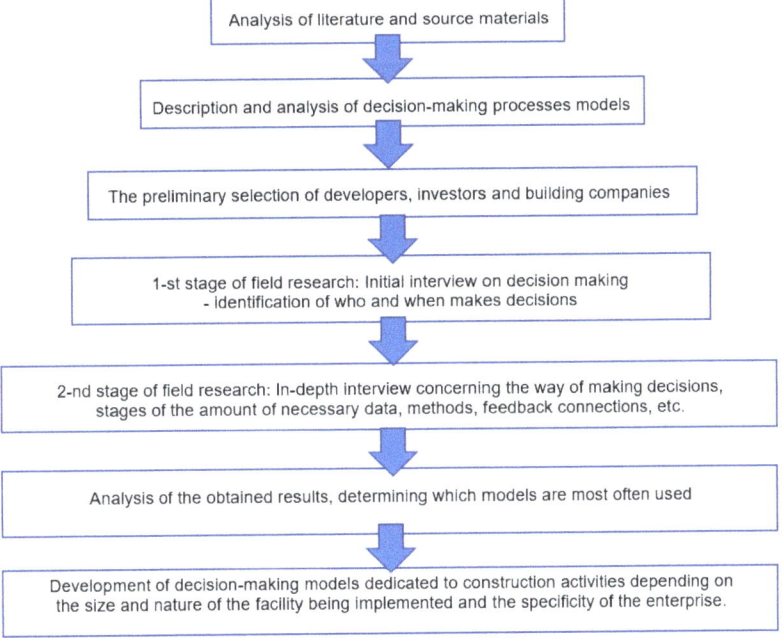

Figure 5. Scheme of research process.

3.2. The Course of the Research

The research has shed light on what problems are most often solved in a systematic manner. First and foremost, these are analyses of variant locations of different development projects. In such cases, decision makers most often use methods based on a multiple-criteria analysis, although the number of steps included may differ slightly, largely depending on the size and complexity of a construction project. Another example of a situation where alternative solutions need to be analysed and selected is a tender, in which the bids are assessed according to previously determined parameters. In a few cases, an analysis of several variants was applied to building material and construction solutions.

Detailed data collected during the study and pertaining to decisions made with respect to possible variants of erection of civil structures are presented in Table 3.

Table 3. Specification of the information gathered about decision models applied to select variants (Erection of civil structures).

No.	Type of Investment (Object Function)	Quantity	No. Object	Size * (S-Small, M-Medium, L-Large)	Number of Sub-Contractors	Applied Variant Assessment Model	Number of Stages
1	Road objects	15	1.1.	S	5	II	7
			1.2.	S	6	II	8
			1.3.	L	12	IV	9
			1.4.	L	15	III	10
			1.5.	M	10	II	7
			1.6.	M	12	II	9
			1.7.	M	10	II	9
			1.8.	M	16	II	8
			1.9.	L	20	IV	9
			1.10.	M	11	IV	8
			1.11.	M	12	II	8
			1.12.	L	16	III	10
			1.13.	S	6	II	7
			1.14.	M	10	II	9
			1.15.	M	14	II	9
2	Bridge structures	5	2.1.	S	5	II	7
			2.2.	S	5	I	7
			2.3.	L	8	III	10
			2.4.	S	4	II	7
			2.5	S	6	II	7
3	Facilities related to environmental protection	7	3.1.	M	10	II	8
			3.2.	M	12	IV	8
			3.3.	S	8	II	8
			3.4.	S	7	II	8
			3.5.	M	10	II	9
			3.6.	L	12	III	10
			3.7.	S	5	II	7
4	Sports facilities	1	4.1.	M	21	II	8
5	School buildings	3	5.1.	S	8	II	7
			5.2.	M	10	II	9
			5.3.	S	9	II	8
6	Healthcare	1	6.1	S	14	II	8
7	Housing estates	2	7.1.	M	12	II	8
			7.2.	S	8	II	8

* dimension was assessed considering the size of a civil structure, the whole construction site, the value of the investment, and the size of the workforce.

Among the analysed civil structures and decision situations connected with their execution, over 40% were road construction projects. Such a large proportion of new roads is the result of the rapidly growing road construction sector stimulated by the EU funds. The second most numerous type of civil engineering projects was the construction of environmental protection facilities. Same as roads, they were mostly built with the contribution of EU funds. Bridges as civil engineering constructions associated with road construction made up 15% of the analysed projects. Each of the remaining types of civil structures corresponded to less than 10% of the total. As many as 4 out of 6 large civil structures represented road structures. Participation in the study each type of objects is presented in the diagram in Figure 6, and the percentage share of various venture in the group of large objects chart in Figure 7.

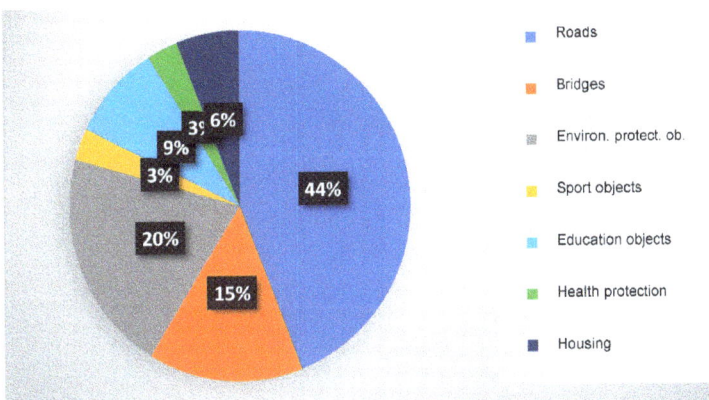

Figure 6. Participation in the study each types of civil structures.

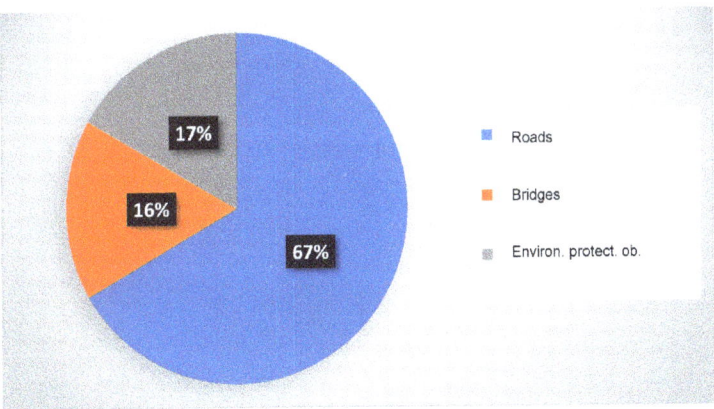

Figure 7. Percentage share of various venture in the group of large civil structures.

The research comprised 6 construction projects classified as large or as highly complex ones. There was the same number of medium-sized and small civil structures (14 civil structures). The percentage of these projects in the total analysed number is illustrated in Figure 8.

The focus in our study was on decision-making processes which occurred during the execution of the analysed construction projects. A review of the documents made available by the companies supplied a wealth of information about these processes. The models used by the decision makers engaged in the analysed cases most often belonged to model II, i.e., more or less advanced methods of multi-criteria analysis. As many as 25 out of 35 analysed cases took advantage of this model. Some other details regarding the applied models are shown schematically in Figure 8.

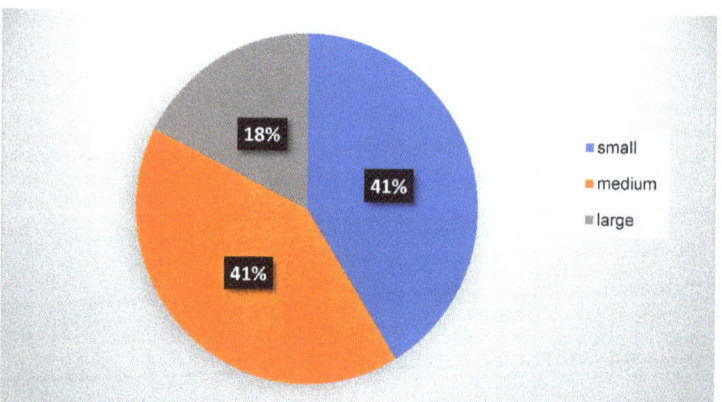

Figure 8. Percentage share of small, medium and large civil structures in research.

From Figure 9, it is clear that Model 5 was not used in any of the analysed cases of construction projects implementation. It is a model based on more advanced mathematical methods and there is no tradition of using these models by the engineering community. Model 5 has found wide application in enterprises when making decisions related to business strategies. This is due to, inter alia, of the fact that construction companies usually employ people with professional training (marketing and management, economics) that allows them to apply operational research in management-related departments. There is a rather shortage of such people when carrying out the works. Models 6 and 7 were not used in both the implementation of projects and business decisions. These methods require much more advanced research tools and are rather used in scientific research on construction activities and not in engineering practice.

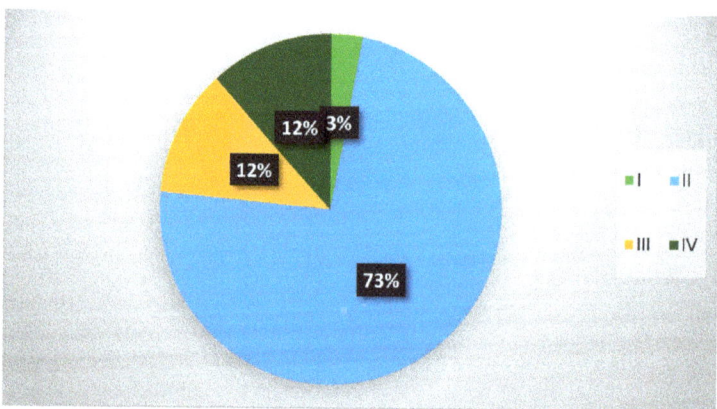

Figure 9. The percentage of different decision-making models used in the studied projects.

Another case associated with making a decision which drew our attention was a change in the profile of operations performed by a given company and the scope of its activities. Such decisions belong to strategic ones, and they are made on the basis of large databases and raw information. A number of building companies operating in the north-eastern part of Poland were analysed in our study. It was found that the construction sector was doing quite well, especially in comparison with other sectors of the economy and despite some symptoms heralding a crisis. Most of the analysed companies did not feel the urge to change the profile of their business activity. Only 15 declared they needed to make amends or widen the scope of operations. Decisions to make such changes were preceded by analyses of the market, area of operations or position of competitors. When making such decisions, it was common to run simulation studies based on analyses of preferences of clients, investors and relative branches of economy. In some cases, an analysis of variant solutions was performed according to the method of multi-criteria analysis. Detailed findings in this regard are presented in Table 4.

Table 4. Specification of the information on strategic decisions (concerning a change in the profile and scope of business).

Company No.	Type of Planned Change	A Kind of Pre-Analysis	Have Options Been Developed?	Used Model	Has the Model Been Verified?
1	Change of business profile	Market research	yes	III	yes
2	Change of business profile	Market research and customer preferences	no	V	yes
3	Offer extension	Study of customer preferences	no	V	yes
4	Extension of the operating area	Customer needs and labour market research	no	III	yes
5	Change of business profile	Needs and production technology research	yes	II	no
6	Offer extension	Study of customer preferences	no	V	no
7	Offer extension	Market research	yes	IV	no
8	Extension of the operating area	Study of customer preferences	no	V	yes
9	Extension of the operating area	Study of customer preferences	no	III	yes
10	Change of business profile	Market research and customer preferences	yes	II	no
11	Offer extension	Market research and customer preferences	no	V	yes
12	Extension of the operating area	Customer needs and labour market research	no	V	yes
13	Offer extension	Market research	no	III	yes
14	Change of business profile	Market research	yes	II	no
15	Change of business profile	Market research	yes	III	no

Decision problems in the companies submitted to our study also pertained to such situations as a change in the profile of business, a change or broadening of the scope and area of business activities. The number of such situations was quite evenly distributed among the types of companies. Details can be seen in Figure 10.

Within this research group, it is possible to notice a much higher frequency of models V—operations studies, and III—Sutherland's model. Both models comprise extended final stages where some feedback is expected in order to improve the developed model. When dealing with strategic decisions, the interviewed decision makers attached much importance to this component of an analysis.

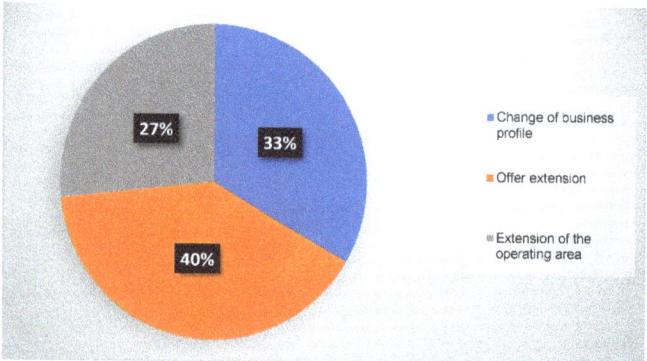

Figure 10. Percentage share of examined decisions.

Detailed information such as the percentage of each model in the total research sample is presented in Figure 11.

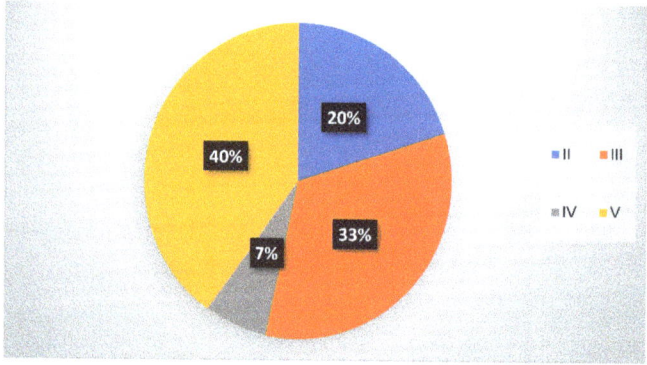

Figure 11. Percentage share of decision models in the research sample.

3.3. Mathematical Methods Used to Support the Analysed Decision-Making Processes

This study also comprised an analysis of the methods for evaluation of variant decisions, applied in all the models. Building investments and civil structures have own specific characteristics, and therefore not all the available methods can be applied in practice. The most important feature expected from decision-support methods employed in the construction business is their capacity for including a large number of criteria of various character. At the stage when variant solutions are compared, it is unnecessary to use a binary assessment; what is actually needed is to determine the degree to which each variant solution satisfies a given parameter. Another essential requirement is to be able to develop descriptive assessment scales that will be readable and easy to use. To use variant evaluation methods in practice, it is also important to have a relatively simple mathematical apparatus and to present the results clearly. Thus, in the construction activity, a certain group of methods tends to prevail. An analysis of an investment for which an evaluation of alternative solutions has been made shows that these are mostly methods where scores are assigned to assessed variants. The formula for calculating an aggregated assessment of variant 'i' is as follows (1):

$$F_i = \sum_{j=1}^{n} c_{ij} \qquad (1)$$

where:

c_{ij}—the assessment of a variant expressed in points scored for criterion c_j by variant v_i,

F_i—total aggregated assessment of variant v_i, ($i = 1, 2, \ldots, m$).

Another equally popular approach is to make an evaluation of solutions that takes into account the varied importance of criteria. This method is sometimes called the weight-score method or the weighted sum method. Here, every criterion is assigned a weight, which facilitates a more effective evaluation of the variants, which takes into account the specific character of the project submitted to evaluation. The sum of weights should always equal 1.0. The problem can be written as (2):

$$F_i = \sum_{j=1}^{n} w_j c_{ij} \qquad (2)$$

where:

c_{ij}—the assessment expressed in points scored for criterion c_j by variant v_i,
w_j—weights.

The following formulas can be used when aggregating mid-term evaluations in scalar methods:

- arithmetic mean (3):

$$F_i = \frac{1}{n} \sum_{j=1}^{n} c_{ij} \qquad (3)$$

- arithmetic weighted mean (4):

$$F_i = \frac{1}{n} \sum_{j=1}^{n} w_j c_{ij} \qquad (4)$$

- sum of arithmetic weighted mean (5):

$$F_i = \frac{\sum_{j=1}^{n} w_j c_{ij}}{\sum_{j-1}^{n} w_j} \qquad (5)$$

Yet another popular method, admittedly more mathematically advanced, is the Analytic Hierarchy Process (AHP) approach, created in the 1980s by T. Saaty. The method owes its popularity to the fact that it approaches the problem of making an assessment by ordering criteria and arranging them in groups of main criteria, after which hierarchical evaluation is performed, which leads to the determination of vectors of the main criteria and subordinate criteria. The mathematical formula relies of a pairwise comparison matrix and proceeds according to the following template:

1. Calculation of the value of a normalized matrix (6):

$$w_{ij} = \frac{a_{ij}}{\sum_{i=1}^{n} a_{ij}} \qquad (6)$$

2. Determination of the value of the vector of sub-priorities (7):

$$w_j = \sum_{j=1}^{n} w_j a_{ij} \qquad (7)$$

where (8):

$$w_j = \frac{\sum_{i=1}^{n} w_{ij}}{n} \qquad (8)$$

In order to verify whether the above-mentioned procedures has been correct, we determine:

- The matrix's own maximum value (9):

$$\lambda \frac{1}{w_i} \sum_{i=1}^{n} a_{ij} w_{j\,max} \qquad (9)$$

- Value of the consistency index (10):

$$C.I. = \frac{\lambda_{max}}{n-1} \qquad (10)$$

- Consistency ratio (11):

$$C.R. = \frac{C.I.}{R.I.} \qquad (11)$$

where:

the C.R. should reach a value < 10%

R.I.—random index, the value of which depends on the "n" number of compared components [19,22].

In addition to selecting and evaluating variant solutions, linear programming methods leading to the optimization of operations are employed in the management of enterprises in the construction activity. A mathematical model of a linear programming problem (12):

$$f(x_1, \ldots x_n) = \sum_{j=1}^{n} c_j x_j \; max \qquad (12)$$

with limitation (13):

$$\sum_{j=1}^{n} \sum_{i=1}^{m} a_{ij} x_j \leq b_j \qquad (13)$$

$(i = 1, \ldots, m)$
$(j = 1, \ldots, n)$
$x_j \geq 0$

An example of the use of one of the methods is presented below. The example mentioned earlier involves the process of making a decision about the location of a production facility in a situation where four variants were available. Six groups of main criteria were used to evaluate each of them, and in each group from three to six factors. In total, the degree of fulfilment of 21 factors was assessed. The evaluation was carried out with the most frequently used weight-score method. Details of the analysis are presented in Tables 5 and 6.

Table 5. List of weights values obtained for the main and sub-criteria.

Criteria	Sub-Criteria	Weights for Main Criteria	Weights for Sub-Criteria	Final Weights
A	A1	0.20	0.18	0.0360
	A2		0.18	0.0360
	A3		0.22	0.0440
	A4		0.16	0.0320
	A5		0.06	0.0120
	A6		0.20	0.0400
B	B1	0.10	0.11	0.0110
	B2		0.55	0.0550
	B3		0.34	0.0340
C	C1	0.25	0.45	0.1125
	C2		0.22	0.0550
	C3		0.33	0.0825
D	D1	0.15	0.11	0.0165
	D2		0.65	0.0975
	D3		0.24	0.0360
E	E1	0.20	0.66	0.1320
	E2		0.17	0.0340
	E3		0.17	0.0340
F	F1	0.10	0.56	0.0560
	F2		0.22	0.0220
	F3		0.22	0.0220

Table 6. List of the results of the variant assessment using the weight-score method.

Criteria	Sub-Criteria	Final Weights	L1	L2	L3	L4	L5
A	A1	0.036	0.072	0.072	0.036	0.108	0.036
	A2	0.036	0.000	0.108	0.036	0.108	0.036
	A3	0.044	0.088	0.088	0.044	0.132	0.132
	A4	0.032	0.096	0.064	0.032	0.096	0.096
	A5	0.012	0.036	0.024	0.024	0.024	0.012
	A6	0.040	0.040	0.080	0.080	0.080	0.080
	Total	0.200	2.200	2.600	1.600	3.200	2.200
B	B1	0.011	0.033	0.011	0.022	0.033	0.033
	B2	0.055	0.165	0.165	0.165	0.165	0.165
	B3	0.034	0.068	0.068	0.000	0.102	0.068
	Total	0.100	0.800	0.600	0.500	0.900	0.800
C	C1	0.113	0.225	0.225	0.225	0.338	0.338
	C2	0.055	0.165	0.110	0.165	0.165	0.165
	C3	0.083	0.165	0.248	0.165	0.248	0.248
	Total	0.250	1.750	1.750	1.750	2.250	2.250
D	D1	0.017	0.050	0.017	0.033	0.033	0.017
	D2	0.098	0.195	0.195	0.098	0.293	0.195
	D3	0.036	0.072	0.072	0.072	0.000	0.108
	Total	0.150	1.050	0.750	0.750	0.750	0.900
E	E1	0.132	0.132	0.396	0.264	0.396	0.132
	E2	0.034	0.034	0.034	0.068	0.102	0.034
	E3	0.034	0.034	0.068	0.068	0.102	0.034
	Total	0.200	0.600	1.200	1.200	1.800	0.600
F	F1	0.056	0.056	0.112	0.056	0.112	0.112
	F2	0.022	0.044	0.044	0.044	0.066	0.066
	F3	0.022	0.066	0.044	0.022	0.066	0.022
	Total	0.100	0.600	0.600	0.400	0.800	0.600
	Sum		7.000	7.500	6.200	9.700	7.350

The results of the calculations after summing up indicate the best-ranked solution.

4. Discussion of the Obtained Results

The research results concerning decisions about which variant solution to choose (Table 3) show that such decision-making situations are evidently dominated by the multi-criteria decision support methods (model II), and this model is particularly preferred in cases of construction projects classified as small and medium-sized ones. This is mostly connected with the character of decisions and specific situations generating numerous assessment criteria. In 25 of the analysed cases, one of the multi-criteria analytical methods was employed. Meanwhile, it could be noticed that the structure of models applied was diverse—the observations show that the eight-step decision support scheme recommended in the literature was employed in 11 cases, while in 8 cases the process was shorter by one stage, hence only seven steps were followed, in contrast to 6 other cases, where the procedure was extended by one stage. The application of the eight-step procedure appeared in construction projects classified as small and medium-sized, while longer procedures appeared in medium-sized projects. Further investigations revealed that the shortening of the procedure in small construction projects affected the final stage. Decision makers skipped the stage of evaluating the effectiveness of the decision they made. They concluded that it was unnecessary in the case of less complicated construction projects. The lengthening of the process in medium-sized construction projects differed in character. In two cases (road construction), the procedure required one more final stage to be added, such as transmitting the feedback in order to correct the decision model adopted at the beginning. In other cases, the stage of input information preparation, intended to help recognise difficulties in executing a given construction project, was expanded to enable identification of problems.

Model III—the Sutherland's model, was applied on four occasions in construction projects classified as large ones. In all the cases, it was reduced to ten steps, and the description of the performed analyses explained that the collection of information and data was combined with empirical research into a single stage (stages 2 and 3).

Model IV—the Holt's model, very close to a multi-criteria model, appeared on 4 occasions. In a form lengthened by adding an additional stage it was used in large construction projects, whereas in the form recommended in the literature, i.e., composed of 8 stages, it was employed in medium-sized civil structures. The lengthening of the process consisted of a stage of obtaining feedback in order to correct the process in the case of large and complicated construction projects.

Model I proved to be far less useful in a decision-making process in construction projects, as it only appeared in one case. A decision supported by a single-criterion model does not allow the user to analyse complicated construction processes.

Table 4 contains the information about the decision-making processes concerning development strategies of construction companies. It shows that other decision support models were applied in this case. Six cases of using model V, 5 cases of using model III, 3 cases of using model II and 1 case of using model IV were observed.

Model V was mainly employed when making a decision to broaden the range of services offered. The application of this model entailed market analysis and studies of clients' preferences. In one case, this model was used to plan a change in the company's business profile, and once it served to make a decision to expand the operation area. The adopted model was verified in all but one cases. This model is based on operations studies, it takes into consideration the use of linear and non-linear programming, and it offers many possibilities with respect to modelling decision problems.

Model III was as just popular as Model V (5 cases). Model III enables the user to carefully prepare and conduct the entire decision-making process. By going through 11 stages, one is able to prepare input data more accurately and to verify the outcome. This model was applied in various situations, but its implementation was associated with an assessment of different variant solutions.

Model II, which was used in three of the analysed cases, is similar in principle but less expanded. It was used only when making a decision to change the company's business profile. No verification of the model was noticed in any of the three cases. Model IV, close in its philosophy to model II, appeared only once, when analysing a possibility of broadening the offer, and it was not verified either. Noteworthy is the fact that the cybernetic model was not used at all.

The research reported in this article showed that the most popular decision support methods employed in the construction industry when making an assessment of variant solutions are multi-criteria analytical methods and models which make use of such analyses. The study comprised problems of making decisions while performing small, medium-sized and large construction projects, characterised by different degrees of complexity. It was found that decision support models were somehow modified by users and the modifications depended on the nature of a construction project. The interviews and analyses of all studied cases enabled us to elaborate and suggest decision support models dedicated to the construction sector, including such aspects as the size of a construction project and its complexity.

Suggested decision support models useful when selecting variant solutions in the construction sector:

I.   A model useful for small and medium-sized, less complicated construction projects:
1. Study the decision situation
2. Define the problem
3. Determine criteria applied to assess solutions to the problem
4. Develop variant solutions to the problem

5. Select a method to assess the solutions
6. Assess and select the best solution
7. Implement the solution

II. A model for medium-sized and large, complicated construction projects:
1. Analyse the environment and collect information
2. Define the problem causing difficulties
3. Determine criteria for the assessment of solutions to the problem
4. Develop variant solutions
5. Select a method for making an assessment of the possible solutions
6. Assess all variant solutions and select one
7. Make a decision
8. Implement the decision
9. Obtain feedback
10. Correct the input data, underlying assumptions

Making strategic decisions is most often supported by models III and V. When a decision to be made concerns the future of a company, the most important stage was to collect a complete set of information about the environment (model III), as well as being able to describe the decision situation using a mathematical language (model V) and obtaining feedback which allows the user to verify the model, and in consequence to improve the decision-making process.

Our suggestion of a decision support model for strategic decisions, based on this research and on experiences gained by building companies looks as follows:

1. Identifying the need to make a decision (define the aim).
2. Collecting information about the environment.
3. Building a model (in some cases, describing the situation with a mathematical language).
4. Generating solutions to the decision problem.
5. Making an evaluation of the variant solutions.
6. Verifying the model.
7. Obtaining feedback to correct the model.
8. Receiving the corrected solution.
9. Implementation.

The models designed on the basis of this study take into account the specific nature of civil engineering practice. The analysed research material led to the identification of three basic decision situations: (I) connected with the execution of large and medium-sized construction projects that are highly complex, (II) smaller and less complicated construction projects, and (III) situations when strategic decisions are made in a construction company. The approach recommended in this article is universal and can be useful in practice.

Apart from the tested models applicable in construction, attention should also be paid to the "desirability function approach" used in industrial engineering. This method has been used, among others. for testing the optimal geometry of tall buildings. Its quantitative, relatively simple nature could be used in decision-making processes related to the implementation of complex construction projects [52–55]. This method deserves attention and will be the subject of further research by the authors.

The analysis of the popularity of mathematical methods showed that the commonly used methods are the score and the weighted score ones, which were used in 82% of the analysed investments, in 85% of road and bridge developments, and in 78% of developments connected with environmental protection, which were the most numerous type of development in the time period analysed. The prevalent methods were weighted score ones, which allow the user to stress the importance of some of the criteria. The AHP approach was used as a decision-support tool in 11% of the total number of investments. The remaining 7% had other assessment methods employed, e.g., a linguistic-descriptive method (schools and other education facilities as well as health care facilities), and in one case, which was a sports facility, the developer used a binary relationship model. Linear and non-linear optimization methods were employed to support business decisions in the management of companies. They were used in over 60% of the analysed cases.

5. Summary and Conclusions

Every business activity is invariably connected with making decisions. There are several decision support systems available in management practice. To improve the decision-making process, various model approaches have been developed that use different mathematical tools.

To diagnose the problem of modelling the decision-making process in construction activities, the author of the article, provided that the research covered 34 construction projects and 15 companies operating in the construction industry. The sections include that the methods and approaches described in the literature are also used in engineering practice, although they are subject to certain modifications due to the specificity of this sector, as well as there are no universal methods and procedures modelling decision making. manufacturing process in construction companies.

In the case of small and relatively simple construction projects, simplified models are usually used, where the use of the last steps of verifying the results and improving the applied model is limited. Decision makers pay special attention to the reliable preparation of input data for analysis in the decision-making process. Large and more complex construction projects were often accompanied by a decision support system consisting of more stages, and in these cases, it turned out to be important to obtain feedback and to refine the decision model accordingly.

Research has shown that in large projects it is important to obtain feedback. This is due to, inter alia, from the fact that the implementation of these projects involves much greater financial resources than in small and medium-sized projects. Decision-makers take much more care to verify the correctness of the model, because the effects of wrongly made decisions can be much more severe than in the case of small and medium-sized enterprises.

If it is necessary to make strategic decisions related to the future of a given company, attention was paid to models in which the starting point was to clearly define the goal and collect a complete set of information about the decision-making environment. Various analytical and research methods were used, but feedback was always needed to improve the final solution.

The observations obtained during the research helped the authors to develop decision support models dedicated to engineering practice that may be useful in the implementation of construction projects.

The results of the conducted research indicate that the problem of decision making in construction is wide, decisions are diverse and made in various situations. This topic is not discussed in the literature and it is difficult to find decision models dedicated to construction activities. The literature review shows a variety of approaches used in various areas of the economy. They can be adapted to various purposes, but do not take into account their specificity. For example, decisions in construction processes are always subject to certain limitations due to preceding events. The main criteria, which may be obvious or hidden, can be grouped into a set of goals, a priority system, a course of alternative actions, the consequences of each alternative solution, and a system of selection criteria.

The research does not exhaust the research issues and will be continued to take into account, inter alia, the conditions of uncertainty and risk accompanying the decisions made. Attention should also be paid to the "desirability function approach" used in industrial engineering. This meth-od has been used, among others. for testing the optimal geometry of tall buildings. Its quantitative, relatively simple nature could be used in decision-making processes related to the implementation of complex construction projects.

Author Contributions: E.S.: conceptualization, methodology, survey research, data analysis, resources study, writing-original draft preparation, writing—review and editing, visualization; J.H.: survey research, data analysis, resources study. All authors have read and agreed to the published version of the manuscript.

Funding: This research received no external funding.

Institutional Review Board Statement: Not applicable.

Informed Consent Statement: Not applicable.

Data Availability Statement: Not applicable.

Acknowledgments: We would like to acknowledge every people from construction enterprises for taking part in the survey research.

Conflicts of Interest: The authors declare no conflict of interest.

References

1. Alexandre, L.A. Gender recognition: A multiscale decision fusion approach. *Pattern Recognit. Lett.* **2010**, *31*, 1422–1427. [CrossRef]
2. Hazir, O. A review of analytical models, approaches and decision support tools in project monitoring and control. *Int. J. Proj. Manag.* **2015**, *33*, 808–815. [CrossRef]
3. Klincewicz, K. (Ed.) *Management, Organization and Organization. Review of Theoretical Perspectives*; Scientific Publishing House of the Faculty of Management at the University of Warsaw: Warsaw, Poland, 2016.
4. Matyas, K.; Nemeth, T.; Kovacs, K.; Glawar, R. A procedural approach for realizing prescriptive maintenance planning in manufacturing industries. *CIRP Ann.* **2017**, *66*, 461–464. [CrossRef]
5. Szafranko, E. Decision problems in management of construction projects. In Proceedings of the 3rd International Conference on Innovative Materials, Structures and Technologies, Riga, Latvia, 27–29 September 2017; Volume 251, p. 012048. [CrossRef]
6. Dachowski, R.; Kostrzewa, P. The use of waste materials in the construction industry. *Procedia Eng.* **2016**, *161*, 754–758. [CrossRef]
7. Radziszewska-Zielina, E.; Śladowski, G. Supporting the Selection of a Variant of the Adaptation of a Historical Building with the Use of Fuzzy Modelling and Structural Analysis. *J. Cult. Herit.* **2017**, *26*, 53–63. [CrossRef]
8. Cook, C.; Brismée, J.M.; Pietrobon, R.; Sizer, P., Jr.; Hegedus, E.; Riddle, D.L. Development of a quality checklist using Delphi methods for prescriptive clinical prediction rules: The QUADCPR. *J. Manip. Physiol. Ther.* **2010**, *33*, 29–41. [CrossRef]
9. Citherlet, S.; Defaux, T. Energy and environmental comparison of three variants of a family house during its whole life span. *Build. Environ.* **2007**, *42*, 591–598. [CrossRef]
10. Verbeeck, G.; Hens, H. Life cycle inventory of buildings: A contribution analysis. *Build. Environ.* **2010**, *45*, 964–967. [CrossRef]
11. Szafranko, E. The choice of variant technologies and materials supported by multicriteria methods and an assessment of variants with graphic profiles of criteria. *Mater. Today Proc.* **2018**, *5*, 2002–2009. [CrossRef]
12. Zavadskas, E.K.; Turskis, Z. Multiple criteria decision making (MCDM) methods in economics: An overview. *Technol. Econ. Dev. Econ.* **2011**, *17*, 397–427. [CrossRef]
13. Beheshti, R. Design decisions and uncertainty. *Design Studies* **1993**, *14*, 85–95. [CrossRef]
14. Sobotka, A.; Sagan, J. Cost-saving Environmental Activities on Construction Site–Cost Efficiency of Waste Management: Case Study. *Procedia Eng.* **2016**, *161*, 388–393. [CrossRef]
15. Liu, A.M.; Leung, M.Y. Developing a soft value management model. *Int. J. Proj. Manag.* **2002**, *20*, 341–349. [CrossRef]
16. Kaklauskas, A.; Zavadskas, E.K.; Trinkunas, V. A multiple criteria decision support on-line system for construction. *Eng. Appl. Artif. Intell.* **2007**, *20*, 163–175. [CrossRef]
17. Marques, G.; Gourc, D.; Lauras, M. Multi-criteria performance analysis for decision making in project management. *Int. J. Proj. Manag.* **2011**, *29*, 1057–1069. [CrossRef]
18. Radziszewska-Zielina, E.; Szewczyk, B. Supporting partnering relation management in the implementation of construction projects using AHP and fuzzy AHP methods. *Procedia Eng.* **2016**, *161*, 1096–1100. [CrossRef]
19. Saaty, T.L. Decision making with the analytic hierarchy process. *Int. J. Serv. Sci.* **2008**, *1*, 83–98. [CrossRef]
20. Juan, Y.K.; Gao, P.; Wang, J. A hybrid decision support system for sustainable office building renovation and energy performance improvement. *Energy Build.* **2010**, *42*, 290–297. [CrossRef]
21. Schabowicz, K.; Hoła, B. Application of artificial neural networks in predicting earthmoving machinery effectiveness ratios. *Arch. Civ. Mech. Eng.* **2008**, *8*, 73–84. [CrossRef]
22. Szafranko, E. Possibilities of application of multi-criteria analysis methods to evaluate material and technological solutions in the design of building structures. *Mater. Today Proc.* **2019**, *19*, 1945–1948. [CrossRef]
23. Trivedi, A.; Singh, A. A hybrid multi-objective decision model for emergency shelter location-relocation projects using fuzzy analytic hierarchy process and goal programming approach. *Int. J. Proj. Manag.* **2017**, *35*, 827–840. [CrossRef]
24. Scherer, R.J.; Schapke, S.E. A distributed multi-model-based management information system for simulation and decision-making on construction projects. *Adv. Eng. Inform.* **2011**, *25*, 582–599. [CrossRef]
25. Svenson, O. Differentiation and consolidation theory of human decision making: A frame of reference for the study of pre-and post-decision processes. *Acta Psychol.* **1992**, *80*, 143–168. [CrossRef]
26. Szafranko, E.; Srokosz, P. Applicability of the theory of similarity in an evaluation of building development variants. *Autom. Constr.* **2019**, *104*, 322–330. [CrossRef]

27. Szymanowski, W. Conditions of using information technologies to model decision-making processes in an enterprise. *Ann. Coll. Econ. Econ./Wars. Sch. Econ.* **2017**, *45*, 145–157.
28. Cheng, M.Y.; Roy, A.F. Evolutionary fuzzy decision model for construction management using support vector machine. *Expert Syst. Appl.* **2010**, *37*, 6061–6069. [CrossRef]
29. Guerry, A.D.; Ruckelshaus, M.H.; Arkema, K.K.; Bernhardt, J.R.; Guannel, G.; Kim, C.K.; Spencer, J. Modeling benefits from nature: Using ecosystem services to inform coastal and marine spatial planning. *Int. J. Biodivers. Sci. Ecosyst. Serv. Manag.* **2012**, *8*, 107–121. [CrossRef]
30. Tamošaitiene, J.; Bartkiene, L.; Vilutiene, T. The new development trend of operational research in civil engineering and sustainable development as a result of collaboration between german-Lithuanian-Polish scientific triangle. *J. Bus. Econ. Manag.* **2010**, *11*, 316–340. [CrossRef]
31. Harasymiuk, J. Analysis of selected environmental procedures for construction investments. *Ecol. Eng.* **2017**, *18*, 79–88. [CrossRef]
32. Bucoń, R.; Sobotka, A. Decision-making model for choosing residential building repair variants. *J. Civ. Eng. Manag.* **2015**, *21*, 893–901. [CrossRef]
33. Bolesta-Kukułka, K. *Managerial Decisions in Management Theory and Practice*; Scientific Publishers of the Faculty of Management at the University of Warsaw: Warsaw, Poland, 2000.
34. Berredo, R.C.; Cruz, E.C.; Ekel, P.Y.; Junges, M.F.D.; Contijo, M.M.; Pereira, J.G., Jr.; Popov, V.A. Monocriteria and multicriteria optimization of network configuration in distribution systems. In Proceedings of the WSEAS International Conference on Power Engineering Systems, Tenerife, Canary Islands, Spain, 4–6 July 2005; pp. 117–122.
35. Kasharin, D.V. Intelligent decision support systems in the design of mobile micro hydropower plants and their engineering protection. In Proceedings of the First International Scientific Conference "Intelligent Information Technologies for Industry"(IITI'16), Sochi, Russia, 16–21 May 2016; pp. 239–248. [CrossRef]
36. Opricovic, S.; Tzeng, G.H. Defuzzification within a multicriteria decision model. *International Journal of Uncertainty. Fuzziness Knowl.-Based Syst.* **2003**, *11*, 635–652. [CrossRef]
37. Barker, T.J.; Zabinsky, Z.B. A multicriteria decision making model for reverse logistics using analytical hierarchy process. *Omega* **2011**, *39*, 558–573. [CrossRef]
38. Cheng, M.Y.; Hsiang, C.C.; Tsai, H.C.; Do, H.L. Bidding decision making for construction company using a multi-criteria prospect model. *J. Civ. Eng. Manag.* **2011**, *17*, 424–436. [CrossRef]
39. Sutherland, J.W. *Administrative Decision-Making: Extending the Bounds of Rationality*; Van Nostrand Reinhold: New York, NY, USA, 1977.
40. Chatterjee, P.; Banerjee, A.; Mondal, S.; Boral, S.; Chakraborty, S. Development of a hybrid meta-model for material selection using design of experiments and EDAS method. *Eng. Trans.* **2018**, *66*, 187–207.
41. Hutchins, M.J.; Sutherland, J.W. An exploration of measures of social sustainability and their application to supply chain decisions. *J. Clean. Prod.* **2008**, *16*, 1688–1698. [CrossRef]
42. Clithero, J.A. Improving out-of-sample predictions using response times and a model of the decision process. *J. Econ. Behav. Organ.* **2018**, *148*, 344–375. [CrossRef]
43. Karimi, S.; Papamichail, K.N.; Holland, C.P. The effect of prior knowledge and decision-making style on the online purchase decision-making process: A typology of consumer shopping behavior. *Decis. Support Syst.* **2015**, *77*, 137–147. [CrossRef]
44. Zhang, X.; Wu, Y.; Shen, L.; Skitmore, M. A prototype system dynamic model for assessing the sustainability of construction projects. *Int. J. Proj. Manag.* **2014**, *32*, 66–76. [CrossRef]
45. Turskis, Z.; Gajzler, M.; Dziadosz, A. Reliability, risk management, and contingency of construction processes and projects. *J. Civ. Eng. Manag.* **2012**, *18*, 290–298. [CrossRef]
46. Vukomanovic, M.; Radujkovic, M. The balanced scorecard and EFQM working together in a performance management framework in construction industry. *J. Civ. Eng. Manag.* **2013**, *19*, 683–695. [CrossRef]
47. Bozeman, D.P.; Kacmar, K.M. A cybernetic model of impression management processes in organizations. *Organ. Behav. Hum. Decis. Process.* **1997**, *69*, 9–30. [CrossRef]
48. Mohammadi, F.; Sadi, M.K.; Nateghi, F.; Abdullah, A.; Skitmore, M. A hybrid quality function deployment and cybernetic analytic network process model for project manager selection. *J. Civ. Eng. Manag.* **2014**, *20*, 795–809. [CrossRef]
49. Adeli, H. Neural networks in civil engineering. *Civ. Infrastruct. Eng.* **2001**, *16*, 126–142. [CrossRef]
50. Zavadskas, K.; Antucheviciene, E.; Adeli, H.J.; Turskis, Z. Hybrid multiple criteria decision making methods: A review of applications in engineering. *Sci. Iran.* **2016**, *23*, 1–20. [CrossRef]
51. Antucheviciene, J.; Kala, Z.; Marzouk, M.; Vaidogas, E.R. Solving Civil Engineering Problems by Means of Fuzzy and Stochastic MCDM Methods: Current State and Future Research. *Math. Probl. Eng.* **2015**, *2015*, 34. [CrossRef]
52. Harrington, E.C. The desirability function. *Ind. Qual. Control* **1965**, *4*, 494–498.
53. Derringer, G.C.; Suich, R. Simultaneous optimization of several response variables. *J. Qual. Technol.* **1980**, *12*, 214–219. [CrossRef]
54. Vera Candioti, L.; De Zan, M.M.; Camara, M.S.; Goicoechea, H.C. Experimental design and multiple response optimization. Using the desirability function in analytical methods development. *Talanta* **2014**, *124*, 123–138. [CrossRef]
55. Lacidogna, G.; Scaramozzino, D.; Carpinteri, A. Optimization of diagrid geometry based on the desirability function approach. *Curved Layer. Struct.* **2020**, *7*, 139–152. [CrossRef]

Article

Experimental Method for Flow Calibration of the Aircraft Liquid Cooling System

Yingjie Zhao [1], Fan Yang [2] and Yijiang Ma [2,*]

1 School of Energy and Power, Jiangsu University of Science and Technology, Zhenjiang 212003, China; lszhaoyingjie@126.com
2 School of Naval Architecture and Ocean Engineering, Jiangsu University of Science and Technology, Zhenjiang 212003, China; fyang@just.edu.cn
* Correspondence: yima@nuaa.edu.cn

Abstract: In the process of aircraft operation, the flow calibration of aircraft liquid cooling system has always been one of the research hotspots in engineering. Based on the principle of the differential pressure method, a new experimental flow calibration method is proposed for the aircraft liquid cooling system in this paper. In the reducer and the square bend of the aircraft liquid cooling system, the pressure difference will be generated. The flowmeter is used to measure the flow of the coolant, and the flow rate coefficient of the aircraft liquid cooling system can be calibrated. The experimental platform is established to conduct the flow calibration of the aircraft liquid cooling system, and the influence of the temperature and imported pressure on the flow will be investigated. Results indicate that the experimental method proposed is very effective, and the flow calibration can be realized without damaging the aircraft liquid cooling system.

Keywords: liquid cooling system; flow calibration; differential pressure; experimental method; aircraft

Citation: Zhao, Y.; Yang, F.; Ma, Y. Experimental Method for Flow Calibration of the Aircraft Liquid Cooling System. *Appl. Sci.* **2022**, *12*, 5056. https://doi.org/10.3390/app12105056

Academic Editors: Mariusz Szóstak, Marek Sawicki and Jarosław Konior

Received: 11 April 2022
Accepted: 11 May 2022
Published: 17 May 2022

Publisher's Note: MDPI stays neutral with regard to jurisdictional claims in published maps and institutional affiliations.

Copyright: © 2022 by the authors. Licensee MDPI, Basel, Switzerland. This article is an open access article distributed under the terms and conditions of the Creative Commons Attribution (CC BY) license (https://creativecommons.org/licenses/by/4.0/).

1. Introduction

With generous applications of the avionics and the continuous improvement of the radar power, the air cooling system cannot fully satisfy the cooling requirements, and the liquid cooling system emerges as the times requires. Compared with the air-cooling system, the liquid cooling system has several advantages, such as the larger refrigeration capacity, the simpler system design, the secondary cooling with fuel or punching air, and no influence on the normal work of the aircraft. At the same time, the heat transfer coefficient [1,2] and specific heat of the liquid are much greater than the air, and the liquid cooling system has higher cooling efficiency and stable working ability. Therefore, the liquid cooling system for high-performance aircraft electronic equipment has become an inevitable trend.

Under normal conditions, the main characteristic parameters of the liquid cooling system are the temperature and pressure. In order to evaluate the cooling performance of the liquid cooling system more accurately, the flow of the coolant becomes another important parameter, which can be obtained through the experimental measurement. Around the 21st Century, many liquid flow measurement methods had been proposed [3–7], but there are several disadvantages among these methods proposed, such as the leakage, restriction of the flight condition, the toxic of the refrigerant and so on, which were very difficult to be applied to the liquid cooling system of the aircraft. Therefore, it is necessary to find a safe and feasible flow measurement method for the aircraft liquid cooling system.

In recent years, the aircraft liquid cooling system has become an important development direction of the aircraft refrigeration system, and researches on domestic aircraft liquid cooling system have just started. Zheng et al. [8] used the experimental method to calculate the flow correction coefficient. Tian et al. [9] introduced the usage, the characteristics and the application scope of common flow counters in industry briefly. Ahrens et al. [10] used a telecentric CCD imaging system to track a moving liquid meniscus inside a glass capillary, and developed a new experimental setup for measuring ultra-low flows. Doihara et al. [11] developed a calibration rig, which consisted of a syringe pump and a weighing tank system, to conduct the flow calibration in the flow range of Ma et al. [12] considered the diversion damper situation and the jet flow velocity profile, and developed a measurement error model of diverter to overcome the error calculation difficulties of the diverter in the liquid flow calibration facilities. Nakada et al. [13] constructed a flow measurement system consisting of an acoustic emission sensor and a signal processing circuit, and transformed the acoustic emission signals into the corresponding flow. Restricted by the limited space of the aircraft cabin, some liquid flow measurement equipment commonly used are difficult to implement modification on the aircraft, which can cause many safety problems, such as the leakage of the coolant and the performance of the system. Therefore, these flow measurement methods above are difficult to carry out on the aircraft.

To overcome this problem, a new flow measuring method of the aircraft liquid cooling system is proposed in this paper. This method does not need to modify the liquid cooling system, so it will not affect the normal work of the aircraft liquid cooling system. According to the principle of the differential pressure method, the experimental platform of the aircraft liquid cooling system is established, and the flow calibration is carried out based on the pressure difference of the reducer and the square bend. Experimental results indicate: the flow calibration method proposed is correct and effective, which can be applied to the flow measurement of the aircraft liquid cooling system.

2. Flow Rate Calibration Scheme

There are many flow measurement methods [14–16] proposed interiorly, such as the differential pressure flowmeter method, the volumetric flowmeter method, the float flowmeter method, the blade flowmeter method, the electromagnetic flowmeter method, the vortex flowmeter method and the ultrasonic flowmeter method. Most of these methods need special flowmeters. Many problems will be encountered during the installation of special flowmeters, interfering with the normal operation [17] and affecting the safety check of the liquid cooling system [18]. A variety of flow calibration methods have been proposed based on the physical properties of geometrical characteristics [19,20], and the most feasible method is the differential pressure flow measurement method.

The differential pressure flow measurement principle, is to fix a throttle with an area less than the section area of the pipe in the liquid filled pipe, and the liquid in the pipe will shrink when it passes through the throttle, then the flow velocity will increase and the static pressure will drop at the contraction; finally a certain pressure drop will generate before and after the throttle.

As shown in Figure 1, two cross-sections are taken on a variable diameter pipe, Sections 1 and 2. Section 1 is the cross-section before the contraction of the stream, and Section 2 is the cross-section after the contraction of the stream. According to the Bernoulli Equation of the incompressible ideal fluid and continuity equation of the incompressible fluid constant flow:

$$\frac{p_1'}{\rho_1} + \frac{u_1^2}{2} = \frac{p_2'}{\rho_2} + \frac{u_2'^2}{2} \tag{1}$$

$$A_1 u_1 = A_2 u_2 \tag{2}$$

where: p_1' and p_2' are the average pressures of the fluid at Sections 1 and 2; ρ_1 and ρ_2 are the density of the fluid at Sections 1 and 2; u_1 and u_2' are the flow velocities at Sections 1 and 2; A_1 and A_2 are the areas of Sections 1 and 2.

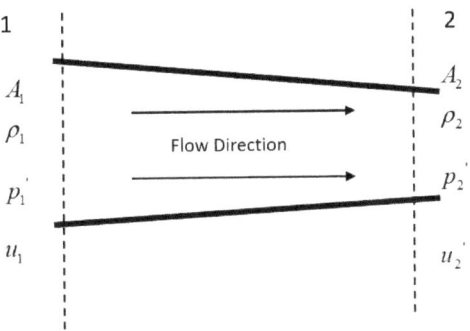

Figure 1. Diagrammatic sketch of a variable diameter pipe.

For the same coolant, $\rho_1 = \rho_2$. Substitute Equation (2) into Equation (1), and the flow velocity at Section 2 can be derived as follows:

$$u_2' = \frac{1}{\sqrt{1-(A_2/A_1)^2}}\sqrt{\frac{2}{\rho}(p_1' - p_2')} \qquad (3)$$

The flow velocity at Section 2 u_2' can be rewritten as follows:

$$u_2' = \frac{1}{\sqrt{1-\mu^2\beta^4}}\sqrt{\frac{2}{\rho}(p_1' - p_2')} \qquad (4)$$

where: μ is the contraction coefficient of the stream; β is the diameter ratio of the throttle device, and $\beta = \sqrt{A_2/A_1}$.

Because p_1' and p_2' are the average pressures of the fluid at Sections 1 and 2, the pressure drop $p_1 - p_2$ is read at the pipe wall according to a certain way of taking pressure in the actual measurement. Therefore, a pressure correction coefficient ψ needs to be introduced, and $p_1' - p_2' = \psi(p_1 - p_2)$. When the taking pressure method is different, the pressure correction coefficient ψ is different from coefficients of other pressure methods. In engineering practice, there is flow velocity loss in the fluid flow, which is different from the assumption of the isentropic constant flow. Therefore, a coefficient ξ is introduced to modify the flow velocity u_2'. The volume flow of the incompressible fluid after modification can be expressed as follows:

$$G = u_2\mu A_0 = \frac{\mu\xi\sqrt{\psi}}{\sqrt{1-\mu^2\beta^4}}A_0\sqrt{\frac{2}{\rho}(p_1 - p_2)} \qquad (5)$$

where: $p_1 - p_2$ is the actual pressure drop measured.

The mass flow of the incompressible fluid [21] after modification can be expressed as follows:

$$G = u_2\mu A_0 = \frac{\mu\xi\sqrt{\psi}}{\sqrt{1-\mu^2\beta^4}}A_0\sqrt{2\rho(p_1 - p_2)} \qquad (6)$$

where: G is the flow of coolant in the pipe.

Assume that a is the flow coefficient, and $a = \frac{\mu\xi\sqrt{\psi}}{\sqrt{1-\mu^2\beta^4}}$. The flow coefficient is related to the form of the throttle, the ratio of the diameter, the way of the pressure withdrawal, the Reynolds number of the flow and the roughness of the tube wall. Because μ, ξ and ψ cannot be measured directly, the flow rate coefficient is generally determined by the experiments.

3. Establishment of Experimental Platform

For the aircraft liquid cooling system, the flow measurement is really difficult. Firstly, the pipeline of the aircraft liquid cooling system cannot meet the length requirements of the straight pipe before and after the throttle; Secondly, the installation of the throttle device on the pipe can affect the flow resistance characteristics of the whole system, which can also increase the possibility of the leakage of the coolant.

In the technical index of a certain type of the aircraft liquid cooling system, the imported pressure of the radar components is 930 kPa, and the configuration of a throttle device may cause 50 kPa or even greater pressure loss. The pressure loss has a great impact on the performance of the whole system, and may even make the liquid cooling system work abnormally. Therefore, for the aircraft liquid cooling system, the flow calibration needs to learn from the flow measurement principle of differential pressure flowmeter, and carried out the ground experiments of flow calibration. Based on the principle of the differential pressure method, design and establish the experimental platform to conduct the flow calibration of the aircraft liquid cooling system, and the corresponding flow rate curve can be obtained through the ground calibration experiments.

According to the flow calibration plan proposed above, the flow calibration experimental platform of the aircraft liquid cooling system is established in Figure 2.

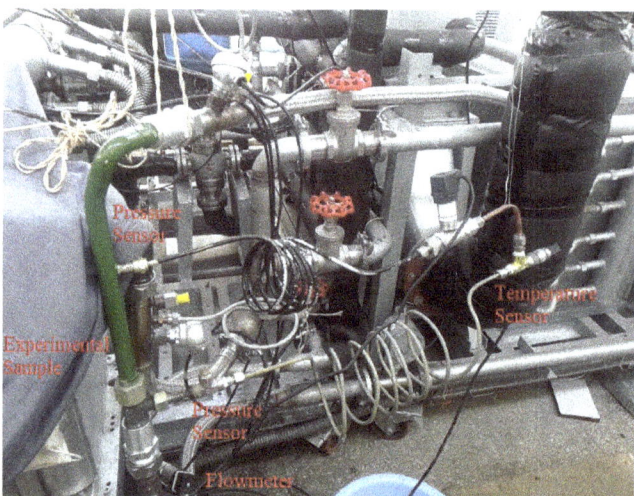

Figure 2. Schematic diagram of the aircraft liquid cooling system.

As shown in Figure 2, the cooling duct of the radar exit is a tube with variable diameters of 16 mm, 18 mm and 20 mm, and there is an elbow bend. Similar to the throttle, the pressure drop will also be caused at the inlet and outlet of the variable diameter pipe and the elbow bend. Therefore, the flow rate coefficient a can be obtained at the variable diameter pipe and the elbow bend by the ground calibration experiments, so as not to destroy the original cooling system layout or increase the additional pressure drop.

The experimental system consists of 7 parts: a pump, a filter, a heater, an imported valve, a flowmeter, an exported valve and an experimental sample. The connection of the flow calibration experimental system is shown in Figure 3.

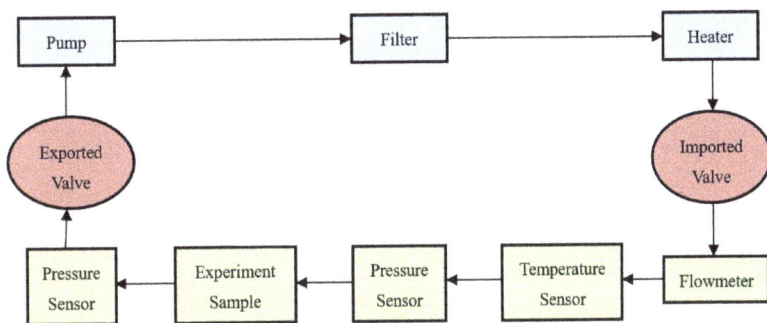

Figure 3. Diagram of connection of experimental platform.

As shown in Figure 3, the pump is used to provide continuous fluid; the heater is mainly used to regulate the temperature of the coolant and heat the coolant in the experimental system; the temperature sensor is used to measure the temperature of the coolant; the flowmeter is used to measure the flow of the coolant in the experiment sample; the function of the inlet valve is to adjust the liquid pressure into the entrance of the experiment sample by adjusting the opening degree of the inlet valve; pressure sensors are used to measure the pressure before and after the experiment sample. The pipe, which needs to be detected, is first connected to the experiment platform. Under different pressures and temperatures, the flow of the ethylene glycol through the pipe is adjusted by the pump in the experimental platform, and the pressure drop at the import and export of the experimental sample is measured by the pressure sensors, which are set before and after the experimental sample.

On the ground experimental platform of the standard flow device, the temperature, pressure and flow of the liquid cooling system are changed according to the experimental requirements, and the relation curve of the pressure drop and flow can be obtained, and the correction curve of the temperature and pressure can be obtained at the same time. The coolant is pressurized in the liquid pump, and filters into the electric heater through the filter. When the coolant is heated to the experimental required temperature, the coolant is flowing into the experimental tube through the regulating value. Then, the coolant flows back to the liquid pump through the flow control valve, and the circulation process is completed. The temperature and the flow in the pipeline remain unchanged, and the relation curve of the flow pressure drop will be obtained by changing the coolant inlet pressure of the system. The pressure drop and flow remain unchanged, and the relation curve of the flow pressure drop will be obtained by changing the coolant temperature.

4. Results and Discussion

4.1. Influence of the Temperature

When the imported valve pressure of the coolant in the pipe is 200 kPa and the coolant temperatures in the pipe are 20 °C, 40 °C and 60 °C, respectively, the relation curve of the coolant flow and the coolant pressure drop are shown in Figure 4. At the same time, we have fitted three functions between the flow and the pressure drop when the temperatures are 20 °C, 40 °C and 60 °C, and the three fitted functions are as follows:

$$\Delta P = -4.26148 + 0.16099G + 0.00106G^2 \quad (T = 20\ °C, P_{imported} = 200\ \text{kPa}) \quad (7)$$

$$\Delta P = 0.67357 - 0.00217G + 0.00224G^2 \quad (T = 40\ °C, P_{imported} = 200\ \text{kPa}) \quad (8)$$

$$\Delta P = 4.70683 - 0.14571G + 0.00329G^2 \quad (T = 60\ °C, P_{imported} = 200\ \text{kPa}) \quad (9)$$

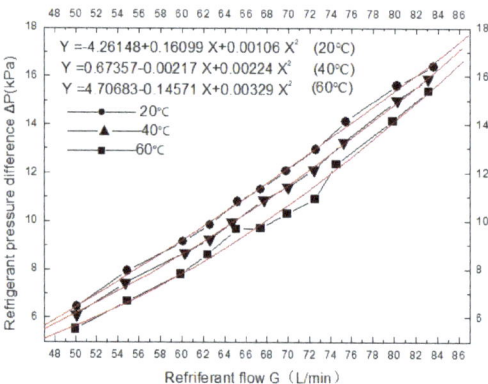

Figure 4. Variation in pressure drop with flow rate (imported pressure 200 kPa).

As shown in Figure 4, when the pressure of the coolant in the pipe is 200 kPa, the influence of the pressure drop on the flow measurement is relatively constant as the flow increases; under different coolant flow conditions, the increasing amplitude of the coolant pressure drop with the flow increasing almost remains the same. At the same time, the influence of the temperature on the flow measurement also cannot be ignored, and the influence of the temperature on the measured value of pressure drop is about 9% under the same coolant flow condition.

When the imported valve pressure of the coolant in the pipe is 350kPa and the coolant temperatures in the pipe are 20 °C, 40 °C and 60 °C, respectively, the relation curve of the coolant flow and the pressure drop is shown in Figure 5. At the same time, we have fitted three functions between the flow and the pressure drop when the temperature is 20 °C, 40 °C and 60 °C, and the three fitted functions are as follows:

$$\Delta P = 0.27686 + 0.00813G + 0.00225G^2 \quad \left(T = 20\ °C, P_{imported} = 350\ \text{kPa}\right) \quad (10)$$

$$\Delta P = -2.79757 + 0.10229G + 0.00140G^2 \quad \left(T = 40\ °C, P_{imported} = 350\ \text{kPa}\right) \quad (11)$$

$$\Delta P = 1.46309 - 0.04899G + 0.00264G^2 \quad \left(T = 60\ °C, P_{imported} = 350\ \text{kPa}\right) \quad (12)$$

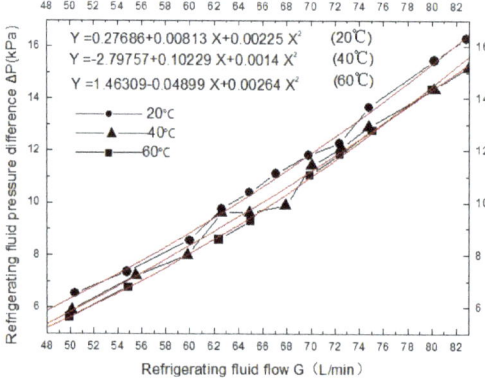

Figure 5. Variation of pressure drop with flow rate (imported pressure 350 kPa).

As shown in Figure 5, the coolant temperature will affect the accuracy of the flow measurement. When the imported pressure of the coolant is 350 kPa, compared with the imported pressure 200 kPa in Figure 3, the influence of the temperature on the flow measurement under two imported pressures above are very close. For the pressure drop measurement, the influence of the coolant temperature is smaller, and the maximum error is less than 7%.

4.2. Influence of the Pressure

When the coolant temperature in the pipe is 20 °C, 40 °C and 60 °C, the imported pressure of the coolant is set to 200 kPa and 350 kPa, and pressure drops under different flow are measured, respectively. As the flow increases, variation curves of the pressure drop are shown in Figures 6–8, respectively.

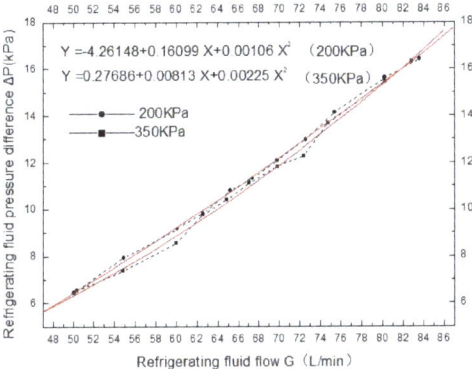

Figure 6. Variation of pressure drop with flow rate (temperature 20 °C).

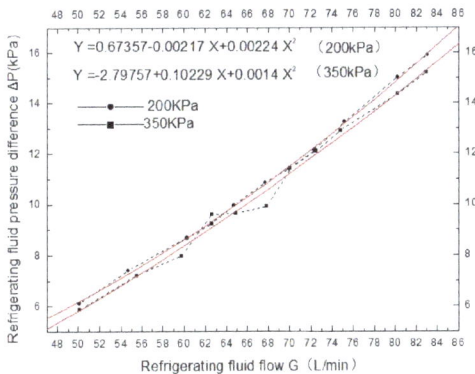

Figure 7. Variation of pressure drop with flow rate (temperature 40 °C).

As shown in Figures 6–8, the influence of the pressure on the flow measurement of the coolant is relatively smaller, and when the temperature increases, the influence of the pressure on the coolant flow gradually increases. From the above analysis, it can be concluded that the influence of the temperature on the coolant flow measurement is relatively larger. This influence has little relation with the flow changing, and gradually decreases as the pressure increases; the influence of the coolant pressure on the coolant flow measurement gradually increases as the coolant temperature increases.

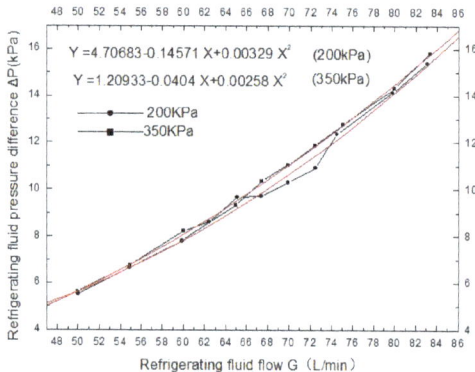

Figure 8. Variation of pressure drop with flow rate (temperature 60 °C).

5. Conclusions

According to the flow characteristics of the fluid in the variable diameter pipe and the elbow bend, a new experimental method of measuring the flow coefficient is proposed in this paper through the differential pressure flow counter, which is developed by the differential pressure flow rate measurement principle.

The volume of the experimental equipment used in the flow calibration experiments proposed in this paper is very small, which can reduce the experimental modification work. It is not necessary to install the throttle device in the pipe of the liquid cooling system, which reduces the risk of the aircraft liquid cooling system leakage. This method does not destroy the original cooling system layout or increase the additional pressure loss. The flow coefficient calibration method proposed in this paper can also be applied to the flight text of other types of aircraft liquid cooling system.

Author Contributions: Conceptualization, Y.Z. and Y.M.; methodology, Y.Z.; validation, Y.Z., F.Y. and Y.M.; formal analysis, F.Y.; investigation, Y.M.; resources, Y.Z.; data curation, Y.M.; writing—original draft preparation, Y.Z.; writing—review and editing, Y.M.; visualization, F.Y.; supervision, Y.Z.; project administration, Y.M.; funding acquisition, F.Y. All authors have read and agreed to the published version of the manuscript.

Funding: This work is supported by the National Natural Science Foundation of China (no. 51605202) and the Natural Science Foundation of Jiangsu Province (no. BK20160550).

Conflicts of Interest: The authors declare no conflict of interest.

Nomenclature

p_1'	average pressure of the fluid at Section 1;
p_2'	average pressure of the fluid at Section 2;
ρ_1	density of the fluid at Section 1;
ρ_2	density of the fluid at Section 2;
μ	contraction coefficient of the stream;
G	flow rate of coolant in the pipe.
u_1	flow velocities at Section 1;
u_2'	flow velocity at Section 2;
A_1	area of Section 1;
A_2	area of Section 2;
β	diameter ratio of the throttle device;

References

1. Zheng, X.; Yang, W. Heat transfer coefficient of film cooling with ellipse-shaped tab. *Trans. Nanjing Univ. Aeronaut. Astronaut.* **2016**, *33*, 155–165.
2. Li, J.; Jiang, Y.; Wang, Y.; Meng, E.; Li, C. Numerical calculation and experimental investigation on airborne three-stream plate-fin condenser. *J. Nanjing Univ. Aeronaut. Astronaut.* **2017**, *49*, 382–388. (In Chinese)
3. Zhang, J. A new way of the flow meter of pressure difference. *Autom. Petro-Chem. Ind.* **2004**, *5*, 87–89. (In Chinese)
4. Li, Y.; Wu, Z. Theoretical analysis and experimental investigation on DIS change measurement by angle pipe. *Irrig. Drain.* **1995**, *14*, 6–11. (In Chinese)
5. Sun, Z.; Zhou, J.; Zhang, H. Numerical simulation and experimental research on measurement characteristics of elbow meter. *Chin. J. Sens. Actuators* **2007**, *20*, 1413–1415. (In Chinese)
6. Li, Y.; Liao, W.; Tian, J. Regression analysis concerning discharge and pressure difference of elbow and calculation of discharge coefficient. *J. Xi'an Univ. Technol.* **1998**, *14*, 373–376. (In Chinese)
7. Silva, F.S.; Velazquez, M.T.; Hernandez, J.R. Experimental study for the use of elbows as flowmeters. *Comput. Stand. Interfaces* **1999**, *21*, 185. [CrossRef]
8. Zheng, J.; Liang, G.; Cheng, Y. Experimental study on the characteristic of elbow flowmeter. *J. China Inst. Metrol.* **1999**, *2*, 41–46. (In Chinese)
9. Tian, Y.; Wang, Y.; Guo, S.; Liu, F.; Hu, Z. Application of common flowmeters. *Contemp. Chem. Ind.* **2011**, *40*, 1294–1296. (In Chinese)
10. Ahrens, M.; Nestler, B.; Klein, S.; Lucas, P.; Petter, H.T.; Damiani, C. An experimental setup for traceable measurement and calibration of liquid flow rates down to 5 nL/min. *Biomed. Tech. Biomed. Eng.* **2015**, *60*, 337. [CrossRef] [PubMed]
11. Doihara, R.; Shimada, T.; Cheong, K.H.; Terao, Y. Liquid low-flow calibration rig using syringe pump and weighing tank system. *Flow Meas. Instrum.* **2016**, *50*, 90–101. [CrossRef]
12. Longbo, M.A.; Zheng, J.; Zhao, J. Research on the flow measurement error model of the diverter in the liquid flow calibration facilities. *Chin. J. Sens. Actuators* **2015**, *28*, 515–520.
13. Nakada, T.; Zheng, Y.; Sakurai, Y. A liquid flow rate measurement method using an AE sensor: Measurement of steady flow rate. *Trans. Jpn. Hydraul. Pneum. Soc.* **2013**, *44*, 49–54. [CrossRef]
14. Wang, Q. The application and the development of flowmeters. *Sci. Technol. Inf.* **2008**, *3*, 32. (In Chinese)
15. Guo, X.; You, X.; Li, L.; Du, W. A new orifice flowmeter with different diameter. *J. Henan Vocat. Tech. Teach. Coll.* **1994**, *22*, 13–15. (In Chinese)
16. Wu, C. *Hydraulics*; Higher Education Press: Beijing, China, 1982. (In Chinese)
17. Feng, W. Coolant flow measurement of aircraft liquid cooling system based on differential pressure method. *ACTA Metrol. Sin.* **2014**, *35*, 248–251. (In Chinese)
18. Cui, Y.; Shi, H.; Chen, C. Flow measurement and calculation of aircraft liquid cooling system. *Value Eng.* **2017**, *36*, 93–96. (In Chinese)
19. Yulang, C.T. *The Flow Measurement Manual*; Chinese Metrology Press: Beijing, China, 1982. (In Chinese)
20. Liang, G.; Cai, W. *The Flow Measurement Technology and Instruments*; China Machine Press: Beijing, China, 2002. (In Chinese)
21. Su, Y. *Flow Measurement and Test*; China Metrology Press: Beijing, China, 1992. (In Chinese)

MDPI
St. Alban-Anlage 66
4052 Basel
Switzerland
Tel. +41 61 683 77 34
Fax +41 61 302 89 18
www.mdpi.com

Applied Sciences Editorial Office
E-mail: applsci@mdpi.com
www.mdpi.com/journal/applsci

www.ingramcontent.com/pod-product-compliance
Lightning Source LLC
LaVergne TN
LVHW070203100526
838202LV00015B/1992